Anastasios Mallios

Modern Differential Geometry in Gauge Theories

Maxwell Fields, Volume I

Birkhäuser
Boston • Basel • Berlin

Anastasios Mallios
University of Athens Panepistimioupolis
Department of Mathematics
GR-157 84, Athens
Greece

Cover design by Mary Burgess.

Mathematics Subject Classification (2000): 53C05, 53C07, 53C80, 18F20, 53D50, 53Z05, 55N30, 19M05, 58A40, 58D27, 58D30, 58K99, 58Z05, 58E15, 81Q70, 81P99, 81T13, 83C45, 83C47, 16D10, 16D40, 16E99, 58C99, 55R05

Library of Congress Control Number: 2005043605

ISBN-10 0-8176-4378-8   e-ISBN 0-8176-4474-1
ISBN-13 978-0-8176-4378-2

Modern Differential Geometry in Gauge Theories:
Yang–Mills Fields, Volume II   ISBN 0-8176-4379-6
Modern Differential Geometry in Gauge Theories:
Volumes I + II (Set)   ISBN 0-8176-4476-8

Printed on acid-free paper.

©2006 Birkhäuser Boston   ***Birkhäuser***

Printed in the United States of America.   (TXQ/SB)

9 8 7 6 5 4 3 2 1

www.birkhauser.com

Contents

General Preface

"What can be said at all can be said clearly."

L. Wittgenstein in *Tractatus Logico-Philosophicus* (Routledge, 1997), p. 3.

"Σοφόν τοί τό σαφές οὐ τό μή σαφές."

Εὐριπ., Ὀρέστης

It is nowadays generally accepted that the theory of *principal fiber bundles* is the appropriate mathematical framework for describing one of the most beautiful, as well as important, physical theories, viz. the so-called *gauge field theory*, or gauge theories, being, in effect, to quote, M. F. Atiyah, *"physical theories of a geometrical character."*

Now, in this context, a *principal fibration* is defined by the (local) *gauge group* (or *internal symmetry group*) of the physical system (*particle field*) under consideration. Yet, the particular physical system at issue is carried by, or lives on, a *"space"* (vacuum) that in the classical case is usually a *smooth* (viz. C^∞-) *manifold*. Within our abstract framework, instead, this is, in general, *an arbitrary topological space*, being also the *base space* of all the *fiber spaces* involved.

Accordingly, *we do not use any notion of calculus* (smoothness) in the classical sense, though we can apply, most of the powerful machinery of the standard *differential geometry*, in particular, the *theory of connections, characteristic classes*, and the like. However, all this is done *abstractly*, which constitutes an *axiomatic treatment of differential geometry* in terms of *sheaf theory* and *sheaf cohomology* (see A. Mallios [VS: Vols I, II]), while, as already noted, *no calculus is used at all!* So the present study can be construed as a further application of that abstract (i.e., axiomatic) point of view in the realm of gauge theories, given, as mentioned before, the intimate connection of the latter theories with (differential) geometry.

Thus, working within the aforementioned abstract set-up, we essentially replace all the previous fiber spaces (viz. principal and/or vector bundles) by the corresponding *sheaves of sections*, the former being, of course, just our model (motivation), while our study is otherwise, as has already explained above, *quite abstract*(!), that is, axiomatic. Of course, *in the classical case the two perspectives are* certainly mathematically speaking (categorically!) *equivalent;* however, the *sheaf-theoretic language*, to which we also stick throughout the present treatment, is even in the standard case, in common usage in the recent physics literature (cf., for instance, Yu. I. Manin [1] or even S. A. Selesnick [1]). Thus, it proves that the same language is at least *physically more transparent*, while, finally, being *more practical*. In addition, *wave functions* are considered as *sections* (i.e., functions whose *domain is varied* as well as their *range*, along *with the point of application*) of appropriate bundles (loc. cit. Vol. II: Chapt. IV; Section 10). Furthermore, it is still very likely that the kind of common base space of the sheaves involved herewith can also be thought of as corresponding to recent aspects of the *"vacuum,"* for instance, " ... *the structure of such spaces is governed by topology, rather than geometry"* (cf. P. J. Braam [1: p. 279]).

On the other hand, a significant advantage of the present abstract formulation of the classical gauge field theory (i.e., the smooth case) lies in the possibility of employing the standard conceptual machinery of the usual (smooth) differential geometry, even for base *spaces* (of the fiber spaces, as above) *that (i) are not smooth enough*, (ii) *include a large amount of singularities* in the classical sense, and (iii) *are not smooth at all* (!), but provide the appropriate framework for the exploitation of the axiomatic theory [VS], as this happens in certain important cases (see concrete examples throughout the sequel). Of course, this potential generalization of the classical theory might very likely be of a particular significance to (mathematical) physicists who long ago were already aware of, as well as tantalized by, the aforesaid type of spaces. Furthermore, the same abstract approach, has certainly theoretical/pedagogical advantages, being, namely, greater perspective, clarity and unification. It is thus more akin to the nowadays generally accepted aspect that *"the basic ideas of modern physics are quite simple"* (see, for instance, H. Fritzcsh [1: p. 211]), or even that " ... *the problems of quantum gravity are much more than purely technical ones; they touch upon very essential philosophical issues"* (cf. G. 't Hooft [1: p. 2]). So it is quite natural to try to manufacture a similar situation pertaining to the mathematics involved; thus, something like this would also be in concord with the apostrophes, as stated in the epigraph of this preface.

Further details about each of the two individual volumes are given by separate prefaces.

Preface to Volume I

The technical aspects of the present volume are as follows: In Chapter I, we are concerned with exactly those basic notions and results of abstract differential geometry that will be of use throughout the rest of this treatise, including both volumes of it. This has been done for the convenience of the reader, who will find here explanations and formal statements of the relevant material used throughout while we refer to [VS] for further details or even complete proofs of the stated results. Yet it may also happen that occasionally we refer to expanded or even new material, in comparison with our previous account on the subject, as given in [VS].

Chapter II deals with the classification of elementary particles according to their spin structure, as is classically the case, however in terms now of sheaf theory; that is, by means of the notion of *vector sheaves* (see Chapter I). In other words, one can identify the states of a (free) elementary particle with sections of an appropriate vector sheaf, where the latter is determined by the spin of the elementary particle at issue. Here, the relevant argument is based, in fact, on a previous work of S. A. Selesnick [1], pertaining to the same classification in terms of vector bundles, which justifies our choice for the title of the chapter (see Chapter II; (6.29)). Among the technical advantages of this sort of classification is, for instance, the resulting cohomological classification of elementary particles on the basis of the standard similar situation one has for vector sheaves (cf. [VS: Section V.2]). The same classification as before helps also to classify in a similar way the so-called *Yang–Mills fields*, of which particular case are the *Maxwell fields* (photons); see Chapter III in the sequel, along with Section I.9, in Volume II of this treatise. An application of the previous type of classification in the case of Maxwell fields is also our considerations in Chapter V, Section 4, concerning geometric prequantization, along with its cohomological classifications.

Chapter III refers to electromagnetism, viz. the electromagnetic field (*photon*), from the point of view of gauge theory, that is, to paraphrase M. F. Atiyah again, of a physical theory within a differential-geometric framework. Of course, the differential geometry that is applied is the abstract one, as has been advocated by [VS], this being also our general perspective throughout the present study. Thus, among other things, one finds here the beginning of a cohomological classification of Maxwell fields (which otherwise is fully presented in the Chapter IV), as well as the formulation within our abstract set-up of the classical action of (the abelian group) $\check{H}^1(X, \mathbb{C}^{\cdot})$, see Section III.4 for the notation on the Maxwell fields (collected into appropriate equivalence classes, thus yielding what we may call the *Maxwell group*) of the space under consideration. As a result, one obtains, for instance, that

(∗)   *two light rays of the same color can differ only, by a "phase factor," viz., by an element of* $\check{H}^1(X, \mathbb{C}^{\cdot})$ *(ibid. (4.55′)).*

In point of fact, by further considering *Hermitian line sheaves*, one can reduce the previous conclusion to the case of the (abelian) group $\check{H}^1(X, S^1)$, which thus corresponds better to the usual physical meaning of the term "phase factor" as applied above (loc. cit., Section 6. In particular, see (6.61.1)). Applications of the preceding are also found Chapters IV and V, pertaining to the (cohomological) classification of the geometric prequantizations of (Hermitian) Maxwell fields.

Thus, continuing in Chapter IV, we systematically consider the aforementioned cohomological classification of Maxwell fields, even of the Hermitian ones; classically, the last adjective is actually referred to the standard *circle group* acting on the corresponding Maxwell fields of the space considered, though in a sheaf-theoretical disguise, according to our general pattern. On the other hand, the cohomology theory employed here is (*Čech*) *hypercohomology*; this extension of the usual sheaf cohomology theory, which is otherwise applied throughout the present treatise, is due in fact to the number of variables involved in dealing with a given Maxwell field. Namely, on the one hand, one has to consider here the *support* (carrier) of the field, while on the other, the field itself, and consequently the use of an appropriate *2-term A-complex* (Section 4). Furthermore, an abstract form of Maxwell's equations (in vacuo) characterizing within the present abstract setting the Maxwell fields is also supplied as a direct consequence of the preceding discussion. Indeed, as an outcome of the language employed, the equations at issue are actually reduced to just one (Sections 6, 7).

Finally, in Chapter V we are concerned, as already said, with the classical theme of geometric prequantization, always in the context of our abstract differential-geometric regime. In this regard, the fundamental result of Weil's integrality theorem has been already discussed in [VS: Chapter VIII; p. 238, Section 11] (See also A. Mallios [7] for an early account of it; yet cf. Chapter III of the present volume, Theorem 3.1, along with the subsequent comments). We further examine, within the aforementioned abstract set-up, the standard classification of prequantizations, in point of fact, those that by definition are referred to a Maxwell (electromagnetic) field. Indeed, this has been already done, within the appropriate context, inChapters III and IV (see, for instance, Sections 5 and 5 respectively), so that our main concern is to further establish the necessary background terminology according to the symplectic sheaves. So, in the end, one gets at the prequantization of elementary particles, in general, of which a particular yet important case is the *graviton* (carrier of the gravitational force). However, this special instance, along with relevant material, is given in Part II of this treatise; cf. Section IV.9 of the second volume of this work. Yet in this connection, as a consequence of our considerations in this chapter and in conjunction with those in Section IV.9, one can further assert that

(∗∗)   *graviton is (pre)quantizable* (!)

as well. For details we refer to Volume II of the present study.

Acknowledgments

The following lines represent only a small part of my indebtedness to all those people who in several ways contributed by their contact or personal communication the present material as well as all of the present consideration of ADG (*abstract* ($\equiv$ modern) *differential geometry*), together with its potential physical applications providing thus an indispensable and corroborative factor of the whole project at issue: Thus it was Elemér Rosinger who some years ago, during one of my visits to the University of Pretoria in South Africa, heard about my intention to present general relativity, the mathematical part of course, e.g., Einstein's equation (in vacuo) in terms of ADG, and in particular using his (sheaf of) algebras of generalized functions; the reaction then was more than enthusiastic, so that project was finally realized in A. Mallios [8]. Somewhat earlier, I had already started to think of the possibility of presenting Yang–Mills theory in terms of ADG, motivated here by the relevant remark (M. F. Atiyah) that the same, being a gauge theory, is in effect of a geometrical character (hence, ADG), yet supported by the common aspect that *"basic ideas of modern physics are quite simple"* (H. Fritzsch) [ADG is, in principle, a *"naive"* theory, viz. *axiomatic* (S. MacLane)]. So the first relevant ideas were already presented in A. Mallios [6], in full details in the same Vol. II of this work, Chapters I–II, thanks, concerning the latter reference, to kind and lively interest in my whole work of K. Iséki and T. Ishihara. So Elemér Rosinger, in that context, post-anticipated me, in point of fact, while supporting me too, at the same time, as concerns the idea of Yang–Mills theory, when he asked for an analogous abstract formulation of the Yang–Mills equations, yet this in his characteristic, for the whole enterprise enthusiastic, stimulating, and always lively manner.

On the other hand, the continued moral and quite definitive support of Steve Selesnick was certainly alive always and perceptible. What I call in this exposition *Selesnick's correspondence* (Vol. I; Chapter II) was the guiding principle, throughout the text, pertaining to its connection with physics, in spite of his usual reservations, referring to the usefulness of that otherwise extremely nice, very convenient and workable (!) idea; later I met an analogous point of view, related with the electromagnetic field, in Yu. Manin's Springer book on *Gauge Field Theory* while quite recently, by that same author, concerning now any other field, in his article in [3] (I owe this last quotation to Yannis Raptis). It was actually also Steve Selesnick who was responsible for a delightful collaboration in the last few years with Raptis, something that has led to an especially fruitful and substantial result, referring in particular to potential physical consequences of ADG for *quantum relativity* and the problem of the so-called *singularities* in general.

The beautiful and very informative recent work of Stathis Vassiliou on the *Geometry of Principal Sheaves*, to appear in the MIA series of Kluwer, came at the right time to vindicate and further extend the scope as well as the applicability of ADG. The ongoing work of Maria Papatriantafillou comes to cover the quite natural *formally categorical* treatise of ADG, both of the aforesaid recent two aspects of ADG being altogether definitive and necessary complements of the whole, thus far, enterprise on the matter. Within an analogous vein of ideas the latest, incomplete, treatise of Elias Zafiris comes already to test the ADG point of view in a *topos-theoretic* environment for the subject, yet with possible applications to *quantum gravity* as well.

During the time of several visits in the last few years to Rabat (Fès, included), Morocco, I had the opportunity to talk about ADG and its potential physical consequences mainly with Mohamed Oudadess and, in effect, with the whole "équipe d'analyse fonctionnelle" that thrives there, in particular, as it concerns topological algebras theory, having thus always an eager and also critical audience, being test, of my own perceptions on the subject. Indeed, a very pleasant atmosphere, still inspiring too, Mohamed Oudadess, at least, being steadily a prompt and critical listener (!), providing me thus with a precious experience of having first reactions of a thoughtful "amateur" (the last denomination is, of course, his own) to the matter, that often led me to greater elaborations of the ideas discussed, to increase understanding.

I have had in similar supporting and inspiring reactions the past from contact with Nelu Colojoără, the late Gerd Lassner, Konrad Schmüdgen, Susanne Dierolf, the late Klaus Floret, Franek Szafraniec, Jan de Graaf, Fredy van Oystaeyen, Roman Zapatrin, and least, with Chris Isham for his incisive corroborative critique, especially concerning our relevant joint work on the subject with Yannis Raptis. The reaction of my Russian editors Vassia Lyubetsky and Sasha Zarelua was supportive, vindicative, and much enlightening, as well.

My special thanks here are due too, for partial financial support during the last few years, to the office of the Special Research Program conducted by the University of Athens and, in particular, to the Vice-Rector Prof. Michael Dermitzakis for his lively and very kind support to my own research work.

The realization and appearance of the material contained in the present two volumes would have not been accomplished, was there the skilful and, really wonderful typing (LATEX) talent of our secretary in the Section of Algebra and Geometry of our Department, Ms. Popi Bolioti. It is a particular pleasure to record here too my wholehearted thanks to her for the excellent job that she has done.

The present two-volume work owes its appearance to the enthusiasm, eager interest, and prompt reaction of Prof. George A. Anastassiou (Univ. of Memphis, USA), *as well as* to the editorial help and extremely kind attention of the executive editor of Birkhäuser, Boston, Ms. Ann Kostant, and her assistant editor Ms. Vaishali Damle.

It is a particular pleasure to express at this place my heartfelt thanks and deep appreciation as well to all the above people for their kindness and the warm attitude they showed toward my work. The same goes also to Ms. Elisabeth Loew, as well as to the rest of the editorial staff at Birkhäuser production.

[It is really amazing that the whole story began simply from one source: the *Math. Z.* article of Stephen Allan Selesnick; see also the Acknowledgments of the first two Volumes on ADG. Then, the enterprise has been continued by pointing out the quite instrumental role the notion of *connection* has had in the whole development of CDG, along with its physical applications.]

Contents of Volume II

Part I

**Maxwell Fields:
General Theory**

1

The Rudiments of Abstract Differential Geometry

"Mathematics aims to understand, to manipulate, to develop, and to apply those aspects of the universe which are formal."

S. Mac Lane in *Mathematics: Form and Function* (Springer-Verlag, 1986). p. 456.

" not ... constructing a building so much as ... having a perspicuous view of the foundations of typical buildings."

L. Wittgenstein in *Culture and Value* (B. Blackwell, 1980). p. 25.

"Geometry [is] a means of turning visual images into formal tools."

S. Mac Lane (loc. cit). p. 257.

The purpose of the first chapter of this treatise is to present the basic ideas and results of *abstract differential geometry*. These results have already been explored in our previous work in [VS], which will be of use throughout the subsequent chapters of the present account. This can also be considered, of course, as a direct consequence of that study, applied, in particular, to the classical set-up of *gauge theories* (the latter being, in effect, *"physical theories of a geometrical character"*; M. F. Atiyah). Strictly speaking, *"geometrical"* here means in terms of a *differential-geometric* point of view. Therefore, for convenience, we recall the necessary issues from [VS] that we are going to employ in the sequel, while for the full details and proofs we refer the reader to this work. Yet, occasionally, the present account contains ameliorated relevant material, or even new material as well, concerning our previous exposition on the subject in [VS]. Thus, we start with explaining the overall fundamental idea of a *differential triad*, the same being, in point of fact, the basis of all ensuing discussion.

1 The Differential Setting

The idea stated in the title of this section is virtually the *fundamental conception* that permeates the whole discipline of differential geometry in any form, that is, the concept of the *basic differential*, which thus becomes the *starting point* for any fundamental notion in differential geometry, as this terminology indicates.

In this connection, one further remarks that the same notion, viz. *the (basic) differential*, refers virtually to *functions, not* to *the space* on which these functions might "live" (i.e., are defined). In point of fact, this is a crucial issue and, indeed, is of particular conceptual importance since this is also a moral of the *abstract differential geometry*. One realizes that

(1.1)   *the functions* (in fact, *sections* of appropriate sheaves) *are of relevance*, and *not the "space" on which these functions "live"* (viz. they are defined).

In turn, this is still of special significance, exactly in the case where the spaces involved are virtually not as "*smooth*" as one might demand (indeed, "*the world around us is far from being smooth enough*"), while, as we shall see, one succeeds in having the desired "*differentials*" between the functions (sections) concerned, which really provide the appropriate machinery to work with, by analogy with the classical (smooth) case.

As a result, *one axiomatizes the situation that appears in the classical case* pertaining to the "*differential machinery*" (viz. the so-called *differential-geometric methods*, determining, at the same time that the latter *powerful technique* may occur in many other cases where the underlying "*space*" (i.e., domain of definition, *source*, of the functions involved) is *very far from being smooth*. Indeed (E.E.Rosinger), one can have *the biggest amount of nonsmoothness* ("*singularities*"), provided the rest is still dense in the space. (See, in particular Volume II of this treatise, Chapter IV, Section 5. It is exactly at this point that one may have a potential application in problems of *quantum relativity*; see loc. cit. for details.)

Finally, from a philosophical point of view, our approach to differential geometry follows the *Leibnizian standpoint* (differentials), *not* the *Newtonian* one (viz. the classical "*geometrical (Cartesian) aspect of the derivative*").

So, to start with, suppose we are given an arbitrary *topological space X* (this will be, throughout the sequel, the *common base space* of the sheaves considered), and let

(1.2) $\mathcal{A}$

be a *sheaf of algebras* on X. By assumption, the algebras involved in $\mathcal{A}$ (thus, local section algebras; hence, fibers too) are

(1.3)   *unital commutative* (linear associative) *algebras over* (the complexes) $\mathbb{C}$.

Thus, henceforward, the pair

(1.4) $(X, \mathcal{A})$,

with X and $\mathcal{A}$ as before, is said to define a $\mathbb{C}$-*algebraized space*.

Of course, the $\mathbb{C}$-*algebra sheaf* $\mathcal{A}$, as above, shares in our case the fundamental rôle, played in the classical theories (real/complex analysis, differential/algebraic geometry), by the so-called ring (algebra) of coefficients, called here *sheaf* (or else *domain*) of *coefficients*. Yet, for obvious reasons that will be made more clear through the ensuing discussion, the same sheaf $\mathcal{A}$, as above, is still termed "*domain of gener-alized coordinates*," or even *structure sheaf*, or just our "*arithmetic*." In this connection, we further remark that, in view of our assumption in (1.3), one obtains

$$(1.5) \qquad\qquad \mathbb{C} \underset{\varepsilon}{\longrightarrow} \subseteq \mathcal{A},$$

the first member of (1.5) denoting the *constant sheaf* (of the complexes) $\mathbb{C}$. That is, one sets, by definition,

$$(1.6) \qquad\qquad \varepsilon(\lambda) := \lambda \cdot \mathbf{1}_\mathcal{A},$$

for any complex number (*constant section*) with $\mathbf{1}_\mathcal{A}$, the *identity section of* $\mathcal{A}$. On the other hand, consider an $\mathcal{A}$-*module* $\mathcal{E}$ on X, and then a *sheaf morphism*

$$(1.7) \qquad\qquad \partial : \mathcal{A} \longrightarrow \mathcal{E}$$

which, in particular, enjoys the two following properties:

$$(1.8) \qquad (i) \quad \partial \text{ is } \mathbb{C}\text{-}linear.$$

Recall that, in view of (1.5), both $\mathcal{A}$ and $\mathcal{E}$ are $\mathbb{C}$- *vector space sheaves* on X, which thus explains (i). Yet, we assume that

$$(1.9) \qquad (ii) \quad \partial \text{ satisfies the } Leibniz \ condition,$$

that is, one has, by definition, the relation;

$$(1.10) \qquad\qquad \partial(s \cdot t) = s \cdot \partial(t) + t \cdot \partial(s),$$

for any (continuous) local sections s, t in $\mathcal{A}(U)$, with U *open* in X.

We recapitulate the two previous conditions by just saying that the sheaf morphism ∂, as in (1.7), which also satisfies (1.9) and (1.10), is our *basic differential*. The same mapping is still defined as a $\mathbb{C}$-*derivation* of $\mathcal{A}$ in $\mathcal{E}$.

Note 1.1 As we shall see, presently below, the above map (*sheaf morphism*) ∂, is *the first* of a sequence of similar "*differentials*" *between* appropriate $\mathcal{A}$-modules, still sharing the proper *Leibniz conditions* (see, for instance, (7.4) in the sequel). Yet, for analogous reasons, *from this point on*, we employ the notation;

$$(1.11) \qquad\qquad \mathcal{E} \equiv \Omega,$$

while later on we shall also set,

$$(1.12) \qquad\qquad \Omega \equiv \Omega^1$$

(cf., for instance, (7.3) in the sequel).

Thus, we call a triple

(1.13) $$(\mathcal{A}, \partial, \Omega),$$

as above, a *differential triad* on X.

As already said, the previous notion is, in effect, the starting-point of all our subsequent discussion in this chapter, that is, in other words, of the whole *abstract differential-geometric* standpoint that is advocated here, or even in [VS]. Indeed, the *sheaf morphism* ∂, as given by (1.7), is *the familiar*

(1.14) $$\text{``}dx\text{''}$$

of the classical theory (smooth manifolds). Thus, the usual *local* (smooth) *coordinate functions* of the standard theory can actually be construed as *local* (continuous) *sections of*

(1.15) $$\mathcal{A} =^{\mathbb{C}} \mathcal{C}_X^{\infty},$$

viz. of the usual *sheaf of germs of*, say, $\mathbb{C}$-valued, *smooth* ($\mathcal{C}^{\infty}$-)*functions* on the smooth manifold X involved; see also [VS: Chapt. VI; p. 9, Section 2, as well as, Chapt. X; pp. 277ff]).

Now, before we proceed further, we still remark that, as an immediate consequence of the very definition of ∂, one gets the familiar property from the classical theory that, namely,

(1.16) $$\partial_{|_{\mathbb{C}}} = 0$$

(see also (1.5)). Yet, cf. [VS: Chapt. VI; p. 3, Lemma 1.1]).

On the other hand, for later use, we also consider the *extension of* (1.7) *to higher dimensions:* That is, one defines, the following *sheaf morphism* keeping, for convenience, the same notation for ∂ (cf. also (1.11)),

(1.17) $$\bar{\partial} \equiv \partial : M_n(\mathcal{A}) \longrightarrow M_n(\Omega) := M_n(\mathcal{A}) \otimes_{\mathcal{A}} \Omega,$$

given, *coordinatewise*, by the relation

(1.18) $$\bar{\partial}(\alpha) \equiv \partial((\alpha_{ij})) := (\partial(\alpha_{ij})),$$

for any

(1.19) $$\alpha \equiv (\alpha_{ij}) \in M_n(\mathcal{A})(U) = M_n(\mathcal{A}(U)),$$

with U *open* in X, such that one has

(1.20) $$\alpha_{ij} \in \mathcal{A}(U), \qquad 1 \leq i, j \leq n.$$

Yet, we further employ occasionally for (1.17) the notation

(1.21) $$\bar{\partial} \equiv M_n(\partial),$$

by also calling it the *nth square matrix extension of* ∂ (see [VS: Chapt. VII; pp. 4ff]). So one obtains the *n-th square matrix extension of* $(\mathcal{A}, \partial, \Omega)$, as now the *differential triad* on X,

$$(1.22) \qquad (M_n(\mathcal{A}), M_n(\partial), M_n(\Omega)).$$

In this regard, we still remark here that the *"algebra of coefficients"* in (1.22) is *no more* a *commutative* $\mathbb{C}$-algebra sheaf, as in (1.13), *unless* $n = 1$ (see also (1.3)).

1.1 Logarithmic Derivation

By considering the *group sheaf of units of* $\mathcal{A}$ (thus, by definition, a sheaf of (abelian) groups on X),

$$(1.23) \qquad \mathcal{A}^{\bullet},$$

given, by the relation,

$$(1.24) \qquad \mathcal{A}^{\bullet}(U) := \mathcal{A}(U)^{\bullet},$$

for any *open* $U \subseteq X$ (the second member of (1.24) denoting the *group of units* (invertible elements) of the $\mathbb{C}$-algebra $(\mathcal{A}(U))$, one defines the *logarithmic derivation* associated with ∂, as above, as the *sheaf morphism*

$$(1.25) \qquad \tilde{\partial} : \mathcal{A}^{\bullet} \longrightarrow \Omega,$$

such that

$$(1.26) \qquad \tilde{\partial}(\alpha) := \alpha^{-1} \cdot \partial(\alpha),$$

for any $\alpha \in \mathcal{A}^{\bullet}(U)$, as in (1.24). In this connection, one actually proves that

$(1.27) \qquad \tilde{\partial}$ is a *morphism of* the (abelian) *group sheaves* as appeared in (1.25).

That is, *one has*

$$(1.28) \qquad \tilde{\partial}(s \cdot t) = \tilde{\partial}(s) + \tilde{\partial}(t)$$

for any s,t *in* $\mathcal{A}^{\bullet}(U) = \mathcal{A}(U)^{\bullet}$; in particular, *one* thus *obtains that*

$$(1.29) \qquad \tilde{\partial}(\alpha^{-1}) = -\tilde{\partial}(\alpha),$$

for any $\alpha \in \mathcal{A}^{\bullet}(U)$. Another easy consequence of (1.28), in conjunction with (1.16), is also the relation,

$$(1.30) \qquad \tilde{\partial}(\lambda \cdot \alpha) = \tilde{\partial}(\alpha),$$

for any $\lambda \in \mathbb{C}^{\bullet}$ ($\equiv \mathbb{C}\backslash\{0\}$), *and* $\alpha \in \mathcal{A}(U)^{\bullet}$, with U *open in* X.

On the other hand, by still referring to (1.22), one also defines;

$$(1.31) \qquad \tilde{\partial} : \mathcal{GL}(n, \mathcal{A}) := M_n(\mathcal{A}^{\bullet}) \longrightarrow M_n(\Omega),$$

such that *one sets* (see also (1.18)),

$$(1.32) \qquad \tilde{\partial}(\alpha) := \alpha^{-1} \cdot \partial(\alpha),$$

for any

$$(1.33) \qquad \alpha \equiv (\alpha_{ij}) \in \mathcal{GL}(n, \mathcal{A})(U) = GL(n, \mathcal{A}(U))$$

(cf. also (1.34) below), while, by virtue of (1.16) one still concludes that

$$(1.34) \qquad \tilde{\partial}|_{\mathcal{GL}(n,\mathbb{C})} = 0.$$

Yet, concerning the notation, that was applied in (1.31), one has

$$(1.35) \qquad \begin{aligned} \mathcal{GL}(n, \mathcal{A})(U) &:= GL(n, \mathcal{A}(U)) \equiv M_n(\mathcal{A}(U))^{\bullet} \\ &= M_n(\mathcal{A})(U)^{\bullet} = M_n(\mathcal{A}^{\bullet}(U)), \end{aligned}$$

for any *open* $U \subseteq X$ (see also [VS: Chapt.IV; p. 285, Section 1.2]). On the other hand, as a consequence of (1.32), *one obtains*

$$(1.36) \qquad \tilde{\partial}(s \cdot t) = Ad(t^{-1}) \cdot \tilde{\partial}(s) + \tilde{\partial}(t),$$

for any $s \equiv (s_{ij})$ *and* $t \equiv (t_{ij})$ *in* $\mathcal{GL}(n, \mathcal{A})(U) = GL(n, \mathcal{A}(U))$, *with* U *open in* X, as above, where we also set

$$(1.37) \qquad Ad(s)(\tilde{\partial}(t)) \equiv Ad(s) \cdot \tilde{\partial}(t) := s \cdot \tilde{\partial}(t) \cdot s^{-1},$$

for any s, t, as before. In particular, in view of (1.36), *one has*

$$(1.38) \qquad \tilde{\partial}(\alpha^{-1}) = -Ad(\alpha) \cdot \tilde{\partial}(\alpha),$$

for any $\alpha \equiv (\alpha_{ij}) \in \mathcal{GL}(n, \mathcal{A})(U)$.

Now, as we shall see presently below, *the* same *map* ∂ as in (1.7) is, in effect, our first basic (*concrete*) *example of what we* are going to *call*, quite generally, in the sequel, an *A-connection*, which here, in particular, has a *zero curvature*, as well (cf. (7.5) and (7.17) below), yielding a "*flat A-connection*" (ibid.; see also (2.6) in the sequel).

2 $\mathcal{A}$-Connections

The notion we are going to discuss here, in brief—its full account having been given already in [VS: Chapts. VI, VII]—is certainly *the most fundamental* idea *in* contemporary *differential geometry*. Still from the time of the very inception of the latter discipline, in the form, of course, which in that time had the said notion in its classi-cal, namely, counterpart. It is thus another more sophisticated (generalized) form of the concept of the usual *differential*, or even *derivative*, being in accord with the idea of taking also into account eventual *variations of the state of the objects*, that we are going to study, through it (cf., for instance, *general relativity* and, in general, *physics* of today. Yet, the same is still of an *analytic/algebraic* nature, *not* of a *geometric* one (cf., for example, the familiar phrase *"geometric meaning of derivative"* (!) viz., as already noted for the *Newtonian point of view*. Furthermore, the fact that this concept is treated in a *"varying" algebraic way* as, for instance, in a *sheaf-theoretic* way (see also Definition 2.1 below), is very much in accord with the aforementioned *physical* standpoint of *"variation"*, which further allows, at the very end, its potential appli-cation on (involvement with) extremely peculiar (*"singular"*) functions, yet, when the latter are defined on rather arbitrary (topological)—however, *not necessarily(!) smooth*—spaces.

So we start with giving the formal definition in our abstract framework. That is, we have the following:

Definition 2.1 Suppose we are given a differential triad on a topological space X (cf. (1.13)), and let $\mathcal{E}$ be an $\mathcal{A}$-module on X. An $\mathcal{A}$-*connection* on $\mathcal{E}$ is a sheaf morphism

$$(2.1) \qquad D : \mathcal{E} \longrightarrow \mathcal{E} \otimes_A \Omega \cong \Omega \otimes_A \mathcal{E} \equiv \Omega(\mathcal{E}),$$

satisfying the following two conditions:

(i) D is $\mathbb{C}$-*linear*, viz. one has

$$(2.2) \qquad D(\lambda s + \mu t) = \lambda \cdot D(s) + \mu \cdot D(t)$$

for any λ, μ in $\mathbb{C}$ and s, t in $\mathcal{E}(U)$, with U open in X.

(ii) D is a *Leibniz map*, that is, it satisfies the relation;

$$(2.3) \qquad D(\alpha \cdot s) = \alpha \cdot D(s) + s \otimes \partial(\alpha),$$

for any s, t, as in (2.2), and $\alpha \in \mathcal{A}(U)$.

We also call (2.3) the *"Leibniz condition"* for D. The preceding definition is, of course, very general and may not have a sense, in general, even in the classical case (e.g., complex analytic vector bundles on complex (analytic) manifolds; M. F. Atiyah [1]). Now, at the other end, and by further referring to the classical case (C^∞-manifolds), we know, as we also explain it in the sequel (see Subsection 2.1 below), that $\mathcal{A}$-connections always *exist in the smooth* (viz. C^∞-) *case*. However, as an outcome of the present *abstract standpoint*, we still realize that

(2.4) $\mathcal{A}$-connections *do exist, even in extremely nonsmooth cases,* being also *of importance for potential applications,* physical or not.

The previous claim, as in (2.4), will be fully clarified, yet, vindicated too, through the subsequent discussion.

Now, as a first *concrete example of an A-connection*, we can consider our *basic differential ∂*, as in (1.7) (cf. also (1.11) for the notation applied), so that one has

$$(2.5) \qquad \partial : \mathcal{A} \longrightarrow \mathcal{A} \otimes_{\mathcal{A}} \Omega = \Omega \equiv \Omega(\mathcal{A})$$

(see (2.1)), the rel. (1.10) being here the desired *Leibniz condition*, as in (2.3). Yet, we still note, in anticipation, that the curvature of ∂ is zero, viz. one has the relation

$$(2.6) \qquad R(\partial) = 0$$

(see (7.5) below). So we also call ∂ the *standard (flat) A-connection* on $\mathcal{A}$; yet, its existence is thus assured from our hypothesis, concerning the differential triad (1.13).

2.1 The Classical Case

By considering the classical example, where

$$(2.7) \qquad \mathcal{A} \equiv^{\mathbb{C}} \mathcal{C}_X^\infty,$$

(see (1.15)), one further sets

$$(2.8) \qquad \Omega_X^1 := \mathcal{S}(\Gamma(^{\mathbb{C}}T^*(X))),$$

that is, the *sheaf of germs of ($\mathbb{C}$-valued) 1-forms* on X. Here $\Gamma(\cdot)$ stands for the *(complete) presheaf of sections of*

$$(2.9) \qquad {}^{\mathbb{C}}T^*(X) = ({}^{\mathbb{C}}T(X))^*,$$

viz. the *complexified cotangent bundle* of X. Therefore,

$$(2.10) \qquad \Omega_X^1, \text{ as given by (2.8), } \textit{is thus, by definition, (isomorphic to) the sheaf (of germs) of sections of the (smooth) } \mathbb{C}\textit{-vector bundle (2.9).}$$

Finally, as a $\mathbb{C}$-*derivation* ∂, between (2.7) and (2.8), one further defines, by analogy with (1.7), the *standard differential*

$$(2.11) \qquad d : \mathcal{C}_X^\infty \longrightarrow \Omega_X^1,$$

such that with each $\mathbb{C}$-valued *(local)* $\mathcal{C}^\infty$-*function* f of X one associates its *differential df*, being thus, by definition, a *(local $\mathcal{C}^\infty$-)1-form* of X. So one gets the triple

$$(2.12) \qquad (\mathcal{C}_X^\infty, d, \Omega_X^1),$$

which thus defines the *standard differential ("smooth") triad*, that can be associated with any given $\mathcal{C}^\infty$-*manifold* X.

Thus, it is the above smooth differential triad on a (smooth) manifold X that actually contributes to the classical notion of a *linear* (or else, *Koszul*) *connection* on a given *smooth vector bundle* on X. On the other hand, within the present abstract point of view, the same connection (alias *"covariant exterior derivation"*) operates, in fact, on the *sections* of the bundle at issue, which is thus the case even in the classical situation, as above (cf. connections applied on *"vector fields,"* viz., in effect, sections of the *tangent bundle*).

More precisely, a $\mathcal{C}_X^\infty$-*connection* (cf. also (2.7), for the notation applied), or just a *linear* (alias *Koszul*) *connection* on the complexified *tangent bundle* of X

$$(2.13) \qquad\qquad {}^{\mathbb{C}}\mathcal{T}(X) \equiv \mathcal{E}$$

is, according to the preceding, a *sheaf morphism*

$$(2.14) \qquad\qquad D : \mathcal{E} \longrightarrow \Omega^1(\mathcal{E}) \equiv \mathcal{E} \otimes_{\mathcal{A}} \Omega^1,$$

where Ω^1 is given by (2.8), while, by analogy with (2.8), we also set (by an obvious *abuse of notation*, in connection with (2.13), as above

$$(2.15) \qquad\qquad \mathcal{E} := \mathcal{S}(\Gamma({}^{\mathbb{C}}\mathcal{T}(X)))$$

(see also (2.10)). Therefore, one here obtains;

$$(2.16) \qquad\qquad \mathcal{E} \otimes_{\mathcal{A}} \Omega^1 = \mathcal{E} \otimes_{\mathcal{A}} \mathcal{E}^* = \mathcal{H}om_{\mathcal{A}}(\mathcal{E}, \mathcal{E}) \equiv \mathcal{E}nd\mathcal{E}$$

so that, by (2.14), one gets at a $\mathbb{C}$-*linear morphism*

$$(2.17) \qquad\qquad D \in Hom_{\mathbb{C}}(\mathcal{E}, \mathcal{E}nd\mathcal{E}),$$

according to our hypothesis for (2.14), which further satisfies the *Leibniz condition*. That is, one obtains

$$(2.18) \qquad\qquad D(s) \in (\mathcal{E}nd\mathcal{E})(U) \equiv Hom_{\mathcal{A}|_U}(\mathcal{E}|_U, \mathcal{E}|_U),$$

for any $s \in \mathcal{E}(U)$, in such a manner that, *for any* $t \in \mathcal{E}(U)$, one has;

$$(2.19) \qquad\qquad D(s)(t) \in \mathcal{E}(U),$$

for every *open* $U \in X$. Thus, one gets at the familiar situation, pertaining to the *"Christoffel functions"* (in fact, *sections*). For details on the notation applied herewith, we refer to [VS: Chapt. VII; Section 5, pp. 123ff].

However, for convenience of the reader, we do give the usual properties of a *"linear connection"* in our case, according to the preceding nomenclature; thus, one has the relations

$$(2.20) \qquad\qquad D(s)(\alpha \cdot t) = \alpha \cdot D(s)(t),$$

by virtue of (2.18), *for any* s, t *in* $\mathcal{E}(U)$ *and* $\alpha \in \mathcal{A}(U)$, with U *open* in X. Moreover, one has

$$(2.21) \qquad\qquad D(\alpha \cdot s)(t) = \alpha \cdot D(s)(t) + \partial(\alpha)(t) \cdot s$$

(*"Leibniz condition,"* in view of our hypothesis for D); see also (2.3) and (2.18), by further taking into account that (cf. (2.9) and (2.13)),

$$(2.22) \qquad \Omega_X^1 := ({}^{\mathbb{C}}\mathcal{T}(X))^* \equiv \mathcal{E}^*,$$

such that we still have

$$(2.23) \qquad \partial(\alpha)(t) \equiv t(\alpha) \in \mathcal{A}(U),$$

which thus corresponds, within our abstract case, to the *familiar formula* of classical differential geometry,

$$(2.24) \qquad (df)(\xi) = \xi(f),$$

for $f \in \mathcal{C}^\infty(X)$ and $\xi \in \mathcal{X}(X)$, a *smooth vector field* on X. In this regard, see also loc.cit., p. 129; (5.41) and p. 130; (p. 45).

Now, this same *formula* (2.24), as above, still *explains* classically *the rôle of the standard differential triad* (2.12) in the definition of a linear connection on a given smooth manifold X, where, of course, our assumption in (2.22), that is, *the very definition of Ω_X^1 is here of fundamental importance.* That is to say, this has to do with the same *definition;* one considers *linear connections in the classical theory,* which thus depends *essentially* on the underlying space. On the other hand the latter space is circumvented by the *abstract theory,* axiomatizing (generalizing) the previous situation through the *"basic differential"* ∂, as in (1.13) above. Yet, the same might have a particular significance for *physical applications* in questions, pertaining, for instance, to *quantum gravity*; see Chapter IV of Part II (Volume II) of this treatise.

Furthermore, it is still well-known that *linear connections always exist* in the classical theory, a fact that will also be further clarified in the sequel, as it concerns its relevance within the *abstract setting,* advocated herewith (see Section 4 below).

Note 2.1 By looking at (2.15), we have considered therein

$$(2.25) \qquad \Gamma({}^{\mathbb{C}}\mathcal{T}(X)),$$

as the *(complete) presheaf of sections* of the *(smooth)* $\mathbb{C}$-*vector bundle*

$$(2.26) \qquad {}^{\mathbb{C}}\mathcal{T}(X) \equiv \mathcal{E},$$

viz. those of the *tangent bundle* of the given (smooth) manifold X. Alias, (2.25) is thus the *(complete) presheaf of (smooth) vector fields* on X, or even, *equivalently,* the *sheaf of germs of sections of* (locally defined smooth) *vector fields* on X (see our notation in (2.24)). Therefore, one has, by the very definitions,

$$(2.27) \qquad \Gamma({}^{\mathbb{C}}\mathcal{T}(X)) \equiv \Gamma(\mathcal{T}(X)) := \{\Gamma(\mathcal{T}(X)|_U)\}_{U \subseteq X, \ open},$$

such that *one* further sets

$$\Gamma(\mathcal{T}(X)|_U) := \Gamma(U, \mathcal{T}(X)) \cong \Gamma(\mathcal{T}(U))$$
$$\equiv \Gamma(U, \mathcal{T}(U)) \equiv \mathcal{X}(U).$$

(2.28)

For convenience, we applied above an obvious *abuse of notation*, which is certainly clear from the context. A similar *abuse of terminology* has been also employed, for simplicity's sake, in (2.9) and (2.15), between (vector) *bundles* and their corresponding (vector) *sheaves of sections*.

2.2 Local Definition of an $\mathcal{A}$-Connection

Given the framework of a *differential triad*, as in (1.13), and an *$\mathcal{A}$-module $\mathcal{E}$* on X (cf. (1.4)), we consider below the way we can restrict a given *$\mathcal{A}$-connection D* of $\mathcal{E}$ on a so-called *"local gauge"* of $\mathcal{E}$, whenever, of course, the latter notion gets a meaning (e.g., if $\mathcal{E}$ is a *vector sheaf* on X; of course, this is not all!), the only case, that a given *$\mathcal{A}$-module* on X has a local gauge, as the latter concept will be determined below. (See (2.29)).

Thus, for convenience, we start with first fixing up the relevant terminology employed herewith: Namely, suppose that we are given an *$\mathcal{A}$-module $\mathcal{E}$* on X, as before. Then, one defines a *local gauge of $\mathcal{E}$*, as an open set $U \subseteq X$, in such a manner that one has the following relation

(2.29)
$$\mathcal{E}|_U = \mathcal{A}^n|_U, \ n \in \mathbb{N},$$

within an *$\mathcal{A}|_U$-isomorphism of* the *$\mathcal{A}|_U$-modules* concerned. That is, in other words, by restricting $\mathcal{E}$ on a local gauge U of it, one gets at a concrete (as well as, complete) *"arithmetization,"* or else *"local scaling,"* of $\mathcal{E}$, by means of $\mathcal{A}$ (our *"arithmetics"*), in the sense that, speaking *sectionwise,* one has that

(2.30)   *any section of $\mathcal{E}$ over an open set $V \subseteq U$ can be expressed* entirely *through* corresponding *sections of $\mathcal{A}$ on V*.

Indeed, by virtue of (2.29), one obtains;

(2.31) $$(\mathcal{E}|_U)(V) = \mathcal{E}(V) = (\mathcal{A}^n|_U)(V) = \mathcal{A}^n(V) = \mathcal{A}(V)^n,$$

within *isomorphisms of* the *$\mathcal{A}(V)$-modules* involved. (See also [VS: Chapt. I; p. 55, (11.40)]).

Note 2.2 As an immediate consequence of the same argument, as in (2.31), one concludes, *more generally,* that, whenever two *sheaves* (of sets) $\mathcal{E}$ and $\mathcal{F}$ on (a *topological space*) X satisfy the relation (*isomorphism* of sheaves)

(2.32)
$$\mathcal{E}|_U = \mathcal{F}|_U,$$

for an *open $U \subseteq X$*, then, *one* still *obtains* (the following *bijections* of sets)

(2.33)
$$\mathcal{E}(V) = \mathcal{F}(V),$$

for any *open $V \subseteq U$*.

Now, based on (2.29) and the hypothesis therein that $n \in \mathbb{N}$, we also refer to the *open set* $U \subseteq X$, at issue, as a *local gauge of $\mathcal{E}$ of finite rank* ($n \in \mathbb{N}$). Furthermore, since $\mathcal{A}^n$ is, by definition, a *free $\mathcal{A}$-module of rank* n, with $n \in \mathbb{N}$, by looking at a *basis of sections of*

$$(2.34) \qquad \mathcal{A}^n(U) = (\mathcal{A}(U))^n \equiv \mathcal{A}(U)^n$$

(one has actually here an *isomorphism of $\mathcal{A}(U)$-modules*), as, for example, at its *canonical (Kronecker) basis*, say,

$$(2.35) \qquad \varepsilon_i := (\delta_{ij}) \in \mathcal{A}^n(U) = \mathcal{A}(U)^n, \ 1 \le i, j \le n$$

(in point of fact, *restriction on U of the* corresponding (homonymous) *global basis of $\mathcal{A}^n$ over X*), one gets at the relation;

$$(2.36) \qquad s = \sum_{i=1}^{n} \alpha_i e_i,$$

for any *local (continuous) section* $s \in \mathcal{E}(U)$, with $a_i \in \mathcal{A}(U)$ such that

$$(2.37) \qquad e_i := \varphi(\varepsilon_i) \in \mathcal{E}(U), \ 1 \le i \le n,$$

where φ denotes the $\mathcal{A}|_U$-*isomorphism in* (2.29). Therefore, we can still say that we actually

$$(2.38) \qquad \text{"replace } \mathcal{E}(U) \text{ by } \mathcal{A}^n(U) = \mathcal{A}(U)^n \text{"}$$

viz., by a (finite, the rank of $\mathcal{E}$ over U) power of $\mathcal{A}(U)$ (cf. also (2.34)). It is just in this sense that we *"arithmetize" $\mathcal{E}$, by means of or (over)* U, as the latter set is appeared in (2.29). Yet, in view of (2.36), we can also assert that

(2.39)   *the* above *replacement ("arithmetization"), as in* (2.38), *is effectuated through the correspondence* (in fact, *bijection*)

$$(2.39.1) \qquad s \longleftrightarrow (\alpha_i) \in \mathcal{A}(U)^n = \mathcal{A}^n(U).$$

Thus, based on the preceding, we still refer to the *open set* $U \subseteq X$, as above, succinctly, as the *local gauge of $\mathcal{E}$ over U*,

$$(2.40) \qquad e^U \equiv \{U; (e_i)_{1 \le i \le n}\},$$

in *such* a manner *that* (2.29), or, *equivalently*, (2.36), *is in force*. Finally, the same isomorphism (2.29) characterizes the given $\mathcal{A}$-module $\mathcal{E}$ on X, as a *locally free $\mathcal{A}$-module*, of (finite) *rank* $n \in \mathbb{N}$, on (the *open*) $U \subseteq X$. (See also (2.30), (2.31), along with Note 2.2 in the preceding).

Now, if a similar isomorphism as in (2.29), holds true for every point $x \in X$, with respect to an appropriate (open) neighborhood U of x in X, one then speaks of $\mathcal{E}$ as a locally free $\mathcal{A}$-module of rank $n (\in \mathbb{N})$ on X, alias a *vector sheaf* on X. Especially, *for* $n = 1$, we speak of a *line sheaf* on X. In this regard, we still write, concerning the latter notion,

(2.41)

$$(2.41.1) \qquad rk_{\mathcal{A}} \mathcal{E} \equiv rk \mathcal{E} = n, \ n \in \mathbb{N}.$$

It is in effect the above two types of $\mathcal{A}$-modules on X that we are actually concerned with throughout the subsequent discussion.

Therefore, by further considering an $\mathcal{A}$-*connection* D, as in (2.1), on a given $\mathcal{A}$-*module* $\mathcal{E}$ on X for which (2.29) is valid, one then obtains, as a local expression of D with respect to that particular given open set $U \subseteq X$, the relations

(2.42)
$$\begin{aligned}
D(s) \in \Omega(\mathcal{E})(U) &\equiv (\Omega \otimes_{\mathcal{A}} \mathcal{E})(U) = (\mathcal{E} \otimes_{\mathcal{A}} \Omega)(U) \\
&= ((\mathcal{E} \otimes_{\mathcal{A}} \Omega)|_U)(U) = (\mathcal{E}|_U \otimes_{\mathcal{A}|_U} \Omega|_U)(U) \\
&= (\mathcal{A}^n|_U \otimes_{\mathcal{A}|_U} \Omega|_U)(U) = (\mathcal{A}^n \otimes_{\mathcal{A}} \Omega)(U) = \Omega^n(U) \\
&= \Omega(U)^n = \mathcal{A}(U)^n \otimes_{\mathcal{A}(U)} \Omega(U) = \mathcal{E}(U) \otimes_{\mathcal{A}(U)} \Omega(U),
\end{aligned}$$

for any $s \in \mathcal{E}(U)$ (cf. also (2.34), (2.38)). Hence, by virtue of (2.36), *one* now obtains

(2.43)
$$\begin{aligned}
D(s) = D\left(\sum_{i=1}^n \alpha_i e_i\right) &= \sum_{i=1}^n D(\alpha_i e_i) \\
&= \sum_i (\alpha_i D(e_i) + e_i \otimes \partial(\alpha_i)) = \sum_i e_i \otimes \left(\partial(\alpha_i) + \sum_j \alpha_j \omega_{ij}\right),
\end{aligned}$$

where we have also set (cf. (2.42))

(2.44) $$D(e_i) = \sum_{j=1}^n e_j \otimes \omega_{ij} \in \Omega(\mathcal{E})(U) = \mathcal{E}(U) \otimes_{\mathcal{A}(U)} \Omega(U), 1 \leq i \leq n,$$

such that *one has*

(2.45) $$\omega \equiv (\omega_{ij}) \in M_n(\Omega(U)) = M_n(\Omega)(U).$$

Of course, the above matrix of *local 1-forms* on U is uniquely defined, by means of (2.44), according to our hypothesis for (2.40). Consequently, one thus infers, through the preceding argument, that

(2.46)   an $\mathcal{A}$-connection D of $\mathcal{E}$ is uniquely defined locally with respect to a given local gauge e^U of $\mathcal{E}$ (cf. (2.40)) whenever we know its values at the elements of any basis (of sections) of e^U, hence, equivalently, if we are given the $n \times n$ matrix (2.45).

In this regard, see also [VS: Chapt. VII; p. 100, (1.8), along with p. 101, Theorem 1.1]. Yet, the same $n \times n$ matrix ω, as defined by (2.44) and (2.45), is called the *local $\mathcal{A}$-connection matrix of D* with respect always to a given local gauge e^U of $\mathcal{E}$, as in (2.40). Furthermore, by virtue of (2.43), one can still say that

(2.47)   the $\mathcal{A}$-connection D of $\mathcal{E}$ as above is locally given on U, where e^U is a local gauge of $\mathcal{E}$ (see (2.40)), by the relation

$$(2.47.1) \qquad\qquad D = \partial + \omega.$$

Here, by an obvious abuse of notation, we have set in (2.47.1)

$$(2.48) \qquad \partial \equiv \partial^n|_U : \mathcal{A}^n|_U \longrightarrow \Omega^n|_U = \Omega(\mathcal{A}^n)|_U$$

for the respective $\mathcal{A}|_U$-*connection* on (the free $\mathcal{A}|_U$-*module*) $\mathcal{A}|_U = (\mathcal{A}|_U)^n$, derived from the given (flat) $\mathcal{A}$-connection ∂ on $\mathcal{A}$, as in (1.13). Thus, one gets through (2.47.1) at a familiar expression in physics by speaking about the above $n \times n$ matrix Ω as a *potential*, so that one can still refer to (2.47.1) by saying that

(2.49) D is locally expressed by the potential ω.

In other words, one concludes the (bijective) correspondence

$$(2.50) \qquad\qquad D|_U \longleftrightarrow \omega \equiv (\omega_{ij}) \in M_n(\Omega(U)),$$

in the sense of (2.47.1), whenever we are given a local gauge e^U of $\mathcal{E}$, as above. Yet, by further referring to

$$(2.51) \qquad\qquad\qquad D|_U,$$

as for instance in (2.50), we also note that the same can be construed as the

(2.52)   restriction, or even *pullback*, of D on U (in the latter case, via the canonical injection $U \underset{\longrightarrow_i}{\subseteq} X$).

More on the second last notion, as above, will also be said in the sequel (see Section 3 below; Section V.5.5, in particular (5.103)).

We consider now the case that the previous situation about $D|_U$ (cf., for instance, (2.50)) is varied throughout X; that is, in the case that we are given a vector sheaf $\mathcal{E}$ on X, then by the very definitions (see, for example, (2.41)), one gets an open covering of X, say

$$(2.53) \qquad\qquad\qquad \mathcal{U} = (U_\alpha)_{\alpha \in I},$$

whose individual open sets $U_\alpha, a \in I$, are local gauges of $\mathcal{E}$. We call such a $\mathcal{U}$ a *local frame of $\mathcal{E}$*. (Of course, the existence of a local frame, as before, for an $\mathcal{A}$-module $\mathcal{E}$ on X characterizes it in turn by the same definitions as a vector sheaf on X).

Therefore, given a vector sheaf $\mathcal{E}$ on X along with a local frame $\mathcal{U}$ of $\mathcal{E}$, as before, by further employing our previous argument connected with (2.45) and a given $\mathcal{A}$-*connection* D of $\mathcal{E}$, one obtains a 0-cochain of local $\mathcal{A}$-connection matrices of D, which is thus associated with the given local frame $\mathcal{U}$ of $\mathcal{E}$ as in (2.53); that is, one has

(2.54)
$$\omega_{\mathcal{U}} \equiv \omega \equiv (\omega^{(\alpha)}) \in C^0(\mathcal{U}, M_n(\Omega)),$$

such that

(2.55)
$$\omega^{(\alpha)} \equiv (\omega_{ij}^{(\alpha)}) \in M_n(\Omega(U_\alpha)), \alpha \in I.$$

Yet by further applying the terminology of (2.49), we also refer to (2.54) as the 0-*cochain of potentials* that is *associated with an $\mathcal{A}$-connection D and a local frame $\mathcal{U}$ of the* given *vector sheaf $\mathcal{E}$* on X. On the other hand,

> a 0-cochain of potentials, is one that is associated with an $\mathcal{A}$-connection D of a given vector sheaf $\mathcal{E}$ on X of rank $n \in \mathbb{N}$ and a local frame $\mathcal{U}$ of X if and only if the following relation holds:

(2.56)

(2.56.1)
$$\omega^{(\beta)} = Ad(g_{\alpha\beta}^{-1})\omega^{(\alpha)} + \tilde{\partial}(g_{\alpha\beta}),$$

for any α, β in I (cf. (2.53)), with

(2.56.2)
$$U_{\alpha\beta} \equiv U_\alpha \cap U_\beta \neq \emptyset.$$

We call the above relation (2.56.1) the *transformation law of potentials*.

We thus have here a criterion of defining an $\mathcal{A}$-connection on a given vector sheaf through local data (cf. (2.47.1), along with (2.50)). Yet, concerning the notation employed in (2.56.1), we denote by

(2.57)
$$(g_{\alpha\beta}) \in Z^1(\mathcal{U}, \mathcal{GL}(n, \mathcal{A}))$$

a *coordinate 1-cocycle of $\mathcal{E}$* with respect to $\mathcal{U}$ (see, for instance, (1.34) as well as Chapter III; (2.14) and (2.25) in the sequel), while we further set

(2.58)
$$Ad(g_{\alpha\beta}^{-1}) \cdot \omega^{(\alpha)} := g_{\alpha\beta}^{-1}\omega^{(\alpha)}g_{\alpha\beta},$$

where one has by definition (cf. (2.57))

(2.59)
$$g_{\alpha\beta} \in \mathcal{GL}(n, \mathcal{A})(U_{\alpha\beta}) = GL(n, \mathcal{A}(U_{\alpha\beta})),$$

with α, β in I as in (2.56.2). In this connection, see also [VS: Chapter VII; p. 112, Theorem 3.2].

Scholium 2.1 By looking at the local definition of an $\mathcal{A}$-connection D of $\mathcal{E}$, as in (2.47.1), in conjunction with (2.50) one realizes that locally (however, see also (2.56)) D *consists of elements (sections) of $\mathcal{A}$*, given the differential triad

(2.60)
$$(\mathcal{A}, \partial, \Omega),$$

whenever

(2.61)

the $\mathcal{A}$-module Ω can be determined by means of $\mathcal{A}$ in point of fact through

$$(2.61.1) \qquad \operatorname{im} \partial = \partial(\mathcal{A}).$$

In this connection, we still note that (2.61) is what in effect classically happens; cf. *Kähler definition of a connection* (see, e.g., [VS: Chapter XI; p. 324, (5.18)], or even N. Bourbaki [3: Chapter III; p. 132, Lemme 1, and p.133, Proposition 18] along with D. Eisenbud [1: p. 384, Definition, p. 388, Section 16.2, and p. 407, Theorem 16.24].

The above still points out the significance of (2.60) in building up the whole machinery of the abstract setting. Yet, the same may have potential applications in physical arguments pertaining to fundamental items that are essentially needed to look at a differential-geometric way of modeling physical theories, where in particular the standard (smooth) manifold concept suffers, as is for instance the case with quantum gravity. Of course,

(2.62)

one is led to an analogous situation with (2.61), pertaining to the description of Ω in terms of $\mathcal{A}$ (although now only locally) in the case that Ω is a vector sheaf on X too (see (2.29)). Yet, this happens also in the classical theory; cf. Section 2.1 above.

2.3 Gauge Transformation

The concept of gauge transformation is among the most fundamental in the theory of $\mathcal{A}$-connections and in general refers to the *way* we understand how things vary. Thus expressed in mathematical parlance, this leads to a change of means by which objects are described, hence to a change of our local (generalized) coordinates (*arithmetical effectuations* of the same objects), that is, to a change of our previously defined local gauges (cf. (2.29), or even (2.40)).

Consequently, by considering a vector sheaf $\mathcal{E}$ on X of rank $n \in \mathbb{N}$, together with a local frame $\mathcal{U}$ of $\mathcal{E}$, as in (2.53), one defines

(2.63)

a *gauge transformation of $\mathcal{E}$, relative to $\mathcal{U}$*, as the following *1-cochain* (of $\mathcal{U}$ with coefficients in $\mathcal{GL}(n, \mathcal{A})$, where $n = rk\mathcal{E} \in \mathbb{N}$),

$$(2.63.1) \qquad \begin{aligned} g &\equiv (g_{\alpha\beta}) \in C^1(\mathcal{U}, \mathcal{GL}(n, \mathcal{A})) \\ &= \prod_{(\alpha,\beta)} \mathcal{GL}(n, \mathcal{A})(U_{\alpha\beta}) = \prod_{(\alpha,\beta)} GL(n, \mathcal{A}(U_{\alpha,\beta})), \end{aligned}$$

where $(\alpha, \beta) \in I \times I$, such that

$$(2.63.2) \qquad U_{\alpha\beta} \equiv U_\alpha \cap U_\beta \neq \emptyset.$$

A (local) gauge transformation of $\mathcal{E}$ entails also a change of the (generalized) local coordinates (*concrete realizations*) of $\mathcal{E}$; see (2.29). On the other hand,

g as in (2.63.1) *effectuates* the given vector sheaf $\mathcal{E}$ itself if and only if g is a *1-cocycle*, that is, whenever one has

(2.64)

$\quad$ (2.64.1) $\qquad g \equiv (g_{\alpha\beta}) \in Z^1(\mathcal{U}, \mathcal{GL}(n, \mathcal{A})) \subseteq C^1(\mathcal{U}, \mathcal{GL}(n, \mathcal{A}))$.

In that case, we still call $g \equiv (g_{\alpha\beta})$, a *coordinate 1-cocycle of $\mathcal{E}$*.

As already said (cf. (2.57)), this is actually the 1-cocycle that appeared in the transformation law of potentials as in (2.56.1). Therefore, one concludes that the same law expresses the fact that

(2.65)
$\quad$ the various local realizations of a given $\mathcal{A}$-connection D of $\mathcal{E}$, through local matrices of "1-forms" (cf. (2.55)), are to each other *gauge equivalent* in the sense that they respect a local change of coordinates, as for instance exhibited by a coordinate 1-cocycle of $\mathcal{E}$ (cf. (2.64.1)). Yet, the above can still be construed as another application (verification) of the (physical) *principle of general covariance* pertaining to the way physical laws (in our case, $\mathcal{A}$-connections) are transformed, hence realized too.

In this connection, see also [VS: Chapt. VII; p. 107, (2.5), as well as, p.108, (2.11), along with the subsequent discussion therein]. Yet the following note is in order, referring to the particular terminology employed in (2.65).

Note 2.3 The *gauge equivalence* of local $\mathcal{A}$-connection matrices of an *$\mathcal{A}$-connection D of $\mathcal{E}$*, as expressed by the corresponding *transformation law of potentials*, is essentially the *local form* of the general formula relating gauge equivalent $\mathcal{A}$-connections with respect to a given $\mathcal{A}$-automorphism of $\mathcal{E}$, viz. via an element

(2.66) $\qquad\qquad\qquad\qquad \phi \in \mathcal{A}ut\mathcal{E}$.

Cf. [VS: Chapt. VII; p.88, (17.7) and (17.10)]. That is, one thus considers here the restriction of the $\mathcal{A}$-connection D on two given local gauges of $\mathcal{E}$, U_α and U_β, satisfying (2.63.2) (cf. also (2.50); see also [VS: Chapt. VII; p. 110, Scholium 2.1]).

Important applications of the above transformation law of potentials, as given by (2.56.1), will be presented in subsequent chapters of this treatise. Thus, by considering, for example, the particular case of a line sheaf $\mathcal{L}$ on X, one gets the aforementioned law in the following familiar form (cf. (2.56.1), for $n = 1$):

(2.67) $\qquad\qquad\qquad\qquad \delta(\theta_\alpha) = \tilde{\partial}(g_{\alpha\beta})$.

Now, as we shall see later (cf. Chapt. III; Lemma 2.1), (2.67) yields, in effect, a local characterization of a given Maxwell field

(2.68) $\qquad\qquad\qquad\qquad (\mathcal{L}, D)$

on X. That is, more precisely,

a given pair

(2.69)   (2.69.1) $((g_{\alpha\beta}), (\theta_\alpha)) \in Z^1(\mathcal{U}, \mathcal{A}^{\cdot}) \times C^0(\mathcal{U}, \Omega)$

characterizes (locally, in terms of a given local frame $\mathcal{U}$ of $\mathcal{L}$)$(\mathcal{L}, D)$, through the law (2.67).

On the other hand, motivated further by (2.67) and setting

(2.70) $\delta(\omega^{(\alpha)}) := \omega^{(\beta)} - Ad(g_{\alpha\beta}^{-1})\omega^{(\alpha)},$

one obtains (2.56.1) in the form

(2.71) $\delta(\omega^{(\alpha)}) := \tilde{\partial}(g_{\alpha\beta}),$

which also will be of use in the sequel. So the last relation will be applied in Part II of the present treatise by further looking at the cohomological classification of Yang–Mills fields (see Chapt. I; Section 9).

Furthermore, see also Chapter III; (2.39) in the sequel for a physical interpretation of (2.67), hence by extension of (2.71) as well. On the other hand, further considerations of the group of gauge transformations of a given vector sheaf $\mathcal{E}$ on X will also be applied in Part II of this study (cf. Chapt. I; Section 5.1) in conjunction with *metric* notions, still referring to Yang–Mills fields.

3 Induced $\mathcal{A}$-Connections

We discuss below the result of applying standard functors of linear and multilinear algebra on $\mathcal{A}$-connections of $\mathcal{A}$-modules on a topological space X, the latter being, as usual, the base space of a given differential triad, as in (1.13) (see also (1.4)).

Thus, by considering a given family

(3.1) $(\mathcal{E}_i)_{i \in I}$

of $\mathcal{A}$-*modules* on X, where each one is endowed with an $\mathcal{A}$-connection $D_i, i \in I$, the corresponding *Cartesian product*, and *Whitney sum*, $\mathcal{A}$-module of the family (3.1).

(3.2) $\prod_{i \in I} \mathcal{E}_i$ *and* $\bigoplus_{i \in I} \mathcal{E}_i,$

respectively, are (canonically) endowed as well with the $\mathcal{A}$-*connections*

(3.3) $\prod_{i \in I} D_i$ *and* $\bigoplus_{i \in I} D_i,$

still called *Cartesian product* and *Whitney sum*, $\mathcal{A}$-connections, respectively. Of course, the latter operators are defined coordinatewise.

Note 3.1 By looking at any one of the two $\mathcal{A}$-modules on X, as in (3.2), we further remark that the same are occasionally vector sheaves on X, but not, of course, of finite rank (just locally free $\mathcal{A}$-modules on X) even if the given $\mathcal{A}$-modules $\mathcal{E}_i$, $i \in I$, as in (3.1), are vector sheaves in the sense employed hitherto, viz. of finite rank. Thus, we have here particular examples of an $\mathcal{A}$-module on X possessing an $\mathcal{A}$-connection without being necessarily a vector sheaf (locally free $\mathcal{A}$-module of finite rank).

Furthermore, one gets at analogous examples as above by taking, for instance, a projective limit of vector sheaves with $\mathcal{A}$-connections (E. Vassiliou), or even an inductive limit of such; therefore,

(3.4)   projective and inductive limits of vector sheaves with $\mathcal{A}$-connections provide $\mathcal{A}$-modules, in general, endowed with $\mathcal{A}$-connections.

Now, the case of a projective limit, in conjunction with our previous example in Section 2.1, has a special bearing on a type of (smooth) "infinite-dimensional" manifold, considered by M. E. Verona [1]; see also [VS: Chapt. X; p.281, Subsection 1.1].

Yet the case of inductive limits of vector sheaves is also related (cf. Subsection 2.1) to the type of infinite-dimensional (smooth) manifolds, (suitable) inductive limits of ordinary (finite-dimensional) C^∞-manifolds), recently considered by J. Bernstein–V. Lunts [1: pp. 96ff.], in studying "smooth models" of classifying spaces of Lie groups (ibid., p. 103).

Another *concrete example* of the above is the *(finite) Whitney sum* of the standard $\mathcal{A}$-connection ∂ of a given differential triad on X (cf. (1.13)). Thus, one gets the *("n-dimensional") Whitney sum* of ∂,

(3.5) $$\partial^n := \underbrace{\partial \oplus \cdots \oplus \partial}_{n\text{-times}}, \qquad n \in \mathbb{N},$$

being, of course, an $\mathcal{A}$-connection on the (free) $\mathcal{A}$-module $\mathcal{A}^n$. We have already used it in (2.48), and shall also employ it in the sequel.

To continue, suppose we are given the pairs

(3.6) $$(\mathcal{E}, D_\mathcal{E} \equiv D) \quad and \quad (\mathcal{F}, D_\mathcal{F} \equiv D')$$

consisting of $\mathcal{A}$-modules and corresponding $\mathcal{A}$-connections on them, as indicated. Hence, one can further consider the $\mathcal{A}$-connections

(3.7) $$D_{\mathcal{E} \otimes_A \mathcal{F}} \quad and \quad D_{\mathcal{H}om_A(\mathcal{E}, \mathcal{F})}$$

on the $\mathcal{A}$-modules

(3.8) $$\mathcal{E} \otimes_A \mathcal{F} \quad and \quad \mathcal{H}om_A(\mathcal{E}, \mathcal{F}),$$

respectively. Precisely, one defines

$$(3.9) \qquad D_{\mathcal{E} \otimes_A \mathcal{F}} := (D_{\mathcal{E}} \otimes 1_{\mathcal{F}}) + (1_{\mathcal{E}} \otimes D_{\mathcal{F}}) \equiv D \otimes 1 + 1 \otimes D',$$

while we still set

$$(3.10) \quad D_{\mathcal{H}om_A(\mathcal{E}, \mathcal{F})}(\phi) := D_{\mathcal{F}} \circ \phi - (\phi \otimes 1_{\Omega}) \circ D_{\mathcal{E}} \equiv D' \circ \phi - (\phi \otimes 1) \circ D$$

such that

$$(3.11) \qquad\qquad \phi \in \mathcal{H}om_A(\mathcal{E}, \mathcal{F})(U) = Hom_{A|_U}(\mathcal{E}|_U, \mathcal{F}|_U)$$

with U *open* in X. In the above relations

$$(3.12) \qquad\qquad\qquad 1_{(\cdot)} \equiv 1$$

stands for the *identity A-automorphism* of the A-module concerned.

In this connection, it is worth noting here the form that (3.10) takes in the particular case $\mathcal{E} = \mathcal{F}$; thus, one obtains

$$(3.13) \quad D_{\mathcal{E}nd\mathcal{E}}(\phi) = D \circ \phi - (\phi \otimes 1) \circ D \equiv D \circ \phi - \phi \circ D \equiv [D, \phi] \equiv L_D(\phi)$$

for any

$$(3.14) \qquad\qquad \phi \in (\mathcal{E}nd\mathcal{E})(U) = Hom_{A|_U}(\mathcal{E}|_U, \mathcal{E}|_U) \equiv End(\mathcal{E}|_U).$$

In this regard, the last term in (3.13) denotes the *"Lie (covariant) derivative of ϕ"* with respect to the given A-connection D of $\mathcal{E}$.

In particular, by taking

$$(3.15) \qquad\qquad\qquad \phi \equiv 1_{\mathcal{E}} \in Aut(\mathcal{E}),$$

that is, the *identity A-automorphism* of $\mathcal{E}$, one gets, by virtue of (3.13),

$$(3.16) \qquad\qquad\qquad L_D(1) = 0.$$

On the other hand, by considering two different A-*connections* D and D' of $\mathcal{E}$, one concludes from (3.10) that

$$(3.17) \qquad\qquad D_{\mathcal{E}nd\mathcal{E}}(1) = D' - D \in \Omega(\mathcal{E}nd\mathcal{E})(X)$$

such that if $\mathcal{E}$ is a vector sheaf on X, one still obtains

$$(3.18) \qquad D_{\mathcal{E}nd\mathcal{E}}(1) = D' - D \in \Omega(\mathcal{E}nd\mathcal{E})(X) = Hom_A(\mathcal{E}, \Omega(\mathcal{E})),$$

that is,

$$(3.19) \quad \begin{array}{l} \text{the difference of two } A\text{-connections of } \mathcal{E} \text{ is actually an } A\text{-morphism} \\ \text{(viz. a ``tensor''), not merely a } \mathbb{C}\text{-linear morphism,} \end{array}$$

a fact that can be verified directly according to the definitions (cf. (2.3)). See also Section 5 below.

Now, given a pair

(3.20) $(\mathcal{E}, D)$,

consisting of an $\mathcal{A}$-module $\mathcal{E}$ on X and an $\mathcal{A}$-connection D of $\mathcal{E}$, we further consider the induced $\mathcal{A}$-connection on the corresponding dual $\mathcal{A}$-module of $\mathcal{E}$,

(3.21) $\mathcal{E}^* := \mathcal{H}_\mathcal{A}(\mathcal{E}, \mathcal{A})$.

Thus, by taking the standard pair (cf. (1.13))

(3.22) $(\mathcal{A}, \partial)$,

one has, by virtue of (3.21) and (3.10),

(3.23) $D_{\mathcal{E}^*}(u) \equiv D_{\mathcal{H}om_\mathcal{A}(\mathcal{E}, \mathcal{A})}(u) \equiv D^*(u) := \partial \circ u - (u \otimes 1) \circ D \equiv \partial \circ u - u \circ D$,

such that (cf. (3.11), along with (3.21))

(3.24) $u \in \mathcal{H}om_\mathcal{A}(\mathcal{E}, \mathcal{A})(U) \equiv \mathcal{E}^*(U)$.

On the other hand, suppose, in particular, that $\mathcal{E}$ is a vector sheaf on X; so, by looking at (2.1), one has by definition (cf. also [VS: Chapt. IV; p. 302, Theorem 6.1])

(3.25) $D^* : \mathcal{E}^* \longrightarrow \Omega(\mathcal{E}^*) \equiv \mathcal{E}^* \otimes_\mathcal{A} \Omega = \mathcal{H}om_\mathcal{A}(\mathcal{E}, \mathcal{A}) \otimes_\mathcal{A} \Omega = \mathcal{H}om_\mathcal{A}(\mathcal{E}, \Omega)$.

Therefore, by taking $U \subseteq X$ as a local gauge of $\mathcal{E}$ (cf. (2.29)) and an element $u \in \mathcal{E}^*(U)$ one obtains, in view of (3.25),

(3.26) $D^*(u) \in \Omega(\mathcal{E}^*)(U) = \mathcal{H}om_\mathcal{A}(\mathcal{E}, \Omega)(U) = Hom_{\mathcal{A}|_U}(\mathcal{E}|_U, \Omega|_U)$,

so that by considering further a local (continuous) section

(3.27) $s \in (\mathcal{E}|_U)(U) = \mathcal{E}(U)$

(cf. also [VS: Chapt.I; p. 55, (11.40)]), one gets, by virtue of (3.26) and (3.23),

(3.28) $D^*(u)(s) = \partial(u(s)) - (u \otimes 1)(D(s)) \in \Omega(U) = (\Omega|_U)(U)$,

where precisely one has (see also (2.43) in the preceding)

(3.29) $D(s) \in \Omega(\mathcal{E})(U) = \mathcal{E}(U) \otimes_{\mathcal{A}\mathcal{A}(U)} \Omega(U)$,

in view of our hypothesis that U is a local gauge of $\mathcal{E}$ (cf. (2.42)). [Of course, one could repeat here the preceding argument, by accepting this hypothesis for U, and thus taking, instead, more generally, an $\mathcal{A}$-module $\mathcal{E}$ on X. However, for convenience, we argued along the previous lines; but see [VS: Chapt.VII; p. 119, Section 4].

On the other hand, by employing an obvious abuse of notation concerning (3.23), we still consider the following relation as a defining formula for D^*, the dual $\mathcal{A}$-connection of D; that is, based on (3.23), one sets

$$(3.30) \qquad \partial(u(s)) = u(D(s)) + D^*(u)(s),$$

for any u, s, as in (3.24) and (3.27), respectively, that is, another, equivalent, expression of (3.23).

Now, looking at a local $\mathcal{A}$-connection matrix of D,

$$(3.31) \qquad \omega \equiv (\omega_{ij}) \in M_n(\Omega(U)),$$

relative to a given local gauge e^U of $\mathcal{E}$ (cf. (2.40), (2.45), or even (2.50)), one concludes that

(3.32)   the corresponding local $\mathcal{A}$-connection matrix of D^*, relative to $(e^U)^*$, the dual of e^U (see (3.33), (3.34) in the sequel), is given by the relation

$$(3.32.1) \qquad \omega^* \equiv (\omega_{ij}^*) = -{}^t\omega \equiv (-\omega_{ji}) = -(\omega_{ji}) \in M_n(\Omega(U)).$$

For convenience we also recall the relevant calculations concerning (3.32.1) (see [VS: Chapt.VII; p. 119, Section 4]): So, by considering the dual (local gauge) of e^U, (cf. also (2.40))

$$(3.33) \qquad (e^U)^* := \{U; (e_i^*)_{1 \le i \le n}\},$$

such that

$$(3.34) \qquad e_i^*(e_j) := \delta_{ij}, \qquad \le i, j \le n,$$

where,

$$(3.35) \qquad \delta_{ij} = 1 \ or \ 0 \ in \ \mathcal{A}(U)$$

The (canonical) local Kronecker gauge of $\mathcal{A}^n$ on the open $U \subseteq X$, one gets, through (3.33), a local gauge of $\mathcal{E}^*$ on U; that is, one has

$$(3.36) \qquad \mathcal{E}^*|_U = \mathcal{A}^n|_U = (\mathcal{A}|_U)^n,$$

within an $\mathcal{A}|_U$-isomorphism of the $\mathcal{A}|_U$-modules involved, determined by (3.34). (See also [VS: Chapt. II; p. 137, (6.23) and (6.22), or even Chapt. IV; p. 298, (5.2.1)]). Yet, in view of our hypothesis for U (see (2.40)), one also obtains, based further on (3.36) and the last part of our previous citation, the relations

$$(3.37) \qquad \mathcal{E}^*|_U = \mathcal{A}_U^n = \mathcal{E}_U = (\mathcal{A}|_U)^n = (\mathcal{E}|_U)^*$$

within $\mathcal{A}|_U$-isomorphisms of the $\mathcal{A}_U$-modules involved, indeed, for any $\mathcal{A}$-module $\mathcal{E}$ and open $U \subseteq X$, being a local gauge of $\mathcal{E}$.

Thus, by looking further at (2.44), one has the one-to-one correspondence

$$(3.38) \qquad (D(e_i))_{1 \le i \le n} \longleftrightarrow \omega \equiv (\omega_{ij})_{1 \le i, j \le n},$$

which uniquely defines the given $\mathcal{A}$-connection D of $\mathcal{E}$. Hence, by analogy, and based on (3.28), one obtains

$$D(e_j^*)(e_i) = \partial(e_j^*(e_i)) - (e_j^* \otimes 1)(D(e_i))$$

$$= \partial(\delta_{ij}) - (e_j^* \otimes 1) \left(\sum_{k=1}^{n} e_k \otimes \omega_{ki} \right)$$

$$= -\sum_k e_j^*(e_k) \cdot \omega_{ki} = -\sum_k \delta_{jk} \cdot \omega_{ki}$$

$$= -\omega_{ji} \equiv \omega_{ij}^*.$$

Accordingly, we set (see (3.32.1))

$$(3.40) \qquad \omega^* \equiv (\omega_{ij}^*) = -{}^t\omega \equiv (-\omega_{ji}) = -(\omega_{ji}) \in M_n(\Omega(U)),$$

obtaining thus the $n \times n$ *local $\mathcal{A}$-connection matrix of D^** with respect to the local gauge $(e^U)^*$ of $\mathcal{E}^*$, as in (3.33). This also explains our previous assertion in (3.32.1). ■ In this connection, we further remark that one actually has here the relation

$$(3.41) \qquad D^*|_U = (D|_U)^*,$$

for any open set $U \subseteq X$; see also [VS: Chapt. VI; p. 29, (6.24)] for the analogous situation that one has concerning the restriction to an open $U \subseteq X$ of an $\mathcal{A}$-connection D of an $\mathcal{A}$-module $\mathcal{E}$ on X in general. Thus, one has

$$(3.42) \qquad (D|_U)(s|_U) := D(s)|_U,$$

such that

$$(3.43) \qquad (D|_U)(s|_U)(x) = D(s)(x),$$

for any $x \in U$ and $s \in \mathcal{E}(U)$ with U and $\mathcal{E}$ as before. In other words, one realizes that we have actually employed, implicitly the notion of the pullback of an $\mathcal{A}$-connection in terms of a given continuous function.

In particular, by referring to (3.42) (or even to (3.41)), we remark that in connection with our previous discussion, one considers here as a continuous function the canonical injection

$$(3.44) \qquad i_U : U \underset{\longrightarrow}{\subseteq} X$$

of the open set $U \subseteq X$ as above. Thus, concerning the relevant differential triads, one sets

$$(3.45) \qquad (\mathcal{A}, \partial, \Omega)_{|U} := (\mathcal{A}|_U, \partial|_U, \Omega|_U),$$

getting thus at the pullback (restriction) of a given differential triad on X on an open set $U \subseteq X$, which in turn defines a differential triad on U as indicated by the second member of (3.45).

More generally, having a continuous map

$$(3.46) \qquad f : X \longrightarrow Y$$

and a differential triad on Y along with a pair on it

$$(3.47) \qquad (\mathcal{E}, D),$$

as in (3.20), one further considers the pullback of D via f, $f^*(D)$ on $f^*(\mathcal{E})$, the pullback on X of the given $\mathcal{A}$-module $\mathcal{E}$ on Y, being thus an $f^*(\mathcal{A})$-module on X. So one sets

$$(3.48) \qquad f^*(D)(f^*(t)) := f^*(D(t)) = D(t) \circ f$$

in such a manner that, one has, in particular,

$$(3.49) \qquad f^*(D)(f^*(t))(x) \equiv f^*(D(t))(x) = (D(t) \circ f)(x) = D(t)(f(x)),$$

for any $t \in \mathcal{E}(V)$, with V an open set in Y and $x \in f^{-1}(V)$. Thus, regarding in particular (3.42), one obtains

$$(3.50) \qquad D|_U = i_U^*(D),$$

while (3.49) still explains (3.43).

Further details concerning the pullback (inverse image) of sheaves in general are given in [VS: Chapt. I; pp. 79ff.]. On the other hand, we shall also consider in the sequel the pullback of Maxwell fields in connection with (geometric) prequantization of elementary particles. See Chapter V, Section 5.5. One finds therein an account of general properties of the "pullback functor" in particular concerning Maxwell fields. For analogous considerations in the case of Yang–Mills fields cf. Part II of the present treatise, Chapt. I; Section 9.

Scholium 3.1 Of course, the notion of the pullback of a differential triad through a continuous map (cf., for instance, (3.45) above, for the particular case of the map (3.44), or even Chapt. V; (5.72) and (5.73), for the general case) is of an obvious importance if one wants to define a differential triad on a given topological space in general. Thus, it suffices to have on the initial space a continuous map whose values belong to a space already endowed with a differential triad, hence, provided, for example, one has on the given space any continuous numerical-valued map.

The previous situation has been recently extended to the case of the direct image ("push-out") of a differential triad as well by M. Papatriantafillou [1] as an outcome of her work pertaining in general, to a study of the category of differential triads as has been advocated in [VS] and further employed throughout the present treatise. Now, by duality with the aforementioned example, the significance of the existence of a continuous curve (map) into a given topological space concerning the possibility of existence of a differential triad on the range of the map at issue is also clear.

We continue in the next section by looking at conditions guaranteeing the existence of $\mathcal{A}$-connections, which, in general, is not always feasible, due to the generality of this notion as it is applied within the present abstract setting. On the other hand, as we shall see in the sequel, $\mathcal{A}$-connections do exist in some very peculiar and important cases as well (see, e.g., Part II, Chapt. IV; Section 5: Rosinger's algebra sheaf).

4 Existence of $\mathcal{A}$-Connections. Criteria of Existence

As already said, one does not expect that within the present abstract form of the classical theory of differential geometry $\mathcal{A}$-connections always exist; of course, in principle, this is due to the generality of the notions that are employed. We do have, however, similar examples in the classical theory too, as is, for instance, the case with the standard result and accompanied criterion of M. F. Atiyah [1] pertaining to the existence of holomorphic connections, the latter being a particular instance of $\mathcal{A}$-connections, where

$$(4.1) \qquad\qquad \mathcal{A} \equiv \mathcal{O}_X,$$

the last term in (4.1) denoting the sheaf of germs of ($\mathbb{C}$-valued) holomorphic functions on a complex (analytic) manifold X. So in the case of a Stein manifold X, every holomorphic vector bundle on X admits a holomorphic ($\mathcal{O}_X$-)connection, as opposed to the case of an arbitrary complex manifold, where the so-called *Atiyah criterion* (cf., for example, (4.12)) is not fulfilled. For the details, see also [VS: Chapt.VI; p.76, Section 15.2].

Within our abstract framework, still guided by the classical case, one obtains analogous criteria as well as, sufficient conditions for the existence of $\mathcal{A}$-connections, which also clarify the essence of the corresponding classical issues: Thus by considering first the smooth case, viz. (finite-dimensional) C^{∞}-manifolds, one then remarks that in that case the corresponding structure sheaf

$$(4.2) \qquad\qquad \mathcal{A} \equiv C_X^{\infty}$$

is a fine sheaf on the paracompact (Hausdorff) C^{∞}-manifold X (take, e.g., X to be 2nd countable),so that one concludes that

(4.3)   there always exist (smooth) connections for every smooth $\mathbb{C}$-vector bundle on X.

In point of fact, we have the previous case, even in our abstract set-up, provided we can guarantee an analogous framework as in (4.3). Indeed, one thus gets the following result:

on a paracompact (Hausdorff) space X endowed with a differential triad,

(4.4)

    (4.4.1) $\qquad\qquad\qquad (\mathcal{A}, \partial, \Omega)$

    (see (1.13)),

    (4.4.2) every fine vector sheaf $\mathcal{E}$ on X admits an $\mathcal{A}$-connection.

See [VS: Chapt. VI; p. 85, Theorem 16.1]; here one employs the existence of a partition of unity, viz. of $1_{\mathcal{E}} \equiv 1$, the identity automorphism of $\mathcal{E}$, guaranteed according to the definitions by virtue of our hypothesis (cf. (4.4.2)) that $\mathcal{E}$ is a fine sheaf on X (see also [VS: Chapt. III, p. 238, Definition 8.1, along with (8.23) and (8.23′)]).

    Furthermore, still based on the aforementioned work of M. F. Atiyah, it is shown that working within the present abstract setting, to provide the existence of an $\mathcal{A}$-connection on a given vector sheaf $\mathcal{E}$ on X, one needs (indeed, one has here a criterion; see (4.11) in the sequel) an appropriate

(4.5)   relation between the local "changes" of an $\mathcal{A}$-connection, (local) fragments, and analogous changes of the (local) "coordinates" of $\mathcal{E}$.

These changes are always determined by means of a given local frame of $\mathcal{E}$, say

(4.6) $\qquad\qquad\qquad \mathcal{U} = (U_{\alpha})_{\alpha \in I}.$

Furthermore, the same frame can be chosen locally finite in the case X happens to be paracompact (Hausdorff); in other words, $\mathcal{U}$, as above, yields then a locally finite open covering of X, which the relation

(4.7) $\qquad\qquad\qquad \mathcal{E}|_{U_{\alpha}} \underset{\eta_{\alpha}}{\cong} \mathcal{A}^{n}|_{U_{\alpha}}$

holds true within an $\mathcal{A}|_{U_{\alpha}}$-isomorphism of the $\mathcal{A}|_{U_{\alpha}}$-modules concerned.

    Thus, by means of $\mathcal{U}$ and the standard flat $\mathcal{A}$-connection ∂ on $\mathcal{A}$ (as extended to $\mathcal{A}^{n}$), one gets a 0-cochain of $\mathcal{A}$-connections on $\mathcal{E}$, restricted to the U_{α}'s, say,

(4.8) $\qquad\qquad\qquad (D_{\alpha}) \in C^{0}(\mathcal{U}, \mathcal{H}om_{\mathbb{C}}(\mathcal{E}, \Omega(\mathcal{E})))$

(*Levi-Civita 0-cochain of $\mathcal{E}$, relative to $\mathcal{U}$*), such that the relation we are looking for (see (4.5)), reads now as follows:

     the Levi-Civita ($\mathcal{A}$-connection) 1-cocycle that corresponds to (4.8), (see also (4.7)),

(4.9)   (4.9.1) $\delta(D_{\alpha}) \equiv (D_{\beta} - D_{\alpha}) \in Z^{1}(\mathcal{U}, \Omega(\mathcal{E}nd\mathcal{E})) \cong Z^{1}(\mathcal{U}, M_{n}(\Omega)),$

     is gauge equivalent, through the 0-cochain

    (4.9.2) $\qquad\qquad (\eta_{\alpha}) \in C^{0}(\mathcal{U}, \mathcal{I}som_{\mathcal{A}}(\mathcal{E}, \mathcal{A}^{n})),$

to the (logarithmic derivative of the corresponding to $\mathcal{U}$ "coordinate") 1-cocycle of $\mathcal{E}$,

$$(4.9.3) \qquad \tilde{\partial}((g_{\alpha\beta})) = (\tilde{\partial}(g_{\alpha\beta})) \in Z^1(\mathcal{U}, M_n(\Omega)).$$

That is, one has

$$(4.9.4) \qquad\qquad\qquad \delta(D_\alpha) \underset{\mathcal{U}}{\sim} (\tilde{\partial}(g_{\alpha\beta})),$$

in the sense that one has, by definition,

$$(4.9.5) \quad \delta(D_\alpha) = D_\beta - D_\alpha = (\eta_\alpha^{-1} \otimes 1)\tilde{\partial}(g_{\alpha\beta})\eta_\alpha \equiv Ad(\eta_\alpha^{-1})\tilde{\partial}(g_{\alpha\beta}).$$

Therefore, one finally gets in cohomolog

$$(4.10) \qquad [(\delta(D_\alpha))] \equiv [\delta(D_\alpha)] = [(\tilde{\partial}(g_{\alpha\beta}))] \in H^1(X, M_n(\Omega)).$$

Thus, one further defines

$$(4.11) \qquad \mathfrak{a}(\mathcal{E}) := [\delta(D_\alpha)] = [(\tilde{\partial}(g_{\alpha\beta}))] \in H^1(X, M_n(\Omega)),$$

calling $\mathfrak{a}(\mathcal{E})$ the *Atiyah class of* $\mathcal{E}$. The corresponding Atiyah criterion for the existence of an $\mathcal{A}$-connection of $\mathcal{E}$ is just the vanishing of the Atiyah class, viz. the condition

$$(4.12) \qquad\qquad\qquad \mathfrak{a}(\mathcal{E}) = 0 \in H^1(X, M_n(\Omega)).$$

In this regard, see also [VS: Chapt. VI; p. 67, Corollary 12.1].

    Furthermore, by still assuming that Ω, as in (4.4.2), is in particular a vector sheaf on X as well, one concludes that

(4.13)   the vanishing of the above Atiyah class of $\mathcal{E}$ amounts to the splitting of appropriate "$\mathcal{A}$-extensions" (short exact sequences of vector sheaves), the latter providing further two more criteria for the existence of $\mathcal{A}$-connections of (the given vector sheaf) $\mathcal{E}$.

See Chapt. VI; p. 73, Theorem 14.1. In this regard, we further note that the so-called *sheaf of $\mathcal{A}$-connection coefficients*

$$(4.14) \qquad\qquad\qquad\qquad M_n(\Omega)$$

(cf., for example, (2.45) and (2.46) concerning the terminology) is also, in point of fact, the structure sheaf of the aforementioned $\mathcal{A}$-extensions, which are further cohomologically determined through their corresponding characteristic classes, the latter taking values from the $\mathcal{A}(X)$-module (cohomology group)

$$(4.15) \qquad\qquad\qquad\qquad H^1(X, M_n(\Omega))$$

(see also (4.11) above, as well as [VS: Chapt. VI; p. 74, (14.13)]).

Precisely the aforesaid characteristic classes are equal to each other as well as to the cohomology classes appearing in (4.11), so that they actually

(4.16)   measure the obstruction of having the given vector sheaf $\mathcal{E}$ an $\mathcal{A}$-connection.

See also the reference following (4.13), along with the same citation, p. 72; (14.4) and (14.6).

5 The Space of $\mathcal{A}$-Connections

Suppose we are given a differential triad

(5.1) $$(\mathcal{A}, \partial, \Omega)$$

on a topological space X (cf. (1.13)) along with an $\mathcal{A}$-module $\mathcal{E}$ on X. Then we denote by

(5.2) $$Conn_\mathcal{A}(\mathcal{E}),$$

the *set of A-connections on* $\mathcal{E}$. Of course, we assume here that

(5.3)   in both the above two cases (5.1) and (5.2) we do not consider the "trivial" example of a zero $\mathcal{A}$-connection.

Thus, suppose that D is a given $\mathcal{A}$-connection of $\mathcal{E}$. Then, the above set (5.2) is given by the relation

(5.4) $$Conn_\mathcal{A}(\mathcal{E}) = D + Hom_\mathcal{A}(\mathcal{E}, \Omega(\mathcal{E})).$$

In this regard, one has

(5.5) $$Hom_\mathcal{A}(\mathcal{E}, \Omega(\mathcal{E})) = \mathcal{H}om_\mathcal{A}(\mathcal{E}, \Omega(\mathcal{E}))(X),$$

so that in the particular case that $\mathcal{E}$ is a vector sheaf on X, one obtains

(5.6) $$\mathcal{H}om_\mathcal{A}(\mathcal{E}, \Omega(\mathcal{E})) \equiv \mathcal{H}om_\mathcal{A}(\mathcal{E}, \mathcal{E} \otimes_\mathcal{A} \Omega) = \mathcal{H}om_\mathcal{A}(\mathcal{E}, \mathcal{E}) \otimes_\mathcal{A} \Omega$$
$$\equiv (\mathcal{E}nd\mathcal{E}) \otimes_\mathcal{A} \Omega \equiv \Omega(\mathcal{E}nd\mathcal{E}).$$

See [VS: Chapt. IV; p. 304, Corollary 6.1]. Therefore, in the case under consideration, based on (5.5) and (5.6), one obtains, in view of (5.4), for the set of $\mathcal{A}$-connections of the vector sheaf $\mathcal{E}$ on X,

(5.7) $$Conn_\mathcal{A}(\mathcal{E}) = D + \Omega(\mathcal{E}nd\mathcal{E})(X),$$

with D a given $\mathcal{A}$-connection of $\mathcal{E}$. That is, in other words, in the case of a vector sheaf $\mathcal{E}$ on X, whenever we are given an $\mathcal{A}$-connection D of $\mathcal{E}$, then any other $\mathcal{A}$-connection, say D' of $\mathcal{E}$, is given by the relation

(5.8)   (5.8.1) $D' = D + u,$

such that

(5.8.2) $u \in \Omega(\mathcal{E}nd\mathcal{E})(X).$

Consequently, as a result of (5.7), or, equivalently, of (5.8), one gets that

whenever we are given a vector sheaf $\mathcal{E}$ on X, then the set of $\mathcal{A}$-connections of $\mathcal{E}$,

(5.9)   (5.9.1) $Conn_A(\mathcal{E})$

(cf. also (5.3)), is an affine space, modeled on the $\mathcal{A}(X)$-module $\Omega(\mathcal{E}nd\mathcal{E})(X)$.

Especially, if we consider a line sheaf $\mathcal{L}$ on X, since in that case one has (see [VS: Chapt. II; p. 139, Lemma 6.2])

(5.10) $\mathcal{E}nd\mathcal{L} = \mathcal{A},$

up to an $\mathcal{A}$-isomorphism of the $\mathcal{A}$-modules concerned, one concludes from (5.7) the following relation, pertaining to the set (in point of fact, affine space, cf. (5.9)) of $\mathcal{A}$-connections of a line sheaf $\mathcal{L}$ on X,

(5.11) $Conn_A(\mathcal{L}) = D + \Omega(X),$

where D stands, of course, for a given $\mathcal{A}$-connection D of $\mathcal{L}$. Thus, one gets the following correspondence (in fact, bijection) determining the previous set (affine space):

(5.12) $D' \longleftrightarrow \omega,$

such that one actually has

(5.13) $D' = D + \omega,$

where $\omega \in \Omega(X)$, that is, a "1-form" on X.

In particular, by considering the given differential triad on X, as in (5.1), thus the "sheaf of coefficients" $\mathcal{A}$ itself as a line sheaf on X, one concludes that

any $\mathcal{A}$-connection d on $\mathcal{A}$ (different from the "standard" (flat) one ∂ as in (5.1)) is given by the relation

(5.14)   (5.14.1) $d = \partial + \omega,$

where ω is a "1-form" on X, that is, an element

(5.14.2) $\omega \in \Omega(X).$

Namely, what also amounts to the same thing,

(5.15)

> any "nonstandard differential" on $\mathcal{A}$ (different from the given one $\partial \equiv$ "dx") is uniquely determined through a 1-form, say ω on X; that is, one has the correspondence
>
> (5.15.1) $d \longleftrightarrow \omega,$
>
> where now $\omega \in \Omega(X)$ is usually given (locally) by its corresponding (local) coordinate functions.

Thus, by still assuming that Ω is a vector sheaf on X as well, one concludes that

(5.16)

> the aforesaid (local) coordinate functions of ω are in effect (local) sections of $\mathcal{A}$. By employing physical parlance, the same are given by the familiar "potentials" (viz. potential functions/sections).

In particular, by considering Ω as a vector sheaf on X as before, such that

$$(5.17) \qquad\qquad rk\Omega = m \in \mathbb{N},$$

then one further obtains, when referring to (5.16),

$$(5.18) \qquad \omega|_U \in \Omega(U) = \Omega|_U(U) = (\mathcal{A}^m|_U)(U) = \mathcal{A}^m(U),$$

for any $\omega \in \Omega(X)$ and U a local gauge of Ω, which can still be considered, as such, for the line sheaf $\mathcal{L}$ at issue, as well. [Of course, the previous argument can be extended to any vector sheaf $\mathcal{E}$ on X, with $rk\mathcal{E} = n$, while Ω is also a vector sheaf on X, with $rk\Omega = m$, as before. Thus, by further considering a common local gauge, say, $U \subseteq X$, of $\mathcal{E}$ and Ω, one gets, for any $\omega \in \Omega(\mathcal{E}nd\mathcal{E})(X)$, locally on U, the relation (cf. also (5.8.2))

$$(5.19) \qquad \begin{aligned} \omega|_U &\in \Omega(\mathcal{E}nd\mathcal{E})(U) = (\Omega(\mathcal{E}nd\mathcal{E})|_U)(U) \\ &= \Omega(U) \otimes_{A(U)} M_n(\mathcal{A})(U) = M_n(\mathcal{A}^m(U)). \end{aligned}$$

That is, again $\omega \in \Omega(\mathcal{E}nd\mathcal{E})(X)$ is always locally expressed (on a given local gauge, as above) through potentials, functions, i.e., true, sections of $\mathcal{A}$; thus, the usual situation one has in practice (including physical applications as well)].

Proofs for all the preceding assertions, as well as further details, can be found in [VS: Chapt. VI; p. 29, Section 7]. On the other hand, frequent use of the above material is made in subsequent chapters of this treatise, including Part II. See the following Section 6 for an important notion concerning physical applications, still based on the space of $\mathcal{A}$-connections as above (cf. 5.2), (5.7)), that one of the so-called *"moduli space"* (see Section 6.1 in the sequel).

6 Related $\mathcal{A}$-Connections. Moduli Space of $\mathcal{A}$-Connections

By referring to the type of $\mathcal{A}$-connections considered already by the title of the present section, we first remark that this actually has, as we shall see presently below, a special bearing, for instance, on the so-called *gauge equivalent ($\mathcal{A}$-)connections*. Indeed, this is a fundamental notion in the theory of ($\mathcal{A}$-)connections, as in particular concerns physical applications of this same concept of a connection.

Classically speaking, the aforementioned gauge equivalence refers virtually to an appropriate change of (local) charts, which of course amounts to a corresponding (local) action of the group of "diffeomorphisms" of (the base manifold) X; finally, this entails, in turn, a similar action of the group of automorphisms of the ($\mathbb{C}$-vector) bundle (vector sheaf) under consideration, domain of definition of the $\mathcal{A}$-connections involved. Now it is virtually the latter issue of the matter that we essentially take into account concerning the present abstract setting. Therefore, within the aforesaid context, the term *related $\mathcal{A}$-connections* refers, in effect, to an interrelation among two $\mathcal{A}$-connections defined on given $\mathcal{A}$-modules through an $\mathcal{A}$-(iso)morphism between the latter objects.

More precisely, given the $\mathcal{A}$-modules $\mathcal{E}$ and $\mathcal{F}$ over X, and an $\mathcal{A}$-morphism

$$(6.1) \qquad \phi \in \mathcal{H}om_A(\mathcal{E}, \mathcal{F}),$$

the $\mathcal{A}$-connections $D_{\mathcal{E}}$ on $\mathcal{E}$ and $D_{\mathcal{F}}$ on $\mathcal{F}$ are said to be ϕ-related whenever the following diagram is commutative:

$$(6.2) \qquad
\begin{array}{ccc}
\mathcal{E} & \xrightarrow{\quad D_{\mathcal{E}} \quad} & \Omega(\mathcal{E}) \equiv \mathcal{E} \otimes_A \Omega \\
\Big\downarrow{\phi} & & \Big\downarrow{\phi \otimes 1} \\
\mathcal{F} & \xrightarrow[\quad D_{\mathcal{F}} \quad]{} & \Omega(\mathcal{F}) \equiv \mathcal{F} \otimes_A \Omega,
\end{array}$$

that is, whenever one has the relation

$$(6.3) \qquad D_{\mathcal{F}} \circ \phi = (\phi \otimes 1_\Omega) \circ D_{\mathcal{E}}.$$

Hence, in the particular case that ϕ is an $\mathcal{A}$-isomorphism, that is, when one has

$$(6.4) \qquad \phi \in \mathcal{I}som_A(\mathcal{E}, \mathcal{F}),$$

(6.3) can be put in the form

$$(6.5) \qquad D_{\mathcal{F}} = (\phi \otimes 1_\Omega) \circ D_{\mathcal{E}} \circ \phi^{-1} \equiv (\phi \otimes 1) \circ D_{\mathcal{E}} \circ \phi^{-1},$$

or even, in *abbreviated form*,

(6.6) $$D_{\mathcal{F}} = Ad(\phi) \cdot D_{\mathcal{E}}.$$

One still employs, concerning the last relation, the following notation:

(6.7) $$D_{\mathcal{E}} \underset{\phi}{\sim} D_{\mathcal{F}},$$

and says that the $\mathcal{A}$-connections $D_{\mathcal{E}}$ and $D_{\mathcal{F}}$ are ϕ-related, or even gauge equivalent, through (the $\mathcal{A}$-isomorphism) ϕ, as in (6.4). As we shall see presently below, the above relation (6.7) restricted to the set of $\mathcal{A}$-connections of a given $\mathcal{A}$-module $\mathcal{E}$ (cf. (5.2)) yields an equivalence relation on that set.

On the other hand, a particular instance of the previous relation (6.6) is the following one, which we shall also use in the subsequent discussion; one has

(6.8) $$D_{\mathcal{E}nd\mathcal{E}} = Ad(\theta) \cdot D_{\mathcal{E} \otimes_{\mathcal{A}} \mathcal{E}^*},$$

where θ denotes the $\mathcal{A}$-*isomorphism*

(6.9) $$\mathcal{E} \otimes_{\mathcal{A}} \mathcal{E}^* = \mathcal{H}om_{\mathcal{A}}(\mathcal{E}, \mathcal{E}) \equiv \mathcal{E}nd\mathcal{E},$$

with $\mathcal{E}$ a given vector sheaf on X (see [VS: Chapt. IV; p. 302, theorem 6.1]). Concerning (6.8), see loc. cit. p. 23; (5.40), along with p. 24 concluding remarks, following (5.44).

6.1 Moduli Space

In the particular case that $\mathcal{E} = \mathcal{F}$ as above, the set appearing in (6.4) becomes the group of $\mathcal{A}$-automorphisms of $\mathcal{E}$; namely, one has

(6.10) $$Aut_{\mathcal{A}}(\mathcal{E}) \equiv Aut\mathcal{E} := \mathcal{I}som_{\mathcal{A}}(\mathcal{E}, \mathcal{E}),$$

which is also the group sheaf of units of the $\mathcal{A}$-algebra sheaf of $\mathcal{A}$-endomorphisms of $\mathcal{E}$,

(6.11) $$\mathcal{H}om_{\mathcal{A}}(\mathcal{E}, \mathcal{E}) \equiv \mathcal{E}nd\mathcal{E},$$

that is, one still obtains, by the definitions (cf. also Lemma 6.1 below),

(6.12) $$Aut\mathcal{E} = (\mathcal{E}nd\mathcal{E})^{\cdot}.$$

See also [VS: Chapt. II; p. 138, (6.29)], concerning the terminology applied herewith, along with loc. cit. Chapt. IV, p. 282, Lemma 1.1, pertaining to the proof of the last relation above (6.12); indeed, based still on the proof of the last quoted result, one gets the following useful (cf. (6.12)) generalized version of it:

Lemma 6.1 Let $\mathcal{E}$ be a unital $\mathcal{A}$-algebra sheaf on a topological space X (see [VS: Chapt. II; p. 138, Definition 6.2]) and

(6.13) $$\mathcal{E}^{\cdot}$$

the sheaf on X generated by the presheaf

$$(6.14) \qquad\qquad U \longmapsto \mathcal{E}(U)^{\cdot}.$$

(Here U varies over the open subsets of X, while the range of (6.14) stands for the group of units of the unital $A(X)$-algebra $\mathcal{E}(U)$). Then (6.14) yields a complete presheaf (of groups) on X, so that one has,

$$(6.15) \qquad\qquad \mathcal{E}^{\cdot}(U) = \mathcal{E}(U)^{\cdot},$$

within a group isomorphism, for any open $U \subseteq X$.

On the other hand, by still considering a unital $\mathcal{A}$-algebra sheaf $\mathcal{E}$ on X as in the previous lemma, and based further on (6.15), one gets the following isomorphism of the group sheaves concerned (cf. also (6.13), along with [VS: Chapt. I; p. 55, (11.40)]):

$$(6.16) \qquad\qquad \mathcal{E}^{\cdot}|_U = (\mathcal{E}|_U)^{\cdot},$$

for any open $U \subseteq X$. Indeed, for any open $V \subseteq U$, one obtains

$$(6.16') \qquad (\mathcal{E}^{\cdot}|_U)(V) = \mathcal{E}^{\cdot}(V) = \mathcal{E}(V)^{\cdot} = (\mathcal{E}|_U)(V)^{\cdot} = \mathcal{E}(V)^{\cdot},$$

which thus proves (6.16)

Within the same vein of ideas, and by restricting ourselves to the (unital) $\mathcal{A}$-algebra sheaf on X,

$$(6.17) \qquad\qquad \mathcal{E}nd\mathcal{E},$$

that can be associated with any given $\mathcal{A}$-module $\mathcal{E}$ on X, as in (6.11) above, one gets, based on the definitions (see loc. cit. Chapt. I, p. 137, (6.23)), the following relation:

$$(6.18) \qquad\qquad (\mathcal{E}nd\mathcal{E})|_U = \mathcal{E}nd(\mathcal{E}|_U)$$

(isomorphism of unital $\mathcal{A}|_U$-algebra sheaves) with U open in X. Hence, in view of (6.12) and (6.16), one has

$$(6.19) \qquad\qquad (\mathcal{A}ut\mathcal{E})|_U = \mathcal{A}ut(\mathcal{E}|_U)$$

(isomorphism of group sheaves on U) for any open $U \subseteq X$ and a given $\mathcal{A}$-module $\mathcal{E}$ on X, as before. On the other hand, still within the previous framework, one concludes that

$$(6.20) \qquad\qquad (\mathcal{A}ut\mathcal{E})(U) = \mathcal{A}ut(\mathcal{E}|_U)$$

(isomorphism of groups), while one still sets, by definition,

$$(6.21) \qquad\qquad \mathcal{A}ut\mathcal{E} := (\mathcal{A}ut\mathcal{E})(X),$$

for any $\mathcal{A}$-module $\mathcal{E}$ on X. [For the proof of (6.20) one can apply either (6.21), for the open set $U \subseteq X$ itself, in conjunction with (6.19) and [VS: Chapt. I, p. 55, (11.40)], or the definition of $(\mathcal{A}ut\mathcal{E})(U)$, according to (6.10) and loc. cit. Chapt. I; p. 134, Definition 6.1].

Now, as we shall see presently, the preceding relations have a special bearing on concrete calculations in terms of the structure sheaf $\mathcal{A}$, being of particular significance for physical applications as well when in the special case considered, $\mathcal{E}$, as before, is a vector sheaf on X, while the open set $U \subseteq$, a local gauge of $\mathcal{E}$.

Thus, we assume henceforth that $\mathcal{E}$ is a vector sheaf on X, with

$$(6.22) \qquad\qquad rk\mathcal{E} = n \in \mathbb{N},$$

and U a local gauge of $\mathcal{E}$, so that according to the definitions, $\mathcal{E}$ restricted to U has the form (cf. (2.29)), along with (2.41))

$$(6.23) \qquad\qquad \mathcal{E}|_U = \mathcal{A}^n|_U = (\mathcal{A}|_U)^n,$$

within an $\mathcal{A}|_U$-isomorphism of the $\mathcal{A}|_U$-modules involved. In this connection, we also remark that

$$(6.24) \qquad \text{the local gauges of a given vector sheaf } \mathcal{E} \text{ on } X \text{ provide a basis of the topology of } X.$$

See also [VS: Chapt. II; p. 125, (4.1) and (4.6)]. ∎

Consequently, starting with (6.18), one has, within the previous context (cf. (6.23)),

$$(6.25) \qquad \begin{aligned} (\mathcal{E}nd\mathcal{E})|_U &= \mathcal{E}nd(\mathcal{E}|_U) = \mathcal{E}nd(\mathcal{A}^n|_U) \\ &= (\mathcal{E}nd(\mathcal{A}^n))|_U \equiv M_n(\mathcal{A})|_U = M_n(\mathcal{A}|_U); \end{aligned}$$

cf. also the last $\mathcal{A}|_U$-isomorphism in (6.23). In particular, based on (6.19), one now obtains

$$(6.26) \qquad \begin{aligned} (\mathcal{A}ut\mathcal{E})|_U &= \mathcal{A}ut(\mathcal{E}|_U) = \mathcal{A}ut(\mathcal{A}^n|_U) \\ &= \mathcal{A}ut((\mathcal{A}|_U)^n) = (\mathcal{A}ut\mathcal{A}^n)|_U \\ &= \mathcal{GL}(n, \mathcal{A})|_U = \mathcal{GL}(n, \mathcal{A}|_U). \end{aligned}$$

On the other hand, based on (6.18), along with [VS: Chapt. I; p. 55, (11.40)], one has

$$(6.27) \qquad \begin{aligned} (\mathcal{E}nd\mathcal{E})(U) &= ((\mathcal{E}nd\mathcal{E})|_U)(U) = (\mathcal{E}nd(\mathcal{E}|_U))(U) \\ &= (\mathcal{E}nd(\mathcal{A}^n|_U))(U) = (\mathcal{E}nd\mathcal{A}^n)(U) \\ &\equiv M_n(\mathcal{A})(U) = M_n(\mathcal{A}(U)). \end{aligned}$$

Therefore, for the particular case of a line sheaf $\mathcal{L}$ on X (for $n = 1$, as above), one gets

$$(6.28) \qquad\qquad (\mathcal{E}nd\mathcal{L})(U) = \mathcal{A}(U)$$

for any local gauge U of $\mathcal{L}$. Thus, in particular (cf. also (6.24)), one obtains

$$(6.29) \qquad\qquad\qquad \mathcal{E}nd\mathcal{L} = \mathcal{A},$$

within an $\mathcal{A}$-isomorphism (of the $\mathcal{A}$-algebra sheaves involved). [Of course, one can still get (6.28) directly for any local gauge U of $\mathcal{L}$, hence (6.29) as well, according to the relations

$$(6.30) \qquad \begin{aligned} (\mathcal{E}nd\mathcal{L})(U) &\equiv \mathcal{H}om_{\mathcal{A}}(\mathcal{L}, \mathcal{L})(U) = Hom_{\mathcal{A}|_U}(\mathcal{L}|_U, \mathcal{L}|_U) \\ &= Hom_{\mathcal{A}|_U}(\mathcal{A}|_U, \mathcal{A}|_U) = \mathcal{H}om_{\mathcal{A}}(\mathcal{A}, \mathcal{A})(U) = \mathcal{A}(U). \end{aligned}$$

See also loc. cit. Chapt. II; p. 135, (6.8), as well as p. 136, (6.18)]. In particular, based on (6.20) and (6.23), one now gets

$$(6.31) \qquad \begin{aligned} (\mathcal{A}ut\mathcal{E})(U) &= Aut(\mathcal{E}|_U) = Aut(\mathcal{A}^n|_U) \\ &= Aut((\mathcal{A}|_U)^n) = (\mathcal{A}ut\mathcal{A}^n)(U) = \mathcal{GL}(n, \mathcal{A})(U) \\ &= GL(n, \mathcal{A}(U)) = \mathcal{GL}(n, \mathcal{A}|_U)(U). \end{aligned}$$

Thus, the upshot of all the preceding calculations and the resulting relations is that, technically speaking, as already noted before, though differently,

$$(6.32) \qquad \begin{array}{l} \text{the various calculations are finally referred to the "structure sheaf" } \mathcal{A}, \\ \text{as, for instance, in (5.1) above, hence, according to our hypothesis, to a} \\ \text{inital commutative } \mathbb{C}\text{-algebra sheaf on } X. \end{array}$$

This, of course, reminds us of the classical relevant utterance of N. Bohr ("Bohr correspondence principle"); cf. Chapt. II: (3.8′) in the sequel.

So, by considering a vector sheaf $\mathcal{E}$ on X (with $rk\mathcal{E} = n \in \mathbb{N}$), one concludes that

$$(6.33) \qquad \begin{array}{l} \text{any local automorphism of } \mathcal{E}, \text{ or local section of (the group) } \mathcal{A}ut\mathcal{E} \text{ (cf.} \\ \text{(6.12)) over a local gauge } U \text{ of } \mathcal{E} \text{ "is" that one of } \mathcal{A}^n, \text{ or of } \mathcal{A}ut(\mathcal{A}^n) = \\ \mathcal{GL}(n, \mathcal{A}), \text{ respectively, over } U. \end{array}$$

Thus, precisely speaking, we first conclude that

$$(6.34) \qquad \begin{array}{l} \text{any local automorphism of a given vector sheaf } \mathcal{E} \text{ over a local gauge } U \text{ of} \\ \mathcal{E} \text{ (see (6.19), along with (6.26), as well as, (6.35) below) is (tantamount,} \\ \text{i.e., modulo an } \mathcal{A}|_U\text{-isomorphism to) a local automorphism of } \mathcal{A}^n \text{ over} \\ U, \text{ where } n = rk\mathcal{E}. \end{array}$$

In this connection, by a local automorphism of $\mathcal{E}$ over U, one means (see thus (6.26)) an element

$$(6.35) \qquad \begin{aligned} \phi \in (\mathcal{A}ut\mathcal{E})|_U &= Aut(\mathcal{E}|_U) = Aut(\mathcal{A}^n|_U) \\ &= (\mathcal{A}ut\mathcal{A}^n)|_U = \mathcal{GL}(n, \mathcal{A})|_U = \mathcal{GL}(n, \mathcal{A}|_U). \end{aligned}$$

On the other hand, by virtue of (6.20) and (6.31), one also gets, as a clarification of the corresponding part of our claim in (6.33), that

(6.36)   any (local) section of $\mathcal{A}ut\mathcal{E}$ over a local gauge U of $\mathcal{E}$, as above, is (virtually identified with) a local section of $\mathcal{A}ut(\mathcal{A}^n) \equiv \mathcal{A}ut\mathcal{A}^n$ over U, with $n \in \mathbb{N}$ the rank of the given vector sheaf $\mathcal{E}$ on X.

In this regard, we still note that (6.34) and (6.36) are, in effect, according to the definitions, equivalent formulations of one and the same fact; viz. of the local "identification" of $\mathcal{A}ut\mathcal{E}$ with $\mathcal{A}ut\mathcal{A}^n$ over any local gauge of $\mathcal{E}$, a consequence, of course, of our hypothesis for $\mathcal{E}$ itself. (See (6.35), (6.31), and (6.26), as above; a sheaf is its sections for that matter, cf. [VS: Chapt. I; p. 12, Section 3]. Finally, concerning the preceding material, see also loc. cit., Chapt. VI; p. 92, (17.30)).

The group (6.12) is usually called the *gauge group of* $\mathcal{E}$, being thus within the present abstract setting a sheaf of groups on X (nonabelian, unless $n = 1$; cf. (6.22)). Furthermore, as an alternative to the preceding (yet occasionally, in practice, even a more convenient point of view), one may consider, in place of the sheaf (of groups), as in (6.12) above, equivalently (loc. cit. Chapt. I, p. 73, Theorem 13.1), its corresponding (complete) presheaf of (local) sections

(6.37) $$\Gamma(\mathcal{A}ut\mathcal{E})$$

(see also (6.31), as well as, (6.24)). Thus, taking (6.31) into account, one still refers to

$\mathcal{GL}(n, \mathcal{A})$ (cf. (1.30)) or its (complete) presheaf of sections

(6.38)   (6.38.1) $$\Gamma(\mathcal{GL}(n, \mathcal{A}))$$

as the gauge group(s) of it associated with a given vector sheaf $\mathcal{E}$ on X.

See also (6.34) and (6.36), or even (6.33).

On the other hand, what is mostly of concern to us in conjunction with the particular applications we are going to consider in Part II of this treatise, the geometry of Yang–Mills fields, is, in effect, the

action of the gauge group of $\mathcal{E}$,

(6.39)   (6.39.1) $$\mathcal{A}ut\mathcal{E}$$

(or in any other equivalent form of it, as before; cf. (6.37), or even (6.38)), on the set (affine space) of $\mathcal{A}$-connections of $\mathcal{E}$,

(6.39.2) $$Conn_{\mathcal{A}}(\mathcal{E})$$

(see (5.2) and/or (5.9) in the preceding).

Indeed, for any element

(6.40) $$\phi \in \mathcal{A}ut\mathcal{E} := (\mathcal{A}ut\mathcal{E})(X)$$

(see (6.21)) and any $\mathcal{A}$-connection D of $\mathcal{E}$, viz. an element

$$(6.41) \qquad\qquad D \in Conn_{\mathcal{A}}(\mathcal{E}),$$

one defines a new $\mathcal{A}$-connection of $\mathcal{E}$,

$$(6.42) \qquad\qquad \phi \cdot D \equiv \phi(D) \equiv \hat{D}(\phi),$$

according to the relation

$$(6.43) \qquad \phi(D) \equiv \phi \cdot D := Ad(\phi) \cdot D \equiv \phi D \phi^{-1} := (\phi \otimes 1) \circ D \circ \phi^{-1}$$

(see also, e.g., (6.5) or (6.6) along with [VS: Chapt. VI; p. 89, (17.15), (17.16)]). In other words, one gets a map

$$(6.44) \qquad \tau : Aut\mathcal{E} \times Conn_{\mathcal{A}}(\mathcal{E}) \longrightarrow Conn_{\mathcal{A}}(\mathcal{E}),$$

given by the relation (cf. (6.43))

$$(6.45) \qquad (\phi, D) \longmapsto \tau(\phi, D) := \phi \cdot D \equiv \phi D \phi^{-1},$$

proven to be a group action, as claimed by (6.39); see also loc. cit., p. 90, (17.17). In this context, see further Chapter II in Part II of this treatise, Section 2, in particular (2.26).

Of course, as follows from the definitions, any group action always defines an *equivalence relation on the "action space,"* that is, for the case at hand, on the set (6.39.2), as in (6.44). Thus, concerning the equivalence relation at issue, we further set

$$(6.46) \qquad\qquad D \underset{\phi}{\sim} D'$$

for any two $\mathcal{A}$-connections D and D' of $\mathcal{E}$ for which there exists an element $\phi \in Aut\mathcal{E}$ such that one has

$$(6.47) \qquad\qquad D' = \phi \cdot D$$

(cf. (6.42), (6.43)). Therefore,

(6.48)
the equivalence relation defined by (6.46), or by (6.47) and (6.43), coincides, in effect, with the one hinted at by (6.7) (see also (6.5)). Accordingly, we still get, by (6.44), or even by (6.46), a "gauge equivalence" of the $\mathcal{A}$-connections concerned.

Consequently, in view of the preceding, the quotient set corresponding to (6.44) is denoted by

$$(6.49) \qquad\qquad \mathcal{M}(\mathcal{E}) \equiv Conn_{\mathcal{A}}(\mathcal{E}) / Aut\mathcal{E}$$

and called the *moduli space of* $\mathcal{E}$ (in point of fact, *of the $\mathcal{A}$-connections of* $\mathcal{E}$). [As already said, we return to the present material in Part II of this study; see Chapter II, in particular, Section 2, as well as subsequent chapters.]

Suppose we are given an $\mathcal{A}$-connection D of $\mathcal{E}$, viz. an element

(6.50)
$$D \in Conn_{\mathcal{A}}(\mathcal{E}).$$

[Here, for convenience, we assume as before that $\mathcal{E}$ is a given vector sheaf on X. However, a more general setting, taking $\mathcal{E}$ as an $\mathcal{A}$-module on X, can still be considered; see [VS Chapt. VI; p. 86, Section 17].] So, by looking at the quotient set (6.49), the equivalence class of D in that set is given by the relation

(6.51) $[D] \equiv \mathcal{O}_D := \{D' \in Conn_{\mathcal{A}}(\mathcal{E}) : D' \underset{\phi}{\sim} D, \phi \in Aut\mathcal{E}\}.$

We call the previous set the *orbit of D in $Conn_{\mathcal{A}}(\mathcal{E})$ under the action of $Aut\mathcal{E}$* on that space, as in (6.44), or just, the *orbit of D*, the rest of the previous terminology being, as usual, understood.

Thus, according to the preceding, one still obtains the following relations pertaining to the orbit of D, as above. That is, one has (cf. also (6.42) and (6.43))

(6.52)
$$\mathcal{O}_D = \{\phi \cdot D : \phi \in Aut\mathcal{E}\} = \{Ad(\phi) \cdot D : \phi \in Aut\mathcal{E}\}$$
$$= \hat{D}(Aut\mathcal{E}) \subseteq Conn_{\mathcal{A}}(\mathcal{E}).$$

Therefore, based on the definitions, one further concludes that

given an $\mathcal{A}$-connection D of $\mathcal{E}$, the orbit of D,

(6.53)   (6.53.1) $\mathcal{O}_D \subseteq Comm_{\mathcal{A}}(\mathcal{E}),$

is the set of those $\mathcal{A}$-connections of $\mathcal{E}$ that are gauge equivalent to D with respect to $Aut\mathcal{E}$ (see (6.46), (6.47) and (6.43)).

On the other hand, by still applying the language of the theory of transformation groups (see, for instance, Ph. Tondeur [1: pp. 13ff.]), one realizes, in view of the preceding, that

the *affine space of $\mathcal{A}$-connections of $\mathcal{E}$*,

(6.54.1) $Conn_{\mathcal{A}}(\mathcal{E}),$

is an $Aut\mathcal{E}$-space partitioned by the corresponding orbits of its elements, each one of which, as, for instance (cf. (6.52)),

(6.54.2) $\mathcal{O}_D = \hat{D}(Aut\mathcal{E}),$ with $D \in Conn_{\mathcal{A}}(\mathcal{E}),$

providing thus, within the same space $Conn_{\mathcal{A}}(\mathcal{E})$, an image of the group (of transformations at issue),

(6.54.3) $Aut\mathcal{E} = (Aut\mathcal{E})(X).$

Thus, one can still write, symbolically,

(6.54)   (6.54.4) $Conn_\mathcal{A}(\mathcal{E}) = \sum_D \mathcal{O}_D \equiv \sum_D \hat{D}(Aut\mathcal{E})$

(set-theoretic sum; we still have here an issue of the general principle of what we may call *Gel'fand duality*).

7 Curvature

Physically speaking, the notion in the title of this section is what one virtually realizes when dealing with a "field," which in our abstract terminology corresponds to an $\mathcal{A}$-connection. Thus,

(7.1)   the curvature appears to be (or is realized as) the result (outcome)of an $\mathcal{A}$-connection.

Indeed, as we shall presently see, in contradistinction to the case of an $\mathcal{A}$-connection, which, as already shown in the preceding, does not in general exist, one always gets the curvature of a given $\mathcal{A}$-connection, provided, of course, one affords the appropriate differential set-up (this latter fact being actually another clearcut consequence of the present abstract approach to the standard differential geometry).

Thus, to formulate the notion of curvature,when an $\mathcal{A}$-connection is given, one needs the classically called, *1st prolongations* of the ("differentials") operators already involved thus far; viz. those of the standard $\mathcal{A}$-connection ∂, as well as of the given one D of an $\mathcal{A}$-module $\mathcal{E}$. However, in point of fact, only the 1st prolongation of ∂ is actually needed, that one of D being otherwise concluded.

Therefore, suppose we have a differential triad

(7.2) $(\mathcal{A}, \partial, \Omega)$

on a topological space X, and let us further assume that we are given a $\mathbb{C}$-linear morphism

(7.3) $d^1 : \Omega^1 (\equiv \Omega) \longrightarrow \Omega^2 := \Omega^1 \wedge \Omega^1$

(see also (1.12) for the notation applied) such that the following relation holds:

(7.4) $d^1(\alpha \cdot s) = \alpha \cdot d^1(s) - s \wedge \partial(\alpha) = \alpha \cdot d^1(s) + \partial(\alpha) \wedge s,$

for any $\alpha \in \mathcal{A}(U)$, and $s \in \Omega^1(U)$, with U open in X. Furthermore we suppose that

(7.5) $d^1 \circ \partial = 0,$

that is, equivalently,

(7.6) $\operatorname{im} \partial \subseteq \ker d^1$.

We call the operator d^1, as above, the *1st exterior derivation*, or even *1st exterior derivative operator*, while, occasionally, we still employ, for convenience, the notation;

(7.7) $\partial \equiv d^0$,   along with $\mathcal{A} \equiv \Omega^0$,

so that one has, up to this point, the following (finite) sequence (of "differentials")

(7.8) $0 \xrightarrow{} \mathbb{C} \xrightarrow{\varepsilon} \Omega^0 \xrightarrow{d^0} \Omega^1 \xrightarrow{d^1} \Omega^2$,

which, of course, a priori, is not exact at any place! (apart from the first one; see also (1.5), (1.6), as well as (1.16)). [The exactness of (7.8) at any place, save the trivial one, viz. at $\mathbb{C}$, is, classically speaking, connected with the Poincaré lemma, something, of course, that is not valid, in general. However, there do exist important particular examples for which the lemma still holds, apart from the classical case of C^∞-manifolds: See thus [VS: Chapts. X, XI], as well as Part II of this treatise, Chapt. IV, Section 5.] On the other hand, we still note that, in view of (7.5), one has

(7.9) $d^1 \circ d^0 \equiv d^1 \circ \partial = 0$,

as well as, by virtue of (1.5),

(7.10) $d^0 \circ \varepsilon \equiv \partial \circ \varepsilon = 0$.

Therefore,

(7.11)   the sequence (7.8), as above, is a complex of ($\mathbb{C}$-) vector space sheaves, however, not in general exact. (See also [VS: Chapt. III; p. 146], concerning the terminology employed).

Now, given an $\mathcal{A}$-module $\mathcal{E}$ on X (cf. (7.2)) and an $\mathcal{A}$-connection D on $\mathcal{E}$, one can further define the *1st prolongation of* D, provided we have the two differential operators

(7.12) $d^0 \equiv \partial$ and $d^1 \equiv d$.

Thus, we set

(7.13) $D^1 : \mathcal{E} \otimes_{\mathcal{A}} \Omega^1 \equiv \Omega^1(\mathcal{E}) \longrightarrow \Omega^2(\mathcal{E}) \equiv \mathcal{E} \otimes_{\mathcal{A}} \Omega^2$,

such that

(7.14) $D^1(s \otimes t) := s \otimes d(t) - t \wedge D(s) = s \otimes d(t) + D(s) \wedge t$,

for any $s \in \mathcal{E}(U), t \in \Omega(U)$, and open $U \subseteq X$. We note here that the operator D^1, as in (7.13), is uniquely defined, as a $\mathbb{C}$-linear morphism satisfying (7.14); see [VS: Chapt. VIII; p. 188, Lemma 1.1].

The previous map D^1, as defined by (7.14), is called the *1st covariant exterior derivation* (or *derivative operator*), or just, as already said, the 1st prolongation of D. In the same point of view, as with (7.12) (see also (7.7)), we also set

$$(7.15) \qquad\qquad D \equiv D^0.$$

On the other hand, by looking at (7.14), (7.4), and (7.7), we realize that we can further consider

$$(7.16) \qquad\qquad d^1 \quad \text{as the 1st prolongation of} \quad \partial \equiv d^0.$$

(See also Volume II of this treatise; Chapt. I; Section 1).

Thus, we now define the map

$$(7.17) \qquad\qquad R(D) \equiv R := D^1 \circ D^0 \equiv D^1 \circ D,$$

which, in view of the preceding, is depicted by the following diagram:

$$(7.18)$$

We call the above map, as defined by (7.17), the curvature of the given $\mathcal{A}$-connection D of $\mathcal{E}$. The sequence (see also (7.12))

$$(7.19) \qquad\qquad (\mathcal{A}, \partial, \Omega^1, d, \Omega^2)$$

or simply the pair

$$(7.20) \qquad\qquad (\partial, d) \equiv (d^0, d^1)$$

is called a *curvature datum* on X, while we also refer then to X as a *curvature space*.

On the other hand, based on the definitions, one easily verifies that

$$(7.21) \qquad\qquad R \in Hom_{\mathcal{A}}(\mathcal{E}, \Omega^2(\mathcal{E})) = \mathcal{H}om_{\mathcal{A}}(\mathcal{E}, \Omega^2(\mathcal{E}))(X);$$

therefore, in the case that $\mathcal{E}$ is in particular a vector sheaf on X, one has

$$(7.22) \qquad\qquad R \in ((\mathcal{E}nd\mathcal{E}) \otimes_{\mathcal{A}} \Omega^2)(X) \equiv \Omega^2(\mathcal{E}nd\mathcal{E})(X).$$

See [VS: Chapt. VIII; p. 192, Lemma 2.1]. The above two last relations show in the case of an $\mathcal{A}$-module in general or even in particular in that of a vector sheaf, that

(7.23)   the curvature of an $\mathcal{A}$-connection is not just a $\mathbb{C}$-linear morphism, as happens with an $\mathcal{A}$-connection, but, in effect, an $\mathcal{A}$-morphism, that is, a sheaf morphism that respects the "sheaf of coefficients" $\mathcal{A}$. Practically speaking, the curvature is thus by definition, a "tensor," hence of a quite "*geometric*" nature.

The preceding comments are in complete contrast to what has already been hinted at in the foregoing concerning the notion of an $\mathcal{A}$-connection (see, e.g., the beginning of the section), the latter being, as already said, of an algebraic/analytic nature. Accordingly,

(7.24)   the outcome of an $\mathcal{A}$-connection (viz. of a "potential" or even of a "field") is "geometric" in character; that is, so appears the corresponding curvature to the $\mathcal{A}$-connection at issue, or (by still applying physical language), the so-called "*field strength*." We have thus again presented the situation one usually has in physical phenomena, that "causality" (viz. the ($\mathcal{A}$-)connection, field, potential) precedes (i.e., causes) the result, "field strength" (curvature), which is still for that matter what we virtually perceive! [Usually, "dynamics" (causality) is of an analytic/algebraic nature, while in turn, the result, viz. "kinematics," expresses the "geometry" (curvature). This is reminiscent of Finkelstein's "flow follows fracture"; see also below, Chapt. III; (2.41).]

A further cohomological justification of the above will be discussed in the sequel too. See, for instance, Chapt. III; (3.56) and (3.57), along with Chapter V; Section 4, pertaining to a relevant aspect in terms of geometric prequantization; ibid (4.21).

7.1 Local Form of the Curvature

For convenience we assume here that we have a vector sheaf $\mathcal{E}$ on X, the latter being a given curvature space (cf. (7.19)). (Of course, one could consider instead more generally, as we already have occasionally in the preceding, an $\mathcal{A}$-module $\mathcal{E}$ on X along with a local gauge $U \subseteq X$ of it; cf. Section 2.2.)

So, assume that

(7.25) $$rk\mathcal{E} = n \in \mathbb{N},$$

and let

(7.26) $$\mathcal{U} = (U_\alpha)_{\alpha \in I}$$

be a local frame of $\mathcal{E}$, while we still suppose that D is an $\mathcal{A}$-connection of $\mathcal{E}$. Hence, by taking any local gauge, say U of $\mathcal{E}$, in general (cf. 2.29)) one gets, in view of (7.25) (see also (7.22), (2.42), (3.14) and (2.29)),

$$R|_U \in \Omega^2(\mathcal{E}nd\mathcal{E})(U) = (\mathcal{E}nd\mathcal{E})(U) \otimes_{A(U)} \Omega^2(U)$$
$$= End(\mathcal{E}|_U) \otimes_{A(U)} \Omega^2(U)$$
$$= End(\mathcal{A}^n|_U) \otimes_{A(U)} \Omega^2(U)$$
$$= M_n(\mathcal{A}|_U) \otimes_{A(U)} \Omega^2(U) = M_n(\Omega^2(U)).$$

(7.27)

That is,

(7.28)   the local form of the curvature of an $\mathcal{A}$-connection D on a vector sheaf $\mathcal{E}$ over X of rank $n \in \mathbb{N}$ with respect to a local gauge $U \subseteq X$ of $\mathcal{E}$, is given by the relation

(7.28.1) $$R|_U \equiv (\omega_{ij}^U) \in M_n(\Omega^2(U)),$$

that is, by an $n \times n$ matrix with entries (local) sections of Ω^2 over U, viz. local "2-forms" on X.

Therefore, by considering now a local frame $\mathcal{U}$ of $\mathcal{E}$ as in (7.26), one gets the following 0-cochain of matrices:

(7.29)
$$R \equiv (\omega^{(\alpha)}) \equiv ((\omega_{ij}^{(\alpha)})) \in \prod_\alpha M_n(\Omega^2(U_\alpha))$$
$$= \prod_\alpha \Omega^2(\mathcal{E}nd\mathcal{E})(U_\alpha) = C^0(\mathcal{U}, \Omega^2(\mathcal{E}nd\mathcal{E}))$$

(see also (7.27) and (7.28) along with [VS: Chapt. III; p. 175, (4.11) and p. 234, Lemma 8.1]). However, based further on (7.22), as well as loc. cit., p. 178, Lemma 4.1, in particular, (4.28), one concludes that

(7.30)   the curvature R, as given by (7.29), is virtually a 0-cocycle of $\Omega^2(\mathcal{E}nd\mathcal{E})$ with respect to $\mathcal{U}$ (ibid.) and not just a 0-cochain of the same. Namely, one actually obtains

(7.30.1) $$R \in Z^0(\mathcal{U}, \Omega^2(\mathcal{E}nd\mathcal{E})) = \Omega^2(\mathcal{E}nd\mathcal{E})(X)$$

(see (7.22)).

On the other hand, by taking into account the local $\mathcal{A}$-connection matrix of D corresponding to U as before, say

(7.31) $$\omega \equiv (\omega_{ij}) \in M_n(\Omega^1)(U) = M_n(\Omega^1(U))$$

(see (2.45)), one proves, concerning the previous relation (7.28), that (cf. also (7.12), for the notation)

(7.32) $$R|_U \equiv R = d\omega + \omega \wedge \omega.$$

The above fundamental relation, which yields the local form of the curvature R in terms of that for the given $\mathcal{A}$-connection D of $\mathcal{E}$ with respect to a fixed (however, arbitrarily given) local gauge U of $\mathcal{E}$ as before is called the ("*second*") *Cartan's structural equation* by extending to our abstract case the corresponding classical terminology. (See also Section 8, concerning the homonymous "first" one, referring, as is classically the case as well, to the (local form of the) torsion of R; cf. (8.41).)

7.2 Transformation Law of Field Strength (Curvature)

We have already seen in the preceding (cf. (2.67), along with (2.70) and (2.71)) the way a given "potential" ($\mathcal{A}$-connection) is transformed under the action of a "gauge," namely of an automorphism (sheaf isomorphism) global or local, of the sheaf, in fact of the $\mathcal{A}$-module, domain of definition of the potential, under consideration. So we consider the analogous phenomenon, referring now to the corresponding curvature (alias field strength), that is, to the upshot of a given $\mathcal{A}$-connection.

Thus, by taking for convenience as before a vector sheaf $\mathcal{E}$ on X (see, e.g., (7.2)), with

$$(7.33) \qquad rk_{\mathcal{A}}(\mathcal{E}) \equiv rk\mathcal{E} = n \in \mathbb{N},$$

let us further consider a local gauge transformation of $\mathcal{E}$, say

$$(7.34) \qquad \begin{aligned} g \in (\mathcal{A}ut\mathcal{E})(U) &= Aut(\mathcal{E}|_U) = Aut(\mathcal{A}^n|_U) = (\mathcal{A}ut\mathcal{A}^n)(U) \\ &= \mathcal{GL}(n, \mathcal{A})(U) = GL(n, \mathcal{A}(U)) = \mathcal{GL}(n, \mathcal{A}|_U)(U) \end{aligned}$$

(see also (6.20) and (6.31) in the preceding). Hence, equivalently, one has a local automorphism (viz. "gauge") *of* $\mathcal{A}^n$, while according to the definitions, we further look at the open set $U \subseteq X$ as above as a local gauge of $\mathcal{E}$; thus, by definition and (7.33), one gets the familiar relation

$$(7.35) \qquad \mathcal{E}|_U = \mathcal{A}^n|_U,$$

valid within an $\mathcal{A}|_U$-isomorphism of the $\mathcal{A}|_U$-modules concerned, in effect, *sine qua non*, for (7.34). Consequently, as we usually say,

> a local gauge transformation of a given vector sheaf $\mathcal{E}$ on X of rank $n \in \mathbb{N}$ is virtually reduced to a similar one of $\mathcal{A}^n$ (viz. by definition of the (local) model of $\mathcal{E}$). That is, by virtue of (7.34), one has
>
> $$(7.36.1) \qquad Aut(\mathcal{E}|_U) = Aut(\mathcal{A}^n|_U),$$

(7.36)

> where the open set $U \subseteq X$ satisfies (7.35). [Of course, the last relation, being otherwise a straightforward consequence of (7.35), is here meant from the point of view of the first relation in (7.34); see also (2.66) in the preceding.]

Note 7.1 The way we have considered the "local gauge transformation" of a given vector sheaf $\mathcal{E}$ on X or, equivalently, of its (local) model $\mathcal{A}^n$ is characteristic of the manner we look, within the present abstract setting, at the objects we are interested in: Indeed, directly at them, and not via the underlying space, here the base space X of the sheaves involved, as is usually the case in the classical theory (anyhow, X is here, in general, simply a topological space). This fact, once more, fits well with the needs and perspectives of the mathematical and theoretical physics of today, namely, avoiding the surrounding and/or underlying space, making thus reference directly to the objects themselves that "live" on the "space."

Now, based on Cartan's structural equation (see (7.32)), as well as on the transformation law of potentials (cf. (2.71) or even (2.56.1)), one now obtains the transformation law of curvature; viz. one has the relation

$$(7.37) \qquad R' = Ad(g^{-1})R \equiv g^{-1}Rg,$$

yielding the local form of changing the curvature $R(D) \equiv R$ of a given $\mathcal{A}$-connection D on a vector sheaf $\mathcal{E}$ on X. By employing here our previous notation, as in (7.29), one can still write, for the "transformation law of field strengths," the relation

$$(7.37') \qquad R^{(\beta)} = Ad(g_{\alpha\beta}^{-1})R^{(\alpha)},$$

for any α, β in I (cf. (7.26) such that $U_{\alpha\beta} \equiv U_\alpha \cap U_\beta \neq \emptyset$. (See also (2.56.1), (2.56.2) in the preceding.) Here we have set

$$(7.38) \qquad \begin{aligned} g &\equiv (g^{(\alpha)}) \in C^0(\mathcal{U}, \mathcal{GL}(n, \mathcal{A})) \\ &:= \prod_\alpha \mathcal{GL}(n, \mathcal{A})(U_\alpha) = \prod_\alpha GL(n, \mathcal{A}(U_\alpha)), \end{aligned}$$

with respect to a local frame $\mathcal{U}$ of $\mathcal{E}$, as in (7.26). Furthermore,

$$(7.39) \qquad Ad : \mathcal{GL}(n, \mathcal{A}) \longrightarrow (\mathcal{E}nd(M_n(\mathcal{E})))^\cdot \equiv \mathcal{A}ut(M_n(\mathcal{E}))$$

stands for the adjoint representation of $\mathcal{GL}(n, \mathcal{A})$ in $\mathcal{A}$, defined by the relation

$$(7.40) \qquad Ad(\alpha) \cdot s := \alpha s \alpha^{-1},$$

for any $\alpha \in \mathcal{GL}(n, \mathcal{A})(U) = GL(n, \mathcal{A}(U))$ and $s \in M_n(\mathcal{E})(U) = M_n(\mathcal{E}(U))$ with U open in X.

In particular, suppose we are given a line sheaf $\mathcal{L}$ on X, which is further endowed with an $\mathcal{A}$-connection D. Thus, by anticipating the terminology we employ from Chapter III on, we assume that we have a Maxwell field

$$(7.41) \qquad (\mathcal{L}, D)$$

on X (loc. cit. (1.4)). Therefore, applying (7.37) for the case at issue, viz. for $n = 1$, we conclude that

$$(7.42) \qquad R' = R.$$

That is,

by considering a Maxwell field

(7.43)   (7.43.1) $(\mathcal{L}, D)$

on a curvature space X (cf. (7.19)), the curvature $R(D) \equiv R$ of D does not change under (the action of) any (local) gauge transformation.

Therefore, what amounts to the same thing,

there is no other "space" concerning a given Maxwell field (take, e.g., a photon, carrier of the electromagnetic field, cf. Chapt. III, Definition 1.1) apart from the "space" that is determined by its carrier (viz. by the photon itself).

Yet, within the same vein of ideas, the same relation (7.42) can still be expressed by saying figuratively that

(7.44)

(7.44.1)   "light travels naturally."

Of course, the same holds true for any boson whatsoever (Maxwell field), for instance, the graviton (cf. Chapter IV in Part II of this treatise); see (7.42) as well as (7.43).

The above comments in (7.44) are to be connected with similar ones in (7.24), as well as with those in [VS: Chapt. VIII; p. 203 (4.32)]. We also refer to the above quotation for further details and proofs of the preceding material; see, in particular, Chapt. VIII: Sections 2–4.

8 Fundamental Identities of the Curvature (Continued). Torsion

We continue in this section by obtaining, within the present abstract setting, further fundamental relations pertaining to the curvature of an $\mathcal{A}$-connection on a given vector sheaf $\mathcal{E}$ on X, the latter being, by assumption, a curvature space (see (7.19)). The same space will still be endowed with further "differential" operators (in point of fact, sheaf morphisms, as was also the case hitherto) as the particular case at issue might demand.

Indeed, to formulate the identities we want, within the abstract framework of our study we need first the notion of the 2nd prolongation of the basic differential operator $\partial \equiv d^0$ (cf. also (7.7)), or else second exterior derivative operator, that is, the following $\mathbb{C}$-linear morphism of the $\mathcal{A}$-modules concerned (see also (7.3)),

(8.1) $d^2 : \Omega^2 \longrightarrow \Omega^3 := \Omega^1 \wedge \Omega^1 \wedge \Omega^1,$

such that one sets

(8.2) $d^2(s \wedge t) := d^1(s) \wedge t - s \wedge d^1(t) \equiv ds \wedge t - s \wedge dt = ds \wedge t + dt \wedge s,$

for any s, t in $\Omega^1(U)$ and U open in X. We further assume that

(8.3)
$$d^2 \circ d^1 \equiv d \circ d = 0.$$

Of course, we assumed above that we were basically given a curvature space (datum, cf. also (7.19))

(8.4)
$$(\mathcal{A}, \partial, \Omega^1, d^1 \equiv d, \Omega^2; X)$$

with respect to a topological space X, where in general, we set

(8.5)
$$\Omega^n := \bigwedge_{i=1}^{n} \Omega^i, \qquad n \in \mathbb{N},$$

as well as, for symmetry (see also (7.7)),

(8.6)
$$\Omega^0 \equiv \mathcal{A} \quad \text{and} \quad d^0 \equiv \partial.$$

Thus, the new (finite) sequence of $\mathcal{A}$-modules and $\mathbb{C}$-linear (sheaf) morphisms involved is now the following:

(8.7)
$$(\mathcal{A}, \partial \equiv d^0, \Omega^1, d^1 \equiv d, \Omega^2, d^2 \equiv d),$$

or, for short,

(8.8)
$$(\partial, d^1, d^2).$$

For convenience, we usually call (8.7) a *Bianchi space*, by referring simply to X, while (8.7) or even (8.8) is, strictly speaking, the corresponding *Bianchi datum*. Of course, in this connection, we always assume that (7.5) and (8.3) are in force, while the operators d^1 and d^2 have been defined by (7.3), (7.4), and (8.2), respectively.

Thus, a Bianchi space is a curvature space (cf. (7.19), (7.20)) that is further equipped with the second exterior derivative operator $d^2 \equiv d$, as given by (8.2), satisfying also (8.3). All told, in a given Bianchi space X, as before (see, for instance, (8.8), along with the previous comments thereon), one has the following:

the 1st prolongation of $\partial \equiv d^0$, viz. d^1, is so defined that (7.5) and (8.3) hold. Thus, summarizing, we assume that

(8.9)

(8.9.1)
$$d^1 \circ \partial = 0$$

and

(8.9.2)
$$d^2 \circ d^1 = 0.$$

Accordingly, we have arrived, supplementing (7.8), at the following *Bianchi complex* (viz. a complex of $\mathbb{C}$-vector space sheaves on X):

(8.10)
$$0 \longrightarrow \mathbb{C} \overset{\varepsilon}{\longrightarrow} \Omega^0 \equiv \mathcal{A} \overset{\partial \equiv d^0}{\longrightarrow} \Omega^1 \overset{d^1}{\longrightarrow} \Omega^2 \overset{d^2}{\longrightarrow} \Omega^3.$$

As a first application of the previously defined operator d^2, we apply it, in view of (8.2), to Cartan's structural equation (cf. (7.29)); in this connection, we recall here that according to (7.25), the curvature R is locally a matrix of 2-forms, so that we can still extend d^2 "coordinatewise" to any matrix-type $\mathcal{A}$-module, say

$$(8.11) \qquad\qquad M_n(\mathcal{E}) := M_n(\mathcal{A}) \otimes_\mathcal{A} \mathcal{E}, \qquad n \in \mathbb{N},$$

for any $\mathcal{A}$-module $\mathcal{E}$ on X. Thus, one obtains

$$(8.12) \qquad\qquad dR = [R, \omega] \equiv R \wedge \omega - \omega \wedge R,$$

where ω stands for the corresponding (local) $\mathcal{A}$-connection matrix of D, the given $\mathcal{A}$-connection on $\mathcal{E}$ (see (2.45), or even (2.46)). Equation (8.12) is called the ("second") *Bianchi identity* (see also (8.27)).

On the other hand, by still referring to (8.12), one concludes that

$$(8.13) \qquad\qquad dR - [R, \omega] = dR + [\omega, R] = DR,$$

where for brevity's sake we set (cf. (8.16) below)

$$(8.14) \qquad\qquad DR \equiv D^2_{\mathcal{E}nd\mathcal{E}}(R).$$

Therefore, one further obtains that

the (second) Bianchi identity, as in (8.12), is equivalent to the relation

(8.15) $(8.15.1) \qquad\qquad DR = 0.$

We call the last relation the *"differential" Bianchi's identity*. (In this connection, cf. also [VS: Chapt. VIII; p. 224, Theorem 7.1.)

Of course, we have here to clarify the notation employed in (8.14): thus, as already indicated in (8.14), for brevity's sake, we have set

$$(8.16) \qquad\qquad D \equiv D^2_{\mathcal{E}nd\mathcal{E}}.$$

That is, formally speaking, one considers the 2nd prolongation, or even 2nd covariant derivative operator (or just covariant exterior derivation), which is associated with the $\mathcal{A}$-connection of the $\mathcal{A}$-module (vector sheaf) $\mathcal{E}nd\mathcal{E}$, the latter being provided by a given $\mathcal{A}$-connection D of $\mathcal{E}$ (cf. also (3.13)).

Thus, given a Bianchi space X (see (8.8)) and an $\mathcal{A}$-module $\mathcal{E}$ on X endowed with an $\mathcal{A}$-connection D, that is, given a pair

$$(8.17) \qquad\qquad (\mathcal{E}, D)$$

as above, one defines the 2nd prolongation of D as the $\mathbb{C}$-linear morphism

$$(8.18) \qquad\qquad D^2 : \Omega^2(\mathcal{E}) \longrightarrow \Omega^3(\mathcal{E}),$$

given by the relation

$$(8.19) \qquad D^2(s \otimes t) := s \otimes d^2(t) + t \wedge D(s) \equiv s \otimes dt - D(s) \wedge t,$$

for any $s \in \mathcal{E}(U)$ and $t \in \Omega^2(U)$, with U open in X. We also write (8.19) in the form

$$(8.20) \qquad D^2 := 1_{\mathcal{E}} \otimes d^2 + 1_{\Omega^2} \wedge D.$$

Therefore, in the case of (8.16), one has the operator (see (8.18))

$$(8.21) \qquad D \equiv D^2_{\mathcal{E}nd\mathcal{E}} : \Omega^2(\mathcal{E}nd\mathcal{E}) \longrightarrow \Omega^3(\mathcal{E}nd\mathcal{E}),$$

while, as already seen (cf. (7.22)), one has

$$(8.22) \qquad R \in \Omega^2(\mathcal{E}nd\mathcal{E})(X) = Hom_A(\Omega^2, \mathcal{E}nd\mathcal{E}),$$

as well as the relation

$$(8.23) \qquad \mathcal{E}nd\mathcal{E} \equiv \mathcal{H}om_A(\mathcal{E}, \mathcal{E}) = \mathcal{E} \otimes_A \mathcal{E}^*$$

in the case that $\mathcal{E}$ is, in particular, a vector sheaf on X. In this connection, see also [VS: Chapt. IV; p. 302, Theorem 6.1, and p. 304, Corollary 6.1].

All told, one thus concludes that concerning the curvature $R \equiv R(D)$ of a given $\mathcal{A}$-connection D on a vector sheaf $\mathcal{E}$ on X, the latter being a Bianchi space, one actually obtains the following:

$$(8.24) \qquad dR = [R, \omega] \text{ if and only if } DR = 0.$$

It is usually the last relation, that is, the relation

$$(8.25) \qquad DR = 0,$$

that is practically referred to as "Bianchi's identity." This same relation will be further applied, as is exactly the case in the classical theory, in our abstract setting, as well as in dealing with Yang–Mills fields; see Chapters I–IV in Volume II of the present work.

Note 8.1 By looking at Bianchi's identity as given by (8.25), one further remarks that the same relation can be conceived, in view of (8.13), as expressing equivalently (see (8.24)) that

$$(8.26) \qquad \text{the 2nd exterior derivative operator is actually reduced, through it, to a type of a "Lie derivative" (operator) with respect to the } \mathcal{A}\text{-connection matrix } \omega \text{ (cf. (2.45), or (2.54)).}$$

That is, one obtains

$$(8.27) \qquad d^2(R) \equiv DR = -\mathcal{L}_\omega(R) := -[\omega, R] = [R, \omega]$$

(see also the comments following (8.10)), or the relation

(8.28) $$d^2 \equiv d = -\mathcal{L}_\omega$$

concerning the evaluation of the curvature (field strength) $R \equiv R(D)$ corresponding to a field ($\mathcal{A}$-connection)

(8.29) $$\omega \longleftrightarrow D$$

(see (2.54)).

8.1 Pullback of Curvature

Suppose we have a curvature space X (cf. (7.19), or (7.20)), and let

(8.30) $$f : Y \longrightarrow X$$

be a continuous map. Based on the definitions, one easily proves that

(8.31)   the pullback, via f of a given curvature datum on X provides a curvature datum on Y.

Based on the second Cartan's structural equation (cf. (7.32)), and on the definitions, one obtains the

(8.32)   commutativity of the pullback functor corresponding to a (continuous!) map f, as in (8.30), and the curvature operator, the latter being applied to a given $\mathcal{A}$-connection D of an $\mathcal{A}$-module $\mathcal{E}$ on X. That is, one has

(8.32.1) $$f^* \circ R = R \circ f^*.$$

Indeed, one gets the relation

(8.33) $$f^*(R(D)) = R(f^*(D)),$$

with $f^*(D)$ being the pullback via f of a given $\mathcal{A}$-connection D of an $\mathcal{A}$-module $\mathcal{E}$ on X (cf. (3.48)), thus an $f^*(\mathcal{A})$-connection on $f^*(\mathcal{E})$, the latter being an $f^*(\mathcal{A})$-module on the topological space Y, the pullback on it, through f, of the given $\mathcal{A}$-module $\mathcal{E}$ on X.

In particular, by considering the (canonical) continuous injection

(8.34) $$U \underset{\scriptscriptstyle i_U}{\overset{}{\subset\longrightarrow}} X$$

of an open subset U of a topological space X, where the latter space is endowed with a curvature datum as before, one gets, by virtue of (8.33), the relations

(8.35) $$R|_U \equiv R(D)|_U := i_U^*(R(D)) = R(i_U^*(D)) \equiv R(D|_U),$$

that is, the usual relation

$$(8.36) \qquad\qquad R(D)|_U = R(D|_U),$$

which also fully explains our previous (local) argument in (7.32), that refers to the restriction of a given curvature datum on the topological space X concerned to an open subset U of X.

On the other hand, within the same vein of ideas, one can further consider the curvature of other "induced $\mathcal{A}$-connections," as, for instance, $\mathcal{A}$-connection, Whitney sum, tensor product, or dual $\mathcal{A}$-connection. For this, however, we refer to the pertinent places of [VS]: see Chapt. VIII; p. 231, Section 9.

8.2 Torsion

Suppose now that we are given a curvature space X, as above (see (7.19)), while we further assume that Ω^1 is a vector sheaf on X whose dual vector sheaf

$$(8.37) \qquad\qquad (\Omega^1)^* := \mathcal{H}om_A(\Omega^1, A) \equiv \mathcal{E}$$

is endowed with an $\mathcal{A}$-connection D. [In this regard, one still remarks that in view of our hypothesis for Ω^1, the latter is reflexive, viz. one has $(\Omega^1)^{**} = \Omega^1$, so that $(\Omega^1)^*$ has an $\mathcal{A}$-connection if and only if this is the case for Ω^1; cf. loc.cit., Chapt. IV; p. 299, Theorem 5.1, and Chapt. VII; p. 122, (4.20).]

On the other hand, suppose that

$$(8.38) \qquad\qquad \theta \equiv (\theta_1, \dots, \theta_n) \in \Omega^1(U)^n$$

is a local section-basis of a given local gauge U of Ω^1, where we assume that

$$(8.39) \qquad\qquad rk\,\Omega^1 = n \in \mathbb{N}$$

(cf., for instance, (2.40)). Finally, suppose that (see also (8.11))

$$(8.40) \qquad\qquad \omega \in M_n(\Omega^1(U)) = M_n(\Omega^1)(U)$$

is the local $\mathcal{A}$-connection matrix of D local gauge of $\mathcal{E} \equiv (\Omega^1)^*$ associated with the corresponding to U, as above (viz. the "dual" of U, or, equivalently, of (8.38), of course, one can consider the open set $U \subseteq X$, as before, as a common "domain of definition" of both the local gauges of $\mathcal{E}$ and Ω^1, the two differing thus by the corresponding local section-bases, defined both on U).

Thus, one now defines the (local) torsion of D over the open $U \subseteq X$, as above, according to the following formula:

$$(8.41) \qquad\qquad \Theta|_U \equiv \Theta := d\theta + \omega \wedge \theta \equiv d^1(\theta) + \omega \wedge \theta.$$

So the previous relation is thus, within our abstract setting, the classical first Cartan's structural equation.

On the other hand, suppose that the above topological (in point of fact, in view of our hypothesis, curvature) space X is, in particular, a Bianchi space (cf. (8.7)). Then, one gets the so-called Ricci's lemma, referring to the derivative of the torsion Θ, as the latter is given by (8.41). That is, one obtains the following relation (first Bianchi's identity):

$$(8.42) \qquad\qquad d\Theta = R \wedge \theta - \omega \wedge \Theta.$$

Indeed, the assertion is a straightforward consequence of applying $d^2 \equiv d$ to (8.41), by still taking into account (8.2) and (8.3), as well as (7.32). ∎ In this regard, see also [VS: Chapt. VIII; p. 236, Section 10].

9 $\mathcal{A}$-Connections Compatible with $\mathcal{A}$-Metrics

We start by first recalling the notion of an $\mathcal{A}$-metric (see [VS: Chapt. IV; Section 8).
Suppose that we are given a (partially) ordered algebraized space

$$(9.1) \qquad\qquad (X, \mathcal{A}),$$

that is, a $\mathbb{C}$-algebraized space, as usual (cf. (1.4)), for which now the underlying $\mathbb{R}$-algebraized space carries a suitable partial order; namely, roughly speaking, an appropriate subsheaf (open subset) of $\mathcal{A}$ is singled out, having the pertinent properties, as, for instance, motivated by the standard notion of a partial order, of course now sheaf-theoretically presented. Moreover, let $\mathcal{E}$ be an $\mathcal{A}$-module on X.

An $\mathcal{A}$-metric on $\mathcal{E}$ is, by definition, a sheaf morphism

$$(9.2) \qquad\qquad \rho : \mathcal{E} \oplus \mathcal{E} \longrightarrow \mathcal{A},$$

which is $\mathcal{A}$-bilinear (relative to the $\mathcal{A}$-modules involved) symmetric, and positive definite. The last property of ρ implies that ($\mathcal{A}$-isomorphism into)

$$(9.3) \qquad\qquad \mathcal{E} \underset{\tilde{\rho}}{\subseteq} \mathcal{E}^* \equiv \mathcal{H}om_{\mathcal{A}}(\mathcal{E}, \mathcal{A})$$

("non-degeneracy" of ρ), where we set

$$(9.4) \qquad\qquad \tilde{\rho}(s)(t) := \rho(s, t)$$

for any s, t in $\mathcal{E}(U)$ and every open $U \subseteq X$. The stronger condition ($\mathcal{A}$-isomorphism)

$$(9.5) \qquad\qquad \mathcal{E} \underset{\tilde{\rho}}{\cong} \mathcal{E}^*$$

specifies the given $\mathcal{A}$-metric ρ as strongly non-degenerate. (Warning! We still recall at this point that in the classical (finite-dimensional) case the previous conditions (9.3) and (9.5) are, in effect, equivalent!)

Thus, a *Riemannian $\mathcal{A}$-metric* ρ on a given $\mathcal{A}$-module $\mathcal{E}$ on X is defined now as a sheaf morphism like (9.2) that is $\mathcal{A}$-bilinear, symmetric, positive definite, and strongly nondegenerate. We call the corresponding pair

(9.6) $(\mathcal{E}, \rho)$

a *Riemannian $\mathcal{A}$-module* on X (see also (9.1)).

On the other hand, in the more general case that a sheaf morphism ρ, as in (9.2), is just $\mathcal{A}$-bilinear, symmetric, and strongly nondegenerate, then one speaks of ρ as a *semi-Riemannian* (or even *pseudo-Riemannian*) *$\mathcal{A}$-metric* on $\mathcal{E}$; we refer to the corresponding pair $(\mathcal{E}, \rho)$, as above, as a *semi-Riemannian* (or else *pseudo-Riemannian*) *$\mathcal{A}$-module* on X. (See, for instance, right below the case of a Lorentz or an Einstein *$\mathcal{A}$-metric*). [Thus, in the case under consideration, we generalize by not necessarily assuming positive definiteness of ρ.]

Now suppose that we are given a Riemannian $\mathcal{A}$-module $\mathcal{E}$ on X, as in (9.6), while D is an $\mathcal{A}$-connection on $\mathcal{E}$. We say then that D is compatible with the (Riemannian) $\mathcal{A}$-metric ρ on $\mathcal{E}$ whenever the following relation holds:

(9.7) $D_{\mathcal{H}om_{\mathcal{A}}(\mathcal{E}, \mathcal{E}^*)}(\tilde{\rho}) = 0.$

Here the $\mathcal{A}$-morphism $\tilde{\rho}$ (in point of fact, $\mathcal{A}$-isomorphism, in view of our hypothesis for ρ, cf. (9.5)) is given by (9.4). In this regard, concerning the notation that is further applied in (9.7), see also our relevant discussion in Section 3 in the preceding.

The basic result is the analogous, in our case, fundamental lemma of Riemannian vector sheaves. That is, one obtains that

> under suitable conditions for the pair $(X, \mathcal{A})$, as in (9.1), every vector sheaf $\mathcal{E}$ on X becomes a Riemannian vector sheaf

(9.8)   (9.8.1) $(\mathcal{E}, \rho)$,

> which is further endowed with an $\mathcal{A}$-connection D, compatible with the Riemannian metric ρ, in the sense of (9.7).

In this regard, see [VS: Chapt. VII; p.168, Theorem 9.1, as well as Chapt. IV; p. 326, Section 8.1, in particular, p. 327, Definition 8.5]. Thus, concerning the pair $(X, \mathcal{A})$ as in (9.1), we assume in particular that X is a paracompact (Hausdorff) space, while $\mathcal{A}$ is a strictly positive fine $\mathcal{A}$-module on X equipped with a Riemannian $\mathcal{A}$-metric (cf. (9.6)). On the other hand, one also refers here, as is actually the case in our abstract setting, to a given differential triad

(9.9) $(\mathcal{A}, \partial, \Omega)$,

which, for the case at issue, underlies the pair (9.1), as above.

We continue by considering now through the next section the Hermitian case.

9.1 Hermitian $\mathcal{A}$-Connections

We assume hereafter in this section that we are given an involutive $\mathbb{C}$-algebraized space

$$(9.10) \qquad\qquad (X, \mathcal{A}, -),$$

in such a manner that the map

$$(9.11) \qquad\qquad - : \mathcal{A} \longrightarrow \mathcal{A}$$

is an involutive automorphism (involution) of $\mathcal{A}$, making the structure sheaf $\mathcal{A}$ an involutive $\mathbb{C}$-algebra sheaf on X (see also [VS: Chapt. IV; p. 330, Section 9]).

Thus, given an $\mathcal{A}$-module $\mathcal{E}$ on X, an $\mathcal{A}$-*valued Hermitian* (or else *sesquilinear*) *inner product* on $\mathcal{E}$ is a sheaf morphism

$$(9.12) \qquad\qquad \rho : \mathcal{E} \oplus \mathcal{E} \longrightarrow \mathcal{A}$$

that is $\mathbb{Z}$-bilinear (or else biadditive) yet skew-$\mathcal{A}$-bilinear; that is, one has

$$(9.13) \qquad\qquad \rho(\alpha s, \beta t) = \alpha \cdot \bar{\beta} \cdot \rho(s, t),$$

for any α, β in $\mathcal{A}(U)$ and s, t in $\mathcal{E}(U)$ with U open in X. Finally, we still assume for ρ, as in (9.12), the property ("skew-symmetry" of ρ),

$$(9.14) \qquad\qquad \overline{\rho(s, t)} = \rho(t, s),$$

for any s, t in $\mathcal{E}(U)$, as before.

Thus, the pair

$$(9.15) \qquad\qquad (\mathcal{E}, \rho),$$

with $\mathcal{E}$ and ρ as above, is called a *Hermitian $\mathcal{A}$-module* on X, while we still refer to ρ as a Hermitian $\mathcal{A}$-metric (or just Hermitian metric) on $\mathcal{E}$.

Now, by analogy with the classical case, one also concludes that whenever we are given an involutive $\mathbb{C}$-algebraized space as in (9.10), where further X is a paracompact (Hausdorff) space and $\mathcal{A}$ a fine $\mathcal{A}$-module on X, then

$$(9.16) \qquad \begin{array}{l} \text{every vector sheaf } \mathcal{E} \text{ on } X \text{ can be endowed with a Hermitian metric,} \\ \text{provided we command a Hermitian } \mathcal{A}\text{-metric } \rho \text{ on } \mathcal{A}. \end{array}$$

We use the above result right below, while we refer for details to [VS: Chapt. IV; p. 333, Theorem 9.1].

To continue, suppose now that we are given a differential triad

$$(9.17) \qquad\qquad (\mathcal{A}, \partial, \Omega)$$

on a topological space X, while we still assume that the latter is an involutive $\mathbb{C}$-algebraized space (see (9.10)).

Thus, by considering a Hermitian $\mathcal{A}$-module

$$(9.18) \qquad\qquad (\mathcal{E}, \rho)$$

on X (cf. (9.15)), together with an $\mathcal{A}$-connection D on $\mathcal{E}$, we say that D is compatible with the Hermitian $\mathcal{A}$-metric ρ on $\mathcal{E}$ whenever the condition

$$(9.19) \qquad\qquad \partial(\rho(s, t)) = \rho(D(s), t) + \rho(s, D(t))$$

is satisfied for any s, t in $\mathcal{E}(U)$ with U open in X. We also refer to the $\mathcal{A}$-connection D, as above, as a Hermitian $\mathcal{A}$-connection on $\mathcal{E}$.

In this regard, we have still to explain our previous notation in (9.19): Thus, one obtains an extension of the given $\mathcal{A}$-metric ρ on $\mathcal{E}$, as in (9.12), according to the sheaf morphism

$$(9.20) \qquad\qquad \rho : \Omega(\mathcal{E}) \oplus \mathcal{E} \longrightarrow \Omega,$$

in such a manner that one sets

$$(9.21) \qquad\qquad \rho(s \otimes t, s') := \rho(s, s') \cdot t,$$

for any s, s' in $\mathcal{E}(U)$ and $t \in \Omega(U)$, with U any open set in X (cf. also (2.1)). For convenience, we retained the same symbol ρ in (9.12) and (9.20)). On the other hand, concerning the above definition in (9.21), we also refer for further details to [VS: Chapt. VII; p. 171, Section 7: particularly, cf. (10.6) and (10.7) therein].

In particular, assuming that $\mathcal{E}$ is a vector sheaf on X, with $rk\mathcal{E} = n \in \mathbb{N}$, and by further considering a local gauge e^U of $\mathcal{E}$ (cf. (2.29), or even (2.40)), one concludes that

$$(9.22)$$
> an $\mathcal{A}$-connection D on $\mathcal{E}$ is Hermitian if and only if one has the relation

$$(9.22.1) \qquad\qquad \omega +^t \bar{\omega} = \tilde{\partial}(\rho).$$

Concerning our previous notation in (9.22.1), we have set

$$(9.23) \qquad\qquad \omega \equiv (\omega_{ij}) \in M_n(\Omega(U)) = M_n(\Omega)(U)$$

for the (local) $\mathcal{A}$-connection matrix of D that corresponds to the given local gauge e^U of $\mathcal{E}$ (we also refer, for convenience, to the "local gauge U" of $\mathcal{E}$); cf. (2.45) or (2.46) in the preceding.

On the other hand, by still looking at (9.22.1), one further sets, in view of (9.23),

$$(9.24) \qquad\qquad \bar{\omega} \equiv \overline{(\omega_{ij})} := (\bar{\omega}_{ij}).$$

That is, one sets here as well

$$(9.25) \qquad\qquad \overline{\omega_{ij}} \in \Omega(U), \qquad \text{with } 1 \leq i, j \leq n = rk\mathcal{E},$$

referring thus, by definition, to the "conjugates" with respect to (9.11) of the "coordinates" (components) of $\omega_{ij} \in \Omega(U)$, relative also to a given local gauge of Ω

again over the open $U \subseteq X$ (we can take, for instance, Ω as a vector sheaf on X too). Accordingly, for convenience, one may assume here that

(9.26)   (9.22.1) is referred to a common local gauge $U \subseteq X$ of both $\mathcal{E}$ and Ω, the latter $\mathcal{A}$-modules being considered as vector sheaves on X.

Thus, by analogy with (9.22), our main conclusion hitherto is that

(9.27)   within a suitable set-up for X and $\mathcal{A}$, as before (see the comments below), we ascertain that

(9.27.1)   every vector sheaf on X admits a Hermitian $\mathcal{A}$-connection.

Now, concerning our phraseology at the beginning of (9.27), we actually assume therein that the topological space considered is, in particular, an enriched ordered involutive algebraized space, all this terminology referring, in fact, to appropriate conditions that we demand to be fulfilled by the structure sheaf $\mathcal{A}$ (cf., for instance, (9.10) above, along with [VS: Chapt. IV; p. 336, Definition 10.1]); furthermore, we assume that X is a paracompact (Hausdorff) space, while $\mathcal{A}$ is a fine $\mathcal{A}$-module, which is also endowed with a Hermitian $\mathcal{A}$-metric. Finally, we still suppose that Ω, as in (9.6), is a vector sheaf on X (see also, e.g., (9.26) above).

9.2 Matrices of $\mathcal{A}$-Metrics

Following the same vein of ideas as before, in particular as concerns (9.23), one is led to consider an analogous matrix, pertaining now to the $\mathcal{A}$-metric ρ on the $\mathcal{A}$-module (in fact, vector sheaf) $\mathcal{E}$ involved. Thus, by considering a local gauge

(9.28) $$e^U \equiv \{U; (e_i)_{1 \leq i \leq n}\}$$

of $\mathcal{E}$ (see also (2.40) in the preceding), one sets

(9.29) $$\rho \equiv (\rho_{ij}), \qquad 1 \leq i, j \leq n = rk\mathcal{E},$$

such that

(9.30) $$\rho_{ij} := \rho(e_i, e_j) \in \mathcal{A}(U), \qquad 1 \leq i, j \leq n,$$

as above, where, in view of (9.28),

(9.31) $$(e_i)_{1 \leq i \leq n} \subseteq \mathcal{E}(U)$$

stands for a (local) section-basis of the $\mathcal{A}(U)$-module

(9.32) $$\mathcal{E}(U) \cong \mathcal{A}^n(U) = \mathcal{A}(U)^n.$$

In this connection, see also (9.12), along with (2.36), as well as [VS: Chapt. IV; p. 320, (8.22)]. Thus, one gets the relation

(9.33) $$\rho \equiv (\rho_{ij}) \in M_n(\mathcal{A}(U)) = M_n(\mathcal{A})(U),$$

so that by taking a local frame, say

(9.34) $$\mathcal{U} = (U_\alpha)_{\alpha \in I}$$

of (the vector sheaf) $\mathcal{E}$ (see (2.53)), one actually obtains

(9.35) $$\rho \equiv (\rho_{ij}^{(\alpha)}) \in Z^0(\mathcal{U}, M_n(\mathcal{A})),$$

such that (cf. (9.33)) one sets

(9.36) $$\rho^{(\alpha)} \equiv (\rho_{ij}^{(\alpha)}) \in M_n(\mathcal{A})(U_\alpha), \qquad \alpha \in I.$$

Thus, one concludes that

(9.37) an $\mathcal{A}$-metric ρ on the vector sheaf $\mathcal{E}$ on X is (uniquely determined by) a 0-cocycle of $\mathcal{U}$, a given local frame of $\mathcal{E}$, with values in the $\mathcal{A}$-algebra sheaf on X

(9.37.1) $$M_n(\mathcal{A}) = \mathcal{E}nd(\mathcal{A}^n).$$

Concerning the terminology applied above, see also [VS. Chapt. II; p. 138, (6.29), along with Definition 6.2 therein].

On the other hand, based further on the nondegeneracy of ρ (cf. (9.3)), one obtains that

(9.38) $$\rho \equiv (\rho^{(\alpha)}) \in Z^0(\mathcal{U}, \mathcal{GL}(n, \mathcal{A})) \underrightarrow{\subseteq} Z^0(\mathcal{U}, M_n(\mathcal{A})),$$

such that one has (see also (9.36))

(9.39)
$$\rho^{(\alpha)} \equiv (\rho_{ij}^{(\alpha)}) \in \mathcal{GL}(n, \mathcal{A})(U_\alpha) = GL(n, \mathcal{A}(U_\alpha))$$
$$= M_n(\mathcal{A}(U_\alpha))^{\cdot} = M_n(\mathcal{A})^{\cdot}(U_\alpha)$$

for any $\alpha \in I$, as in (9.34). Hence, by further specifying our previous conclusion in (9.37), one finally obtains

(9.40) an $\mathcal{A}$-metric ρ on a vector sheaf $\mathcal{E}$ on X is (given by) a 0-cocycle of $\mathcal{U}$ (cf. (9.34)) with values in the group-sheaf on X

(9.40.1) $$\mathcal{GL}(n, \mathcal{A}) = M_n(\mathcal{A})^{\cdot},$$

where $n = rk\mathcal{E} \in \mathbb{N}$.

Now, by still taking into account our hypothesis for X and $\mathcal{A}$ (see the comments after (9.27)), one can also apply the analogous, to our case, classical ortho-normalization procedure to specify further our previous conclusions according to standard results. We explain all this by the following.

Scholium 9.1 By considering an ordered algebraized space X with "square root" (alias an enriched ordered algebraized space, cf. [VS: Chapt. IV; p. 336, Definition 10.1]), along with a vector sheaf $\mathcal{E}$ on X, one can locally apply (i.e., by restriction to a local gauge U of $\mathcal{E}$, see (9.28)) the standard Gram–Schmidt orthonormalization process, so that $\mathcal{E}$ acquires then locally an orthonormal gauge (loc. cit. p. 337f., and p. 340, Theorem 10.1). Indeed, the preceding is essentially rooted in the basic assumption that

$$(9.41) \qquad\qquad (\mathcal{A}, \rho)$$

is a Riemannian $\mathcal{A}$-module (ibid.). Thus, by further assuming that $\mathcal{A}$ is, in particular, a strictly positive fine sheaf and X paracompact (Hausdorff), the Riemannian $\mathcal{A}$-metric ρ, as in (9.9), can be transferred to $\mathcal{E}$ (see also (9.8) and subsequent comments therein concerning the terminology applied).

The corresponding matrix of ρ is actually constant, viz. one has

$$(9.42) \qquad \rho \in \mathcal{GL}(n, \mathbb{C})(X) = GL(n, \mathbb{C}) \underset{\longrightarrow}{\subset} \mathcal{GL}(n, \mathcal{A})(X) = GL(n, \mathcal{A}(X))$$

(see also (9.38) above, as well as, loc. cit. p. 340; (10.37)). Therefore, by referring to (9.22.1), one obtains

$$(9.43) \qquad\qquad \tilde{\partial}(\rho) = 0,$$

so that concerning the Riemannian case, one has the relation

$$(9.44) \qquad\qquad {}^{t}\omega = -\omega,$$

while for the Hermitian case one obtains

$$(9.45) \qquad\qquad {}^{t}\omega = -\bar{\omega}.$$

The last two relations characterize, in effect, the compatibility of a given $\mathcal{A}$-connection on $\mathcal{E}$ (having local $\mathcal{A}$-connection matrix ω, cf. (9.23)) with the corresponding $\mathcal{A}$-metrics at issue. Concerning the classical situation, see, for instance, M. Postnikov [1: p. 174, Proposition 4].

On the other hand, we further remark that

$$(9.46)$$
(9.42) is characteristic of the existence of a flat $\mathcal{A}$-connection on $(\mathcal{E}, \rho)$, as above, i.e., of the relation

$$(9.46.1) \qquad\qquad R(D) = 0.$$

See, for instance, concerning the classical case, C.T.J. Dodson–T. Poston [1: p. 313, Corollary 2.06]; in this connection, cf. also [VS: Chapt. VIII; p. 204: (5.4), (5.5), as well as, Chapt. VII; p. 168f., Theorem 9.1 and its proof].

We continue by examining, always within our abstract set-up, other important metrics of the classical theory, some of which will also be considered in the sequel (cf., for instance, Part II; Chapter IV).

9.3 Kähler $\mathcal{A}$-Metrics

Suppose we are given a Hermitian $\mathcal{A}$-module

$$(9.47) \qquad\qquad (\mathcal{E}, \rho)$$

(cf. (9.18)). Furthermore, assume that we are also supplied with an $\mathcal{A}$-endomorphism of $\mathcal{E}$, say

$$(9.48) \qquad\qquad \mathcal{J} \in \mathcal{H}om_{\mathcal{A}}(\mathcal{E}, \mathcal{E}) \equiv \mathcal{E}nd\mathcal{E},$$

relative to the $\mathbb{R}$-algebraized space

$$(9.49) \qquad\qquad (X, \mathcal{A}),$$

which underlies the initially given $\mathbb{C}$-algebraized space, as in (9.10), in such a manner that one has the relation

$$(9.50) \qquad\qquad \mathcal{J}^2 = -id_{\mathcal{E}} \equiv -1.$$

We call (9.48) for which (9.50) is valid a complex structure on $\mathcal{E}$.

We shall say that a given Hermitian $\mathcal{A}$-metric ρ as in (9.47) is in particular a *Kähler $\mathcal{A}$-metric* whenever it preserves the complex structure $\mathcal{J}$, as above, that is, any time one has

$$(9.51) \qquad\qquad \rho(\mathcal{J}(s), \mathcal{J}(t)) = \rho(s, t),$$

for any s, t in $\mathcal{E}(U)$, and open U in X. We still write (9.51) in the form

$$(9.52) \qquad\qquad \rho \circ (\mathcal{J}, \mathcal{J}) = \rho,$$

"invariance of ρ, relative to $\mathcal{J}$" (see also (9.2)).

On the other hand, suppose now that we are given on X, as before, a differential triad

$$(9.53) \qquad\qquad (\mathcal{A}, \partial, \Omega).$$

Thus, by taking, for convenience, the $\mathcal{A}$-module $\mathcal{E}$, as in (9.47) above, a vector sheaf on X, let us assume further that D is a Hermitian $\mathcal{A}$-connection on $\mathcal{E}$. This same $\mathcal{A}$-connection on $\mathcal{E}$, under suitable assumptions for the $\mathbb{C}$-algebraized space

$$(9.54) \qquad\qquad (X, \mathcal{A}),$$

as well as for (9.53), can already be supplied by the $\mathcal{A}$-metric ρ, as in (9.18). (See (9.26) and (9.27.1), along with the subsequent comments therein.)

We shall say that

$$(9.55) \qquad\qquad \mathcal{E} \equiv (\mathcal{E}, \rho, \mathcal{J}; D),$$

as above, is a *Kähler vector sheaf* on X whenever the complex structure $\mathcal{J}$ (cf. (9.48), (9.50)) is parallel with respect to the Hermitian $\mathcal{A}$-connection D; viz. one has

$$(9.56) \qquad\qquad D(\mathcal{J}) \equiv D_{\mathcal{E}nd\mathcal{E}}(\mathcal{J}) = 0$$

(cf. also (9.48), along with (3.13) in the preceding).

On the other hand, (9.56) is equivalent to an analogous relation, concerning a symplectic 2-form ω on X (see also Chapter V in the sequel, Definition 1.1) that is naturally associated with $\mathcal{J}$ by means of the given $\mathcal{A}$-metric ρ on $\mathcal{E}$ (cf. (9.47)). For the corresponding situation in the classical case, see, for instance, W.A. Poor [1: p. 262, Theorem 8.40, along with p. 251, Theorem 8.13]; in this regard, and by further mimicking the standard situation as it concerns the entangled 2-form ω, as above, we also assume that the $\mathcal{A}$-module Ω, as in (9.53), is given by the relation

$$(9.57) \qquad\qquad \Omega := \mathcal{E}^* \equiv \mathcal{H}om_A(\mathcal{E}, \mathcal{A})$$

(in this connection, see also loc. cit., p. 245, Definition 8.4, as well as, [VS: Chapt. IV; p. 305, Corollary 6.2]). However, see also (9.5).

9.4 Einstein $\mathcal{A}$-Metrics

The (Riemannian) $\mathcal{A}$-metrics considered so far were, by definition, symmetric and positive definite, hence, nondegenerate, as well; thus, one has in that case the relation

$$(9.58) \qquad\qquad \mathcal{E} \underset{\xrightarrow{\tilde{\rho}}}{\subseteq} \mathcal{E}^* \equiv \mathcal{H}om_A(\mathcal{E}, \mathcal{A})$$

(see, for instance, (9.2) and (9.3)). In point of fact, we assumed that the following stronger condition than (9.58) is in force:

$$(9.59) \qquad\qquad \mathcal{E} \underset{\tilde{\rho}}{\cong} \mathcal{E}^*,$$

within an $\mathcal{A}$-isomorphism of the $\mathcal{A}$-modules concerned.

On the other hand, in the case of the $\mathcal{A}$-metrics, as in the title of the present section (this also happens in the next one), we actually release the positive definiteness of the $\mathcal{A}$-metric concerned, and just assume that

$$(9.60) \qquad \begin{array}{l} \text{the } \mathcal{A}\text{-metric considered is symmetric and strongly nondegenerate. (For} \\ \text{finite-dimensional } \mathbb{C}\text{-vector spaces (classical theory) the last condition} \\ \text{is already an outcome of the metric being positive definite, or even just} \\ \text{nondegenerate; cf. (9.58).)} \end{array}$$

Thus, an *Einstein $\mathcal{A}$-metric* on a given vector sheaf $\mathcal{E}$ on X is an $\mathcal{A}$-metric

$$(9.61) \qquad\qquad \rho : \mathcal{E} \oplus \mathcal{E} \longrightarrow \mathcal{A},$$

i.e., a sheaf morphism, as indicated that is $\mathcal{A}$-bilinear, symmetric, and strongly nondegenerate, such that the following relation ("Einstein's condition") is in force:

$$(9.62) \qquad \mathcal{R}ic(\mathcal{E}) = \alpha \cdot \rho,$$

where $\alpha \in \mathcal{A}(X)$.

Concerning the notation applied in (9.62), the first member therein stands for the so-called *Ricci operator of* $\mathcal{E}$, where we further assume that

$$(9.63) \qquad (\mathcal{E}, D)$$

is a Yang–Mills field on X, that is, a vector sheaf $\mathcal{E}$ on X followed by an $\mathcal{A}$-connection D. However, the related material here is presented in detail in Part II of this treatise, Chapter IV; Sections 1.1, 1.3.

9.5 Lorentz $\mathcal{A}$-Metrics

The $\mathcal{A}$-metrics in the title of this subsection are, except for (9.62), of the same nature as in the previous subsection, viz. also symmetric and strongly nondegenerate (cf. (9.59)). Thus again, we do not assume the $\mathcal{A}$-metrics concerned to be positive definite. For the case considered, we suppose that they satisfy the following "Lorentz condition":

$$(9.64) \qquad \rho(e_i, e_j) \equiv \eta_{ij} := \begin{cases} 0, & i \neq j, \\ -1, & i = j = 0, \ 0 \le i, j \le n, \\ +1, & i = j \neq 0. \end{cases}$$

Here we assume that we are given a local gauge

$$(9.65) \qquad e^U \equiv \{U \subseteq X, \ \text{open}; \ (e_i)_{0 \le i \le n} \subseteq \mathcal{E}(U)\}$$

of a vector sheaf $\mathcal{E}$ on X (see, for instance, (9.28)) such that on the open $U \subseteq X$, one actually has, by definition,

$$(9.66) \qquad \mathcal{E}|_U = \mathcal{A}^{n+1}|_U = (\mathcal{A}|_U)^{n+1},$$

within $\mathcal{A}|_U$-isomorphisms of the $\mathcal{A}|_U$-modules concerned.

Now, $\mathcal{A}$-metrics of the above type, named *Lorentz $\mathcal{A}$-metrics,* will be considered in Part II of this work, Chapter IV, on general relativity, viewed as a gauge theory (Section 2). Thus, we relegate the discussion to that part of our exposition for any further details.

In this connection, by analogy with the classical theory, we still apply the previous type of $\mathcal{A}$-metrics in the formulation, for instance, of Einstein's equation (*in vacuo*), viz. of the relation (ibid. (3.11))

$$(9.67) \qquad \mathcal{R}ic(\mathcal{E}) = 0.$$

See, e.g., (9.62) in the preceding for the notation applied in (9.67). Thus, the Ricci operator $\mathcal{R}ic(\cdot)$, as above, is employed to a *"Lorentz vector sheaf"*

$$(9.68) \qquad (\mathcal{E}, \rho)$$

on X, the latter topological space being suitably defined according to our general set-up. However, for details we refer, as before, to Part II Chapt. IV; Section 2.

10 The Hodge ∗-Operator. Volume Form

In this final section of the present chapter we discuss the classical *Hodge ∗-operator*, within the abstract framework of $\mathcal{A}$-modules. Thus, we first assume that

(10.1) $$(X, \mathcal{A})$$

is an enriched ordered algebraized space (cf. Scholium 9.1). Furthermore, suppose that

(10.2)
> $(\mathcal{A}, \rho)$ is a Riemannian $\mathcal{A}$-module and $\mathcal{E}$ a free $\mathcal{A}$-module on X, of rank $n \in \mathbb{N}$, while
>
> (10.2.1) $$(s_i)_{1 \leq i \leq n} \in \mathcal{E}(X)^n = \mathcal{E}^n(X)$$
>
> is a given (global) gauge of $\mathcal{E}$.

Then,

> there exists an orthonormal gauge of $\mathcal{E}$, say
>
> (10.3.1) $$(\tilde{s}_i)_{1 \leq i \leq n} \in \mathcal{E}^n(X),$$

(10.3)
> such that
>
> (10.3.2) $$\tilde{\rho}(\tilde{s}_i, \tilde{s}_j) = \delta_{ij}, \qquad 1 \leq i, j \leq n.$$

Here we denote by

(10.4) $$\tilde{\rho} := \underbrace{(\rho \oplus \cdots \oplus \rho)}_{n-times} \circ (\phi \oplus \phi)^{-1}$$

the Riemannian $\mathcal{A}$-metric on the $\mathcal{A}$-module

(10.5) $$\mathcal{E} \underset{\phi}{\cong} \mathcal{A}^n,$$

which thus is (canonically) associated with the given $\mathcal{A}$-metric ρ on $\mathcal{A}$, in view of (10.2).

We can next obtain the corresponding

> volume (element) of the $\mathcal{A}$-metric $\tilde{\rho}$ of $\mathcal{A}^n$ ($\cong \mathcal{E}$, see (10.4), (10.5)), say
>
> (10.6.1) $$\omega,$$

(10.6)
> which is associated with the orthonormal gauge of $\mathcal{E} \cong \mathcal{A}^n$, according to (10.3.1).

That is, *one has,*

(10.7) $\qquad \omega = \tilde{s}_1 \wedge \cdots \wedge \tilde{s}_n \in (\det(\mathcal{A}^n))(X) = \mathcal{A}(X).$

We still refer to ω, as above (thus, a global section of $\mathcal{A}$), as the

(10.8) $\qquad$ volume (form) of the given $\mathcal{A}$-metric ρ of $\mathcal{A}$, by virtue of (10.2).

Of course, the same global section ω of $\mathcal{A}$, as before, is also defined by the relation

(10.9) $\qquad\qquad\qquad\qquad \omega := \sqrt{|\tilde{\rho}|} \cdot \varepsilon_1 \wedge \cdots \wedge \varepsilon_n,$

such that

(10.10) $\qquad\qquad\qquad\qquad (\varepsilon_i)_{1 \leq i \leq n} \subseteq \mathcal{A}^n(X) = \mathcal{A}(X)^n$

is the *canonical Kronecker gauge of* $\mathcal{A}^n$ ($\cong \mathcal{E}$, cf. (10.5)), while one still sets

(10.11) $\qquad\qquad\qquad\qquad |\tilde{\rho}| := |\det(\tilde{\rho}(\varepsilon_i, \varepsilon_j))|.$

In this connection, see also [VS: Chapt. IV; p. 342, (11.9)] for further details on the preceding terminology.

On the other hand, by virtue of our hypothesis for the $\mathcal{A}$-metric ρ (cf. (10.1), (10.2) and (10.5)), one obtains

(10.12) $\qquad\qquad\qquad\qquad \mathcal{E} \underset{\tilde{\rho}}{\cong} \mathcal{E}^* \equiv \mathcal{H}om_{\mathcal{A}}(\mathcal{E}, \mathcal{A})$

within an $\mathcal{A}$-isomorphism of the $\mathcal{A}$-modules concerned. So we come next to the definition of the Hodge ∗-operator, according to the $\mathcal{A}$-isomorphism of $\mathcal{A}$-modules

(10.13) $\qquad\qquad\qquad * : \overset{p}{\bigwedge} \mathcal{E}^* \longrightarrow \overset{n-p}{\bigwedge} \mathcal{E}^*, \qquad 1 \leq p \leqq n,$

in such a manner that one sets;

(10.14) $\qquad\qquad (*\alpha)(\beta) := \omega \cdot (\alpha \wedge \beta^{\#}) \equiv \langle \alpha \wedge \beta^{\#}, \omega \rangle \in \mathcal{A}(X)$

for any

(10.15) $\qquad\qquad\qquad\qquad \alpha \in \left(\overset{p}{\bigwedge} \mathcal{E}^* \right)(X) = \overset{p}{\bigwedge} \mathcal{E}(X)^*$

and

(10.16) $\qquad\qquad\qquad\qquad \beta \in \left(\overset{n-p}{\bigwedge} \mathcal{E} \right)(X) = \overset{n-p}{\bigwedge} \mathcal{E}(X),$

while, based on (10.12), one further sets

(10.17) $\qquad\qquad\qquad\qquad \# := \overset{n-p}{\bigwedge} \tilde{\rho}, \qquad 1 \leq p \leq n,$

such that one has

$$(10.18) \qquad \beta^{\#} = \left(\bigwedge^{n-p} \bar{\rho} \right)(\beta) \in \bigwedge^{n-p} \mathcal{E}^{*}(X) = \left(\bigwedge^{n-p} \mathcal{E}(X) \right)^{*}.$$

Therefore, one finally obtains, pertaining to the notation applied in (10.14),

$$(10.19) \qquad \alpha \wedge \beta^{\#} \in \mathcal{A}(X),$$

so that the notation "·" in the same relation stands, in view also of (10.7), for the usual (ring) multiplication in the $\mathbb{C}$-algebra $\mathcal{A}(X)$.

Furthermore, concerning the application of the operator $*$, as in (10.13), on the volume element ω (see (10.7)) one gets the relations

$$(10.20) \qquad *\omega = 1 \quad \text{and} \quad *1 = \omega,$$

where $1 \in \mathcal{A}(X)$ stands for the identity (global) section of $\mathcal{A}$ (see also (10.13) for $p = n$, as well as (10.14)).

Thus, one gets, by virtue of (10.13), an $\mathcal{A}$-isomorphism of the (free) $\mathcal{A}$-module

$$(10.21) \qquad \bigwedge \mathcal{E}^{*},$$

that is, of the exterior algebra of $\mathcal{E}^{*}$ onto itself, so that one finally concludes that (cf. also (10.12))

$$(10.22) \qquad * \in \mathcal{A}ut_{\mathcal{A}}\left(\bigwedge \mathcal{E}^{*} \right) \cong \mathcal{A}ut_{\mathcal{A}}\left(\bigwedge \mathcal{E} \right).$$

Thus, by extending here the classical terminology, we call the previous map "$*$", as, for instance, in (10.22), the $*$-operator, or the *Hodge operator* on the given free $\mathcal{A}$-module $\mathcal{E}$ on X (see (10.2), (10.5); however, cf. also (10.23) below).

Now, it is quite clear that

(10.23)   one can transfer the previous situation pertaining to the $*$-operator as in (10.13) or (10.22) locally to any given vector sheaf $\mathcal{E}$ on X. (One can further globalize, following a standard argument, by assuming X, paracompact (Hausdorff) and a suitably chosen $\mathcal{A}$; see, for example (9.41) in the preceding, along with subsequent comments.)

Scholium 10.1 By looking at the Hodge operator "$*$", as given by (10.22), one can further remark, by analogy with the classical situation (cf. L. Conlon [1: p. 80ff. and p. 212; Ex. (1)]), that

(10.24)   the only volume elements on X (cf. (10.20)) for a suitable space X (see the ensuing comments) arise in the following manner:

Suppose we are given an ordered algebraized space

$$(10.24) \qquad\qquad (X, \mathcal{A}),$$

with X paracompact (Hausdorff) and $\mathcal{A}$ a strictly positive fine sheaf on X (see [VS: Chapt. IV, p. 318, Definition 8.2, and p. 327, Definition 8.5] for the terminology). On the other hand, consider the group sheaf

$$(10.25) \qquad \mathcal{GL}(n, \mathcal{A})_+ \subseteq \mathcal{GL}(n, \mathcal{A})$$

on X generated by the complete presheaf

$$(10.26) \qquad GL(n, \mathcal{A}(U))_+ \subseteq GL(n, \mathcal{A}(U)),$$

with U open in X, whose individual elements are matrices with positive determinant (see also [VS: Chapt. IV, p. 94, Section 4, in particular, p. 295; (4.8)]). Thus, by further taking a vector sheaf $\mathcal{E}$ on X, and a local frame

$$(10.27) \qquad \mathcal{U} = (U_\alpha)_{\alpha \in I}$$

of $\mathcal{E}$, one gets, in view of the hypothesis for $\mathcal{A}$, a strictly positive partition of unity of $\mathcal{A}$, subordinate to $\mathcal{U}$, say

$$(10.28) \qquad (\phi_\alpha) \subseteq End\mathcal{A} = \mathcal{A}(X).$$

Accordingly, one can further define

$$(10.29) \qquad \tilde{\omega} := \sum_\alpha \phi_\alpha(e^{(\alpha)}) \in \mathcal{A}(X)$$

such that

$$(10.30) \qquad e^{(\alpha)} \equiv \{U_\alpha; (e_i^{(\alpha)})_{1 \le i \le n = rk\mathcal{E}}\}$$

stands for a local gauge of $\mathcal{E}$ on $U_\alpha \in \mathcal{U}$, so that $e_i^{(\alpha)} \in \mathcal{E}(U_\alpha)$, $1 \le i \le n$, while we still set in (10.29)

$$(10.31) \qquad \phi_\alpha \cdot (e^{(\alpha)}) := \phi_\alpha \cdot (e_1^{(\alpha)} \wedge \cdots \wedge e_n^{(\alpha)}), \qquad \alpha \in I$$

(cf. also (10.7), or [VS: Chapt. IV; p. 314f. and p. 341, Section 11]). Therefore, one gets

$$(10.32)$$
$$\omega \equiv [(e^{(\alpha)})] \in \mathcal{B}(\mathcal{U}) := C^0(\mathcal{U}, \mathcal{GL}(n, \mathcal{A})/\mathcal{GL}(n, \mathcal{A})_+)$$
$$\subseteq \mathcal{B} \equiv \varinjlim_{\mathcal{U}} \mathcal{B}(\mathcal{U}) \equiv \sum_\mathcal{U} \mathcal{B}(\mathcal{U}),$$

the "limit" in the last relation being taken with respect to the (upward directed) set of local frames of $\mathcal{E}$. Accordingly, one concludes that

$$(10.33)$$
for every element $\omega \in \mathcal{B}$, as in (10.32), one can consider a uniquely defined volume element on X, say $\tilde{\omega} \in \mathcal{A}(X)$, as given by (10.29). (This also explains our assertion in (10.24).)

Applications of the previous formalism will be considered in Part II of this treatise; see, e.g., Chapter I; Section 4: Yang–Mills equations, in particular, self-dual gauge fields (Section 4.5), as well as Sections 7, 8.

2

Elementary Particles: Sheaf-Theoretic Classification, by Spin-Structure, According to Selesnick's Correspondence Principle

" ... find a purely algebraic theory for the description of reality."

A. Einstein in *The Meaning of Relativity* (Princeton Univ. Press, 1988). p. 166.

" ... The methods of sheaf theory are algebraic."

H. Grauert and R. Remmert in *Coherent Analytic Sheaves* (Springer-Verlag, 1984). p. vii.

Our aim in this chapter is to express the well-known classification of elementary particles, based on their spin-structure, by means of sheaf-theoretic notions; the latter refer, in particular, to our preponderant, throughout this treatise, sheaf structure, as it actually is, an $\mathcal{A}$-module, especially, that one of a vector sheaf (see Chapter I). In this connection, the argument, that is employed is, in effect, the transcription in our case of previous considerations on the same matter by S.A. Selesnick [1]. So, among other things (see also, for instance, Chapt. IV; Section 9), one is thus able to circumvent the so-called *correspondence principle*, hence coming directly to the second quantization (quantum field theory; see Section 3 in the sequel), a fact that is very convenient indeed in geometric quantization theory (cf., for instance, D.J. Simms–N.M.J. Woodhouse [1: p. 86]). In this connection, to recall what happens in the classical situation, we may, for example, quote here H. Goldstein [1: p. 370] according to whom, " ... *if we want to quantize a field, we have first to describe it in the language of mechanics*." Following the point of view presented here, one gets another algebraic way of looking at the (states of) "elementary particles," a fact that, as we shall see (cf. Chapter V), helps also in coping with general (geometric) quantization problems (loc. cit. (5.73), in conjunction with Chapt. IV; Section 9).

1 Preliminaries. Basic Notions

To start with, it seems most appropriate to refer first to the notion of "elementary particles," or else of the "ultimate constituents of matter," according to the common usage of the term "elementary" (see, for instance, R. Levi Setti [1: p. 1]); of course, this has obviously to do with the actual application of the predicate "ultimate constituent," the latter being proved to be, in effect, quite a matter of definition, depending, in most cases, on the particular chapter of physics, under consideration (compare, for example, the corresponding use of the term in physical chemistry, or even in atomic and/or nuclear physics). On the other hand, by referring to elementary-particle physics, whose mathematical description of certain parts of which, via gauge-theoretic jargon, concerns us throughout the present study, we still remark that even in that case, one agrees to consider certain special particles as being "more elementary" than others (e.g., photons, leptons, and the like). Thus, here again, one is actually faced with the definition of the notion at issue, depending, so to say, just on the particular status quo of the theory itself.

Accordingly, by referring in the sequel to an "elementary particle," one actually means something that, literally speaking, is usually accepted as such, without knowing, in effect, that this indeed represents an ultimate constituent of matter (a fact that is reduced to a *Utopia*, rather, as it concerns human knowledge!).

Now, a possible way out of this indeterminable situation, or at least a more convenient set-up within which one can work, is to express the previous notion through those of an appropriate "field" along with concomitant "physical observables" (see below), which finally can be formulated via (sheaf, in point of fact, de Rham) cohomology notions. In this connection, see, for instance, C. von Westenholz [1: pp. 321ff.], as well as Section 7 in the subsequent discussion here. So one essentially arrives, by the previous argument, at an indirect, at first sight, viz. mathematical, way out of the aforementioned inconvenience of the terminology.

Yet another, still mathematical, way of getting around the same inconvenience as before (E. Wigner; see also, for instance, G.G. Emch [1: p. 503; Frobenius–Wigner–Mackey theorem]) is to relate the "elementary particle" under consideration to an irreducible representation of the (internal) symmetry group (e.g., the proper Poincaré group; see, for example, loc. cit. p. 500) that is naturally associated with the physical system (object) at issue (cf. also the subsequent discussion, as well as the introduction to this chapter). In this regard, we also note that the aforementioned Wigner's classification of elementary (quantum) systems refers, of course, to special relativity. An analogous extension of the same point of view to more general situations is certainly clear.

Of course, one further applies in quantum mechanics the notion of a "quasi-particle" or even "excitation," which is rather bending to a better physical interpretation, though we are not going, strictly speaking, to apply it in the sequel. So we actually restrict ourselves to an equally contiguous notion to the preceding, relative to the framework that we are going to apply in the ensuing discussion, namely to that one of a "field" (see below).

Thus (cf. also the subsequent Section 2), a (physical) particle, in our terminology, either elementary or not, will be identified with a corresponding particle field or, equivalently, with its states, the latter being simply a section of an appropriately defined vector sheaf, which finally will represent our particle. (In point of fact, as we shall see, one actually considers pairs consisting of a vector sheaf, whose rank depends on the particular spin of the particle, along with an $\mathcal{A}$-connection of the vector sheaf at issue: the Yang–Mills field; see Chapter III, as well as, Chapt. I; (4.13) in Volume II). So, to repeat it again, particle field, or even, for simplicity, just particle, will mean some particular vector sheaf, that is to be specified (see the next section, or the last comments above), and in effect (viz. equivalently: "a sheaf is its sections," after all, see e.g. [V.S: Chapt. I; Section 3]), its (sheaf of germs of) sections! In this context, by referring to the previous terminology, we actually mean, by definition, *free* (or even *bare*) *particles*, namely, those being in "interaction" with no others (however, see also Section 9.2 in the sequel).

Scholium 1.1 By further referring to the same notion of "field" as above, it is worth noticing here that nowadays, this is considered in physics an "independent, not further reducible fundamental concept" (see A. Einstein [1: p. 140]); indeed, as we shall also realize by the subsequent discussion, the same is really a very convenient way, thus far, of looking at (physical) reality, being essentially of an "algebraic nature" (cf. Chapters III and I, Vol. II). By employing the previous notion under the form of a "particle field" when it is further cohomologically interpreted (see Section 7 below, in particular (7.32)), this still contributes, for instance, to the quantum conception of gravity (cf. Chapt. IV; Section 9 in Vol. II, in particular, (9.19) and (9.20)).

In this connection, see also A. Einstein-N. Rosen [1], as well as, S. Weinberg [1].

2 Classification of Elementary Particles, Through Vector Sheaves, According to Their Spin-Structures

For convenience, we first comment, in brief, on the classical aspect of the intended characterization, as alluded to in the title of this section, which one usually considers for elementary particles, based on their *spin*. We then turn to the corresponding classification followed by S.A. Selesnick [1], which, as already said, is further adopted in the present discussion always within our sheaf-theoretic set-up (see, e.g., Section 6, in particular (6.2)).

2.1 Standard Classification of Elementary Particles by Spin Number

It is generally accepted today to classify all the known elementary particles into two broad subfamilies as to whether their spin (intrinsic angular momentum) is an integer or half-integer. Thus, according to this classification, and denoting that structure constant of the particle by s, one has

(2.1) $$\text{either } s \in \mathbb{Z}, \text{ or } s \in \tfrac{1}{2} \cdot \mathbb{Z}_{odd},$$

where $\mathbb{Z}_{odd}$ stands for the odd integers; hence, one can also set simply

(2.1′) $$s \in \frac{1}{2} \cdot \mathbb{Z}.$$

See also, for instance, R. Levi Setti [1: p. 6]; in this connection, we still remark that in physics one usually describes the above situation by saying that the "spin is quantized" according to the classical Stern–Gerlach experiment (see, e.g., E. Prugovečki [1: p. 9f.]).

On the other hand, it is also standard that one has an equivalent way of classifying "elementary particles" in terms of quantum-statistical mechanics: This refers to the case of many identical particles (cf., for instance, P.A.M. Dirac [1: p. 207f.]), where then one is faced with the invariance of the respective Hamiltonian, under any permutation of the particles, in effect of their characteristics appearing in the Hamiltonian (e.g., spin, position). Thus, one distinguishes particles that present symmetric states (that remain unchanged by permutations of the particles in the sense indicated above), their statistics being named (after the physicists who first studied the phenomenon)

(2.2)   Bose–Einstein statistics

and the respective particles

(2.2′)   bosons,

while those with antisymmetric states are said to obey the

(2.3)   Fermi–Dirac statistics

(in honor of E. Fermi and P.A.M. Dirac, who considered the latter case), and named

(2.3′)   fermions.

In this connection, cf. also, for instance, W. Pauli [2: p. 197]. Thus, in conclusion, one obtains, according to the definitions, as above, that

(2.4)   *bosons* are (elementary) particles with integer spin-number, while *fermions* are those with half-integer spin-number.

In this regard, see also, for example, W. Pauli [1: p. 116 ftn 1, and p. 125]. Furthermore, it is worth noticing here that the above conclusion (2.4) was based, for a certain time, just on "experimental evidence" (loc. cit.), while within the framework of relativistic quantum field theory, one can already refer to it as a consequence (or even an equivalent formulation) of the so-called *"spin-statistics theorem"*; see R.F. Streater–A.S. Wightman [1: p. 150, Theorem 4.10], or N.N. Bogolubov et al. [1: p. 532, Theorem 20.2]. See also the ensuing discussion, in particular (3.14) and (4.1) of Sections 3 and 4, respectively.

Of course, there is still the classical interpretation of spin, in terms of the theory of group representations, Clifford algebras, and their automorphism groups (spinor groups). In this context, see, for instance, R. Deheuvels [1: p. 285ff. and p. 401f.], as well as I.M. Gel'fand et al. [1]. See also A.I. Kostrikin–Yu.I. Manin [1: pp. 164ff.].

Thus, based on the preceding, we come next, as already said, to the classification of elementary particles in terms of vector sheaves, strictly speaking of their (local) sections.

2.2 Classification of Elementary Particles Through Module-Structures (à la Selesnick)

To consider a free (alias "bare") elementary particle is, of course, quite an idealized situation, representing, for theoretical convenience, a hypothetical state of the physical system, at issue, the latter coming to (viz. perceived by) us, in effect, in a "dressed" form that it has acquired after all the occasional interactions that it has undergone from its creation until the time of its observation, that is, of our experimental measurement! Indeed, it is exactly the second word of the last expression that mostly interests us here: Namely, once we refer to a "measurement," we mean, of course, automatically the intervention of some "arithmetic" on the basis of which our measurements are conducted, the same arithmetic being also used, at a higher level, in the formulation of the physical laws (differential equations) that govern our experiment (thus, our arithmetic is in the form of differential calculus, or even differential geometry, and the like).

At the very end, and speaking mathematically, one employs coordinates in terms of this arithmetic, which amounts to the fact that one essentially considers free modules ("coordinates," e.g., vector spaces, namely free $\mathbb{C}$-modules in the classical case) with respect to this algebra (or field, classical situation) arithmetic. Accordingly, still arguing within the same vein of ideas and referring now to the standard "state (Hilbert) space" of quantum mechanics, that corresponds to the particular physical system under consideration, one, assumes, in view of the preceding analysis, that the following relation holds, pertaining to the state space at issue:

$$(2.5) \qquad \check{H}_{phys} = \check{H}_{bare} \oplus \check{H}_{etc}.$$

Concerning (2.5), we first have to explain the notation employed therein, and second to provide more physical grounds for the same assumption. (As already said, we essentially follow the treatment of S.A. Selesnick [1]): Thus, the first member of (2.5) stands for the actual state space of the system at issue, which, in effect, is attained at the moment of measurement, being the result of its state (space) when no interaction was present ("bare state", $\check{H}_{bare}$), along with the state (space) that the system acquires after that particular (hypothetical) moment during the experiment, denoted in (2.5) by $\check{H}_{etc}$, the latter two spaces constituting by assumption the "true" (state) space of the system. Of course, based on usual physical considerations (cf., for instance, "Hellinger–Toeplitz theorem"; see E. Prugovečki [1: p. 195, Theorem 2.10, along with comments on the top of p. 193 therein]), one actually assumes in (2.5) dense subspaces (the notation "ˇ" stands for exactly that). Furthermore, the direct sum decomposition of the (pre-) Hilbert spaces, entangled in the same relations, is vindicated by the exploitation of standard arguments of quantum scattering theory, pertaining here to the so-called *Møller* (*wave*) *operators* that define the *scattering* (or *S-)operator*. It is actually the latter, in effect a unitary operator of the Hilbert (state) space involved, that by definition describes in scattering experiments the transformation of "in-states" (prepared, hence, controlled ones) into the "out-states" (uncontrolled ones).

A further analysis of (2.5) along with a corresponding physical interpretation of it within the framework of the so-called *"second quantization"* will finally supply the desired classification. However, this will be our subject matter in a few of the subsequent sections.

3 Quantum State Modules

Our aim in the present section is to explain how the direct sum ((pre-) Hilbert space) decomposition in (2.5) can actually be construed, when working within quantum field theory (second quantization), as an analogous decomposition of $\mathbb{A}$-modules with respect to an appropriate *algebra of coefficients*, say, $\mathbb{A}$, to be defined presently. Still following here the idea of S.A. Selesnick [1], let us consider, as a guiding example, the formula that gives us the "operator field" that corresponds to the second quantization of the Dirac field (electromagnetism, cf. Section 5 in the sequel) in terms of the one-particle wave functions; thus, one has

$$(3.1) \qquad \phi(x, t) = \sum_{i=1}^{k} u_i(x, t) a_i$$

(see also, for instance, J.D. Bjorken–S.D. Drell [2: p. 49; (13.18)]). Here the u_i's stand for the aforementioned *single-particle wave functions* (solutions of the single-particle Schrödinger equation, ibid., p. 45, (13.3)), while the a_i's represent *annihilation* (alias *ladder*) *operators* (see also, e.g., A. Böhm [1: p. 19]).

Note 3.1 By looking at (3.1), we still remark that one has expressed here, in an analytic form what we call a *light ray*, that is, an element

$$(3.1') \qquad [(\mathcal{L}, D)] \in \Phi^1_{\mathcal{A}}(X) \simeq H^1(X, \mathcal{A}^{\cdot}).$$

See Chapt. III; (1.16), (2.3) in the sequel. In other words, a "number" of Maxwell fields having the same field strength (viz. a "light ray of a certain color").

So, according to the preceding (3.1), one concludes that

(3.2)   the coefficients of the (Hilbert (state) space, cf. (2.5)) operators involved in the second quantization are now functions and not just (complex) scalars, as is the case in the first quantization.

On the other hand, our previous claim in (3.2) can still be considered simply as a consequence of the definition of the (operator) $*$-algebra that corresponds to the physical system at issue (namely, of that generated by the respective ladder operators; see, for instance, A. Böhm [1: pp. 10, 19]). Therefore, as a result of (3.2), one obtains that

(3.3) the state spaces occurring in (2.5) are in fact (pre-Hilbert state) A-modules, called henceforth *quantum state modules*, with respect to the (function) algebra of coefficients involved, that is, in (3.1). (See (3.4) below; in this regard, cf. also Note 3.2, as well as Scholium 3.1 in the sequel.)

The algebra in question is first a unital commutative (linear associative) algebra over the complexes (complex number field $\mathbb{C}$), that is, a *unital commutative $\mathbb{C}$-algebra* $\mathbb{A}$. We further set

$$(3.4) \qquad\qquad \mathbb{A} \equiv \mathcal{C}^\infty(X);$$

viz. we consider the $\mathbb{C}$-algebra in the previous sense (with pointwise defined operations), of $\mathbb{C}$-valued smooth (that is, $\mathcal{C}^\infty$-) functions on a smooth ($\mathcal{C}^\infty$-)manifold X (of finite dimension), this being the spacetime continuum.

Note 3.2 By referring to the abstract setting that we are going to adopt, it is instructive to emphasize in anticipation that X, as in (3.4), will at the final stage be an arbitrary topological space that occasionally (due to cohomological reasons) will be decreed to be paracompact (Hausdorff). On the other hand, the same algebra $\mathbb{A}$, as before, will finally be replaced by an appropriate ($\mathbb{C}$-)algebra sheaf, say $\mathcal{A}$, on the topological space X, that will thus constitute our standard domain of coefficients, alias our (generalized) arithmetic. In that sense, one has

$$(3.4') \qquad\qquad \mathbb{A} = \mathcal{C}_X^\infty \equiv \Gamma(X, \mathcal{C}_X^\infty)$$

(see also Chapter I; (1.15) for the notation applied above). Thus $\mathbb{A}$ as in (3.4) is the algebra of global sections of the $\mathbb{C}$-algebra sheaf $\mathcal{C}_X^\infty$ of germs of smooth functions on (the $\mathcal{C}^\infty$-manifold) X.

On the other hand, the physical grounds for our choice in (3.4) might be the following (cf. also S.A. Selesnick [1]): Of course, the motivation for such a choice is always (3.1), referred to the relativistic quantum field theory. Thus, working on a space-time manifold X (either Minkowskian or Lorentzian), it is always advantageous, pertaining to the differential geometry of X, to have the functions involved smooth enough, so that for convenience they are supposed to be $\mathcal{C}^\infty$. Furthermore, for technical reasons, due to experimental physical experience (thus, one usually considers bound states, in other words, laboratory constraints), the same functions as above are often appropriately smeared out; technically speaking, this actually that the functions at issue are multiplied by (Schwartz) test functions ($\mathcal{C}^\infty$-functions on X with compact support). Now, by still looking at (3.4), one obtains

$$(3.5) \qquad\qquad \mathbb{C} \underset{\longrightarrow_\varepsilon}{\subset} \mathbb{A},$$

expressing the fact that our $\mathbb{C}$-algebra $\mathbb{A}$, as in (3.4), is unital, or else it contains the constants, that is, the scalars $\mathbb{C}$. Therefore, according to the preceding,

(3.6)   the domain of coefficients for the second quantization is an extension of (contains) that of the first quantization (the scalars $\mathbb{C}$). In other words, the usual coordinates (i.e., the complexes $\mathbb{C}$), as applied in the 1st quantization, are now replaced, concerning the 2nd quantization, by the generalized coordinates, or matrix elements of the respective field operators (cf. (3.1)), as given by the elements of (3.4).

In this connection, we still remark that our new extended domain of coordinates, as in (3.4), contains, apart from the complexes $\mathbb{C}$, the aforementioned smearing fields (test functions). We thus attain, through the preceding, a more natural (mathematical) interpretation of looking at the (relativistic)

(3.7)   quantum field theory (second quantization) as the quantum mechanics of infinite systems, or of systems with an infinite number of degrees of freedom; yet, as that one in which "the [wave] field [itself], at each point of the space [-time manifold] is [now] considered, as an independent generalized coordinate.

Thus one realizes the fact that

(3.8)   it is typical of the second quantization that the domain of coordinates (coefficients of the field equations involved) is the algebra $\mathbb{A}$, as in (3.4), and not simply the scalars $\mathbb{C}$, as is the case in the first quantization.

     At this point, it is also of a special significance to remark that here too (as happens with $\mathbb{C}$ as well) the extended domain of coordinates (cf. (3.5)), alias our arithmetics as applied in the second quantization is still commutative! Accordingly, macroscopic measurements (coordinates, generalized or not) appear always to be performed within a commutative domain.

In this connection, it is still of special interest to recall at this point the relevant remarks of N. Bohr,

(3.8′)   " ... *the description of our own measurements of a quantum system must use classical, commutative c-numbers* ... "

The above also constitutes the so-called *Bohr's correspondence principle*. In this regard, see also, for instance, F. Strocchi [1: p. xi], J.D. Bjorken–S.D. Drell [2: p. 11], or L.I. Schiff [1: p. 492; (54.1)]. Furthermore, within the same spirit, as with our previous comments in (3.8), cf. also, for example, S.A. Selesnick [1: p. 1283, as well as p. 1278].

     Now, as already said, the space

$$(3.9) \qquad\qquad \check{H}_{phys}$$

in (2.5) corresponds to the actual state space of the physical system under consideration. The same space, as already explained, by our previous discussion in (3.8), is in effect an $\mathbb{A}$-module with respect to the (unital commutative) $\mathbb{C}$-algebra $\mathbb{A}$, as the latter is given by (3.4). Thus, since our measurements are (canonically) entangled with numbers (elements of our "arithmetic"; see, e.g., our relevant terminology employed in Note 3.2), it is natural to suppose (see also our comments in Section 2.2) that

(3.10)   the space $\check{H}_{phys}$, as defined by (2.5), being in effect an $\mathbb{A}$-module according to our previous discussion, is a *free* $\mathbb{A}$-*module*.

Our previous assumption in (3.10), along with that in (2.5), whose physical/mathematical grounds seems to be quite sound (!), constitutes the basic argument upon which all the ensuing discussion in this and the following few sections of the present chapter rests. See also the subsequent scholium.

Scholium 3.1 Remaining still within the context of second quantization, we should further remark that measurements, being canonically related to the coefficients of the field operators involved (see (3.1)), that is, with elements of the $\mathbb{C}$-algebra $\mathbb{A}$, as in (3.4), are naturally associated with numbers too, appearing in this framework as values of the elements ($\mathbb{C}$-valued functions). Consequently, the usual shuffle of reference that results when ascribing numbers to our calculations (in effect, c-numbers(!); see also (3.8′)), due essentially to (3.5). In this regard, cf. also S.A. Selesnick [1: p. 32, starting remarks of Section 3].
    As already remarked, in view of (3.10) and (2.5), one thus concludes that

(3.11)   the (quantum) state ($\mathbb{A}$-)module (see (3.3) and (3.8)) of a bare particle, $\check{H}_{bare}$ (cf. (2.5)), is a projective $\mathbb{A}$-module, where $\mathbb{A}$ is given by (3.4).

Indeed, according to (2.5), $\check{H}_{bare}$ is a direct summand of a free $\mathbb{A}$-module, by virtue of our assumption for the $\mathbb{A}$-module $\check{H}_{phys}$ (cf. (3.10)). Hence, our claim in (3.11) follows by the characterization of projective modules (see, for instance, S. Mac Lane [1: p. 21, Proposition 5.5], or W.A. Adkins–S.H. Weintraub [1: p. 136, Theorem 5.1]). ∎

On the other hand, it is still to be noticed that the $\mathbb{A}$-module, which contributes to the defining property of a projective $\mathbb{A}$-module, that is, of being the latter a direct summand of a free $\mathbb{A}$-module, can be, according to the essence of the aforementioned characterization, absolutely arbitrary, under the proviso, of course, that the said condition holds. So, still referring to (2.5), we further infer something of a physical significance within the previous framework. That is, one concludes that

(3.12)   the nature of the $\mathbb{A}$-module $\check{H}_{etc}$ (cf. (2.5)) does not virtually affect $\check{H}_{bare}$ of being, in view of (2.5) and (3.10), a projective $\mathbb{A}$-module.

It is our next objective to indicate that the same $\mathbb{A}$-module $\check{H}_{bare}$ as in (3.11) is in effect finitely generated: This means that there exists a finite set of generators (subset) of our $\mathbb{A}$-module such that every element of the module can be expressed as a (finite $\mathbb{A}$-)linear combination of generators with coefficients from the algebra $\mathbb{A}$. Thus, the minimal number of such generators, when it exists (cf., e.g., (3.13) below), is dubbed the rank of the $\mathbb{A}$-module concerned (see, for instance, W.A. Adkins–S.H. Weintraub [1: p. 115]). In this context, we further note that the (natural) number in question (0 is included here) concerns a $\mathbb{Z}_+$-valued rank function defined on the prime spectrum (set of prime ideals) spec$\mathbb{A}$ of our algebra $\mathbb{A}$, which is here decreed to be constant (see also S.A. Selesnick [1: p. 32, Section 3], along with H. Bass [1: p. 127, Theorem 7.1], or N. Bourbaki [4: p. 109, Theorem 1, and p. 111, Definition 1, along with the ensuing comments on the top of p. 112]). To put it formally,

(3.13)   we assume henceforth (for simplicity)! that the rank, say n, of the $\mathbb{A}$-modules $\check{H}_{bare}$, as above, is constant (thus, one has $n \in \mathbb{Z}_+$; however, as we shall see below, one actually has $n \in \mathbb{N}$).

We evaluate next the possibility of the previous rank being finite; that is, we try to sound out those physical grounds that can support such an assumption: Thus, in principle, the so-called *symmetry group* of a physical system, that is, the group labeling (or else parametrizing) the internal structure (states) of the system is usually at the classical level a compact (matrix) Lie group (see, for instance, R.W.R Darling [1: p. 223]). On the other hand, based further on the symmetry axiom, we assume that the same symmetry group acts also at the underlying quantum-mechanical level as well (see, e.g., D.J. Simms–N.M.J. Woodhouse [1: pp. 21, 86 and 150]). Consequently, the particle states under discussion may be included in a finite-dimensional subspace of the representation (Hilbert (state)) space, associated with a (unitary) irreducible representation of the compact symmetry group, as above ("finiteness theorem"; see A. Robert [1: p. 46, Corollary 5.8, or p. 69, Corollary 7.9, along with p. 14, Proposition 2.2]. See, for instance, M.A. Naĭmark [1: p. 439, Theorem 2, and p. 442, Theorem 4]). Hence, by analogy and based on the definition of the (pre-)Hilbert (state) spaces, as in (2.5), otherwise the quantum (state $\mathbb{A}$-) modules as in (3.3), we can now

(3.14)   assume that the rank of the projective $\mathbb{A}$-module $\check{H}_{bare}$, describing a bare particle (cf. (3.11)), is finite.

Therefore, a compilation of (3.11) and (3.14) now entails the following basic assumption:

(3.15)   the quantum state module $\check{H}_{bare}$ (cf. (3.3)), describing a field of bare particle states is a finitely generated projective $\mathbb{A}$-module with respect to the algebra $\mathbb{A}$, as defined by (3.4).

Of course, an $\mathbb{A}$-module when considered as a vector space over the complexes (see (3.5)) may equally well be an infinite-dimensional $\mathbb{C}$-vector space, hence, a great profit by "extending the scalars" (loc. cit., along with (3.6)), modulo, of course, our experience in that; we are still in the era of "scalar mathematics" and not yet fully in that one of "functional mathematics"! In other words, our "functional arithmetic" is still in the beginning!

It is now our final purpose to give to the same Theorem-Axiom, as in (3.15), a sheaf-theoretic version, this being in many respects more flexible, at least, as we shall see (see Section 6 in the sequel). However, before this, we are going to give a closer, more concrete, correspondence between finitely generated projective $\mathbb{A}$-modules and (bare) bosons and fermions (we call it *"Selesnick's correspondence"*; see Section 6 below). Thus, in a sense, we are going to show that our previous relation (2.5) can be construed as the physical counterpart of the classical "Serre–Swan theorem" (cf. Section 5 in the sequel), this being also the quintessence of the relevant study (classification, as alluded to above) in S.A. Selesnick [1]. So, to this end, we start the next section.

4 Free Bosons and Fermions in Terms of Finitely Generated Projective Modules

Based on our previous Theorem-Axiom, as in (3.15), the main objective of the ensuing discussion is to provide sound evidence to the claim that

(4.1)   (fields of states of bare) bosons are described by projective $\mathbb{A}$-modules of rank 1 (singly generated $\mathbb{A}$-modules), where $\mathbb{A}$ is given by (3.4). On the other hand, (bare) fermions (states) correspond to projective $\mathbb{A}$-modules of (finite) rank greater than 1.

Indeed, a *rank-1 projective $\mathbb{A}$-module*, say M, is "locally" equivalent to the algebra $\mathbb{A}$ itself, when the latter is similarly "localized": We employ here the terminology and fundamental results of (topological) commutative algebra; see, for instance, N. Bourbaki [4: Chapt. II; p. 112, Theorem 2], or pertaining in particular to the topological algebra-theoretic counterpart of the same (cf. also the comments presently below), A. Mallios [TA]. Of course, the algebra $\mathbb{A}$ here may be any unital commutative $\mathbb{C}$-algebra, not necessarily a "function algebra," as is, in particular, the case for (3.4). In this connection, it is still to be noticed that the entanglement here of the previous type of terminology (localization theory of commutative (topological) algebras) is also intimately connected, according to the standard point of view, with the sheaf-theoretic aspect of the matter (loc. cit.; see also A. Grothendieck–J. Dieudonné [1: p. 207, Corollaire 1.4.4]), which will be our final target.

For the particular case of our algebra $\mathbb{A}$, as defined by (3.4), we further remark that this is a (very nice and important too, nonnormed) topological algebra (see A. Mallios [TA: p. 131; (4.19)]) whose closed maximal ideals (constituting the so-called Gel'fand space or spectrum of $\mathbb{A}$) may be identified with the points of X, this identification being, in effect, a topological one (viz. a homeomorphism; loc. cit., p. 227, Theorem 2.1, along with Scholium 2.1). Thus, the aforementioned localization refers to the elements of M, which can now be viewed as functions on X "locally defined" only at each point of X (see also A. Mallios [12], [13]).

Consequently, the elements of M (due to the definition of M, as a projective $\mathbb{A}$-module) are (locally defined) functions (emanating) from $\mathbb{A}$, so that, based further on the field-operator representation, as considered in the preceding (see (3.1)), the operators at issue are essentially commuting ones, being in effect identified with elements of $\mathbb{A}$.

Accordingly, the corresponding wave functions are symmetric with respect to the operators involved, e.g., in (3.1). (We thus remark that everything, for instance, in (3.1) is commutative, due to our hypothesis for the algebra $\mathbb{A}$, as in (3.4), a situation that has been explained just above relative to the rank-1 projective $\mathbb{A}$-modules; see also Note 4.1 below.) This exactly entails that the respective particles (cf. (3.15)), whose (quantum) state modules are those we have just considered, their states being, in particular, represented by the previous functions, obey Bose–Einstein statistics, alias the particles at issue are bosons (cf. (2.2) and (2.2′)); in this regard, see also, for example, E. Prugovečki [1: p. 353].

Note 4.1 By still referring, for further clarification, to the symmetry of the corresponding wave functions as above, we also remark that these functions may be viewed (cf. (2.5)) as elements of M ($\equiv \check{H}_{bare}$, for the case at issue) $\cong \mathbb{A}$, "locally"(!), as explained in the preceding; thus, as alluded to before,

(4.2)   one actually transfers the symmetry of the said functions (pertaining, in effect, to the behavior of the latter, relative to an appropriate transformation of their argument, e.g., position-momentum, or "position-spin" variable) to a similar one of the respective functions-operators.

See the aforementioned "local" identification of M, as above, with (an $\mathbb{A}$-submodule of) $\mathbb{A}$. In this connection, see also J.M. Ziman [1: p. 214f; (7.1), (7.4)].

Thus, the preceding fully explains, so far, our assertion in the first part of (4.1).

On the other hand, (finite) exterior powers (over $\mathbb{A}$) of a given (finitely generated) $\mathbb{A}$-module, being always $\mathbb{A}$-modules of the same type as the given one, represent (uniquely) antisymmetric behavior (correspond to alternating maps) of the elements (local functions, as above) of the initial $\mathbb{A}$-module. (See also N. Bourbaki [3: Chapt. III; p. 82, Remarque], or S. Lang [1: p. 588, §9]).

Thus, as a result now of our previous argument, one further concludes that

(4.3)   finitely generated projective $\mathbb{A}$-modules of rank greater that 1 can appear only as (quantum) state modules, in the sense of (3.3), of antisymmetric wave functions.

In this connection, we still note that the aforesaid (wave) functions are actually those that may be associated with (bare) fermions (Pauli exclusion principle; see E. Prugovečki [1: p. 307], or J.M. Ziman [1: p. 32f]). Therefore, one thus finally infers that (bare) fermion states can be represented, within the previous set-up, only through finitely generated projective $\mathbb{A}$-modules, of rank greater that 1. (In this regard, we refer to Selesnick's treatment in [1: p. 34ff, Section 4], which was, as already mentioned, our motivation to the preceding as well as the ensuing discussion on several issues of the present chapter). So the above completely settles now our claim in (4.1).

∎

The foregoing enables us to relate the previous description of the states of (bare) elementary particles to sections of finite-dimensional (smooth) vector bundles, according to a classical identification (viz. equivalence of categories) of such bundles with modules of the above type, where $\mathbb{A}$ is given by (3.4).

This will be our subject matter for the next two sections, in order to come, as already alluded to in the preceding (cf. Section 3 above), to a sheaf-theoretic type of classification of the above. For convenience, we thus first explain through Section 5 the aforementioned identification.

5 Finitely Generated Projective Modules and Vector Bundles (Serre–Swan Theory)

The sort of correspondence hinted at by the title of this section (Serre–Swan correspondence) is a well-known theme, conceived in the late fifties or early sixties of the last century independently of one another by J.-P. Serre [1] and R.G. Swan [1]: We start with a key definition, which still explains, though very succinctly, the relevant situation.

(5.1)

Suppose we are given a topological space X and a unital commutative $\mathbb{C}$-algebra $\mathbb{A}$. Moreover, let

$$(5.1.1) \qquad \xi \equiv (E, \pi, X)$$

be a continuous n-dimensional ($\mathbb{C}$-)vector bundle over X. (See, for instance, K. Jänich [1: p. 117, Definition], or F. Hirzebruch [1: p. 45], concerning the terminology). We say that ξ is *algebraic relative to* $\mathbb{A}$ whenever one has;

$$(5.1.2) \qquad \xi = \ker(\alpha)$$

such that

$$(5.1.3) \qquad \alpha \in M_n(\mathbb{A}), \qquad \text{with } \alpha^2 = \alpha,$$

that is, whenever E is the kernel of an $n \times n$ idempotent matrix with entries from $\mathbb{A}$ (or even that of a projector in $L_\mathbb{A}(\mathbb{A}^n) \equiv L_\mathbb{A}(\mathbb{A}^n, \mathbb{A}^n) \equiv End(\mathbb{A}^n)$). In this connection, we should also remark here that (5.1.2) is only a convenient abuse of notation, the same relation referring to the finitely generated projective $\mathbb{A}$-module, say M, that is (uniquely) associated with ξ; thus, for any $\alpha \in M_n(\mathbb{A})$ as in (5.1.3), one actually defines a (continuous) map

$$(5.1.4) \qquad \check{\alpha} : X \times \mathbb{A}^n \longrightarrow X \times \mathbb{A}^n$$

such that

(5.1.5) $\check{\alpha}(x, z) := (x, \alpha(z))$

for any $(x, z) \in X \times \mathbb{A}^n$, so that one has

(5.1.6) $[\ker(\check{\alpha})] \cong [M] \in K(X)$,

with M the $\mathbb{A}$-module, as above. See also (5.12) in the sequel for the notation; A. Mallios [2: p. 462, Lemma 2.1 and p. 482, Scholium 4.1], or A. Mallios [11] for further details , along with Remark 6.2. By abusing notation, we simply write (5.1.2).

Thus, the classical Serre–Swan theorem asserts the following;

(5.2)   given a compact (Hausdorff) space X, every continuous (finite-dimensional complex) vector bundle over X is algebraic, relative to $\mathcal{C}(X)$ (the $\mathbb{C}$-algebra of $\mathbb{C}$-valued continuous functions on X).

On the other hand,

the same matrix

(5.3.1) $\alpha \in M_n(\mathbb{A})$,   with $\mathbb{A} \equiv \mathcal{C}(X)$,

(5.3)   such that (the second relation of) (5.1.3) is in force defines an n-generated projective $\mathcal{C}(X)$-module (here, $n \in \mathbb{N}$, stands for the "dimension" of the (continuous) $\mathbb{C}$-vector bundle, under consideration, in point of fact, for that of each of its fibers), and conversely (viz. any such module is actually (i.e., modulo an isomorphism of $\mathbb{A}(\equiv \mathcal{C}(X))$-modules) the kernel of a projector (idempotent matrix), as in (5.3.1) (cf. also (5.1.3)).

Thus the Serre–Swan theorem (cf. (5.2)), when we apply categorical language, can now be expressed through the relation

(5.4) $\mathcal{E}(X) = \mathcal{P}(\mathcal{C}(X))$,

valid within a category equivalence supplied by the (global) section functor. Here,

(5.5) $\mathcal{E}(X)$

stands for the *category of continuous (complex) vector bundles* over the compact space X, while

(5.6) $\mathcal{P}(\mathcal{C}(X))$

denotes the *category of finitely generated projective $\mathcal{C}(X)$-modules*. (In this regard, see also M. Karoubi [1: p. 32, Theorem 6.18, along with p. 113, §15], or J. Rosenberg [1: p. 34, Theorem 1.6.3].)

On the other hand, based on topological (algebraic) K-theory of X (loc. cit.), as well as on topological algebras theory (cf., for instance, A. Mallios [TA]), one proves that

(5.7)   every continuous complex n-plane bundle (alias continuous n-dimensional $\mathbb{C}$-vector bundle) over a compact (Hausdorff) space X is algebraic (see (5.1)) relative to any unital commutative complete locally m-convex Q-(topological) algebra (or else, to a Waelbroeck algebra) $\mathbb{A}$ whose spectrum $\mathfrak{M}(\mathbb{A})$ is homotopic to X.

In this connection, cf. A. Mallios, loc. cit., for the terminology applied above, together with A. Mallios [2: p. 485; (5.6)], concerning the proof of our previous claim in (5.7); see also (5.12) in the sequel. ■ Especially, by taking the algebra

(5.8) $\mathbb{A} \equiv \mathcal{C}^{\infty}(X)$

as in (3.4), one concludes that

(5.9) $\mathcal{C}^{\infty}(X)$, with X a compact (Hausdorff) $\mathcal{C}^{\infty}$-manifold, is a topological algebra of the type considered in (5.7), viz. a Waelbroeck algebra.

We further note that the $\mathbb{C}$-algebra $\mathcal{C}^{\infty}(X)$ is equipped with the so-called topology of uniform convergence on compacta (or topology of compact convergence) of the functions and all of their (partial) derivatives, alias $\mathcal{C}^{\infty}$-topology, or Schwartz topology; see A. Mallios [TA: p. 131; (4.19), along with p. 134, Scholium 4.1], as well as [VS: Chapt. XI; p. 371, (11.39)].

Thus, by employing an analogous notation to (5.4), one obtains the following equivalence of categories:

(5.10) $\mathcal{E}^{\infty}(X) = \mathcal{P}(\mathcal{C}^{\infty}(X))$,

with X being a compact (Hausdorff) $\mathcal{C}^{\infty}$-manifold. What amounts to the same thing, one concludes that

(5.11)   every (finite-dimensional) smooth ($\mathcal{C}^{\infty}$-) $\mathbb{C}$-vector bundle over a compact (Hausdorff) smooth manifold X corresponds bijectively (through the category equivalence (5.10) established via the (global) section functor) to a finitely generated projective $\mathcal{C}^{\infty}(X)$-module.

Scholium 5.1 Connected with the preceding comments, referring to the proof of (5.7), we further remark that the relevant argument in A. Mallios [2] is based on what one may call a topological-algebra analogue of Grauert's theorem by extension of a similar result of O. Forster [1: p. 10, Satz 6] in the case of Banach algebras; see also A. Mallios [3: p. 298, Theorem 1.1, and p. 305, Appendix]. So, by further employing topological-algebraic K-theory, one expresses (5.7), hence in particular (5.10), in view of (5.9), through the corresponding Grothendieck groups, according to the following relations:

(5.12) $K(\mathcal{C}(X)) = K(X) = K(\mathfrak{M}(\mathbb{A})) = K(\mathcal{C}(\mathfrak{M}(\mathbb{A}))) = K(\mathbb{A}),$

within isomorphisms of the abelian groups concerned (cf. A. Mallios [3: p. 299; (1.9)]).

Thus, the above constitutes the "$\mathcal{C}^{\infty}$-analogue" of the Serre–Swan theorem, which will also interest us in the sequel. (In this connection, see also K. Lønsted [1: p. 203, comments following Lemma 3.2] as well as C.J. Mulvey [1: p. 65, Corollary 4.2] for another proof of (5.11), generalization of Swan's theorem.)

We proceed now in Section 6 to the aforementioned sheaf-theoretic classification of (the states of bare) elementary particles (cf., in particular, (6.26) in the sequel).

6 Vector Sheaves and Elementary Particles (Continued: Selesnick's Correspondence)

We start with the necessary relevant definitions connecting the preceding with the pertinent sheaf-theoretic terminology applied in the present section.

Thus, suppose that we are given a topological space X and a $\mathbb{C}$-algebra sheaf $\mathcal{A}$ on X, i.e., a sheaf $\mathcal{A}$ on X, whose stalks $\mathcal{A}_x$, $x \in X$, are by definition unital commutative (linear associative) $\mathbb{C}$-algebras. (For a detailed and systematic account of the notions that employed here, we refer to A. Mallios [VS: Chapts. I–II; cf., in particular, Chapt. II, p. 106; (1.76), and p. 104; (1.67), along the ensuing comments].) In this connection, we usually refer to the corresponding pair

(6.1) $(X, \mathcal{A})$

as a $\mathbb{C}$-algebraized space.

A classical and important example of a $\mathbb{C}$-algebraized space that will also concern us in the sequel is the pair

(6.2) $(X, \mathcal{C}_X^{\infty}),$

where X is a smooth (viz. $\mathcal{C}^{\infty}$-)manifold and

(6.2′) $\mathcal{C}_X^{\infty}$

the ($\mathbb{C}$-algebra) sheaf of germs of $\mathbb{C}$-valued smooth $\mathcal{C}^{\infty}$-)functions on X (see also [VS: Chapt. III; p. 239; (8.29)]) and the following Section 6.1.

On the other hand, within our abstract set-up, as in (6.1), suppose we have a sheaf of $\mathcal{A}$-modules, or just an $\mathcal{A}$-module, on X, say $\mathcal{E}$ (see Chapt. I for the terminology employed). Thus, we shall say that the given $\mathcal{A}$-module $\mathcal{E}$ on X, as before, is locally free of finite rank, say $n \in \mathbb{N}$, whenever there exists an open covering of X, say

(6.3) $\mathcal{U} = (U_{\alpha})_{\alpha \in I},$

such that one has

(6.4) $$\mathcal{E}|_{U_\alpha} = \mathcal{A}^n|_{U_\alpha}, \qquad \alpha \in I,$$

within $\mathcal{A}|_{U_\alpha}$-isomoprhisms of the $\mathcal{A}|_{U_\alpha}$-modules concerned. Of course, it is equivalent to assume that

> for every point $x \in X$, there exists an open neighborhood U of x in X such that

(6.5)   (6.5.1) $$\mathcal{E}|_U = \mathcal{A}^n|_U$$

> within an $\mathcal{A}|_U$-isomorphism of the $\mathcal{A}|_U$-modules, under consideration. (See also [VS: Chapt. II; Section 4].)

As we shall see, it is the above property of $\mathcal{E}$ as in (6.5.1), viz. the property of $\mathcal{E}$ of being it locally free, that is of especial importance for subsequent applications (see also Note 6.1 below). In this connection, an open covering of X for which (6.4) or, equivalently, (6.5) holds is called a coordinatizing open covering, or local choice of basis, or a local frame of $\mathcal{E}$ (it is actually the latter term that we usually employ in the sequel).

Note 6.1 The previous terminology that has been applied to the open covering $\mathcal{U}$ of X (see (6.3)) satisfying (6.4) indicates in effect the meaning that sheaves of the above type might have for physical applications, connected in particular with second quantization (or with applications in quantum gravity, cf. Vol. II; Chapter IV in the sequel): Indeed, such sheaves supply, through condition (6.4), equivalently (6.5), the possibility of having local generalized coordinates in the sense that the coordinates (in fact, as we shall presently see, (local) sections of appropriate sheaves, as before) are taken now (locally(!), viz. in a certain neighborhood of each point of X) from our extended arithmetic

(6.6) $$\mathcal{A} \underset{\varepsilon \leftarrow}{\supset} \mathbb{C};$$

see (6.1) above, along with Chapt. I; (1.5).

Within the same vein of ideas,

(6.7)   an open set $U \subseteq X$ for which (6.5.1) holds will be called a *local gauge* of the $\mathcal{A}$-module $\mathcal{E}$ under consideration.

The previous terminology will be further justified by concrete relevant applications in the sequel. Now, to fix its use throughout the rest of our discussion, we still set the following definition, motivated, as we shall see, by standard particular instances. Thus, we have (cf. also A. Mallios [VS: Chapt. II, p. 127, Definition 4.3]) the following definition.

Definition 6.1 Given a $\mathbb{C}$-algebraized space $(X, \mathcal{A})$ (see (6.1)), a locally free $\mathcal{A}$-module of finite rank (usually, greater than one) over X will be called a *vector sheaf* on X. In particular, a locally free $\mathcal{A}$-module of rank 1 (viz. for $n = 1$ in (6.4), or, equivalently, in (6.5.1)) is said to be a *line sheaf* on X.

Example 6.1 Consider the $\mathbb{C}$-algebraized space

$$(6.8) \qquad\qquad (X, \mathcal{C}_X),$$

where $\mathcal{C}_X$ stands for the ($\mathbb{C}$-algebra) sheaf of germs of $\mathbb{C}$-valued continuous functions on a given arbitrary topological space X (see, for instance, loc. cit.; Chapt. I, p. 18, Section 4.2). On the other hand, let

$$(6.9) \qquad\qquad E(\xi) \equiv \xi = (E, \pi, X)$$

be an n-dimensional continuous $\mathbb{C}$-vector bundle on X (cf., for instance, D. Husemoller [1: p. 23, Definition 1.1]), and

$$(6.10) \qquad\qquad \mathcal{E} \equiv \mathcal{S}(\Gamma(E))$$

the corresponding sheaf of germs of sections of E.

For convenience, we first explain the terminology just applied in (6.10): Thus, the sheaf $\mathcal{E}$ under consideration is by definition the sheafification of (alias, the sheaf generated by) the presheaf of (continuous) sections of the given $\mathbb{C}$-vector bundle E, the presheaf in question being defined by means of the section functor Γ; in this connection, see also A. Mallios [VS: Chapt. I; p. 41, Theorem 9.1]. It is proved, in effect, that

(6.11) $\mathcal{E}$, as given by (6.10), is a vector sheaf on X, of rank n (= the (finite) dimension of E), viz. a locally free $\mathcal{C}_X$-module on X of the said rank.

A complete proof of this standard result, even in the more general case that the coefficients (instead of the complexes $\mathbb{C}$) are taken to be from a (suitable) topological algebra $\mathbb{A}$, one considers the so-called (continuous) $\mathbb{A}$-vector bundles on X, as above (take, e.g., $\mathbb{A}$ as in (3.4)), can be found in A. Mallios [5: p. 406, Theorem 1.1], where we also refer for details.

Thus, the above correspondence

$$(6.12) \qquad\qquad E \longleftrightarrow \mathcal{E} \equiv \mathcal{S}(\Gamma(E)),$$

as indicated by (6.10), is in effect a bijection as given by the following relation (loc. cit.; p. 406, (1.23)):

$$(6.13) \qquad\qquad \mathcal{V}ect_{\mathbb{C}}^n(X) = \Phi_{\mathcal{A}}^n(X).$$

In other words, there exists a

(6.13′)   one-to-one and onto correspondence between the set of isomorphism classes of n-dimensional continuous $\mathbb{C}$-vector bundles over X and the set of isomorphism classes of vector sheaves, viz. locally free $\mathcal{A}$-modules, with $\mathcal{A} = \mathcal{C}_X$ (cf. (6.8)), of (finite) rank $n \in \mathbb{N}$ over X.

On the other hand, it is also a standard fact (see, for instance, A. Mallios [VS: Chapt. I; p. 12, Section 3, cf., in particular, p. 16, (3.17)]) that

(6.14)   every sheaf is in effect (uniquely determined by) its sections.

In practice, and since the relevant notion that is included in the previous claim is of particular importance for all that follows, we wish to emphasize at this point the special benefit that is obtained when

(6.15)   any time we have to cope with a vector sheaf (vector bundle, cf. (6.13)), we argue instead, by virtue of (6.14),in terms of its sections, something that is still of paramount importance in physics (see also the subsequent comments in (6.16) below).

Indeed, to comment a bit more on our last sentence in (6.15) above, we understand that nowadays it is standard that

(6.16)   in physics we always strive for fields from the study of which we try to understand their origin (nature), which we do not actually know! However, fields (in effect, their "states") are maps, and in fact, sections of suitable vector bundles (see also Sections 8, 9 in the sequel), hence, in view of (6.13), sections of vector sheaves. Thus, to paraphrase here S. Mac Lane [1: p. 357],

(6.16.1)   "any important function should be construed, as a (continuous) section of some particular sheaf."

In this regard, see also A. Mallios [VS: Chapt. I; p. 22, (4.33)].

The preceding, in particular our previous comment in (6.16.1), might also be viewed as another vindication of the point of view advocated by this section, thus by the present chapter as well (see also (6.14), along with (6.29) in the sequel).

   On the other hand, by further commenting on the technicalities of (6.14) (see thus also loc. cit., p. 10; (2.10), (2.12)), we still remark that when dealing with a (vector) sheaf, say $\mathcal{E}$, one concludes that this is locally determined at every point $x \in X$ by a (continuous) local section, viz. by a continuous map, say

(6.17) $$s : U \longrightarrow \mathcal{E},$$

with $U \subseteq X$ an (open) neighborhood of x such that

(6.18) $$\pi \circ s = id_U,$$

that is, equivalently,

(6.18′) $$\pi(s(x)) = x, \quad x \in U.$$

In this regard, one denotes here by the triple

(6.19) $$(\mathcal{E}, \pi, X)$$

the given (vector) sheaf, or equivalently, the local homeomorphism (sheaf) π determining $\mathcal{E}$ (ibid., Chapt. I; Section 1).

Now, the above, pertaining especially to (6.17)–(6.19), are in force for any sheaf on X in general (loc. cit.). Thus, when looking in particular at a vector sheaf $\mathcal{E}$ on X, as in (6.12), and by further restricting ourselves to a local gauge U of $\mathcal{E}$ (cf. (6.7), for $\mathcal{A} \equiv \mathcal{C}_X$, as above), something that we certainly may always do, according to the definition of $\mathcal{E}$ (cf., for example, loc. cit., Chapt. II; p. 127; (4.9)), one obtains, for both E and $\mathcal{E}$, as in (6.12), the following relations:

$$(6.20) \qquad \Gamma(U, \mathcal{E}) \equiv \mathcal{E}(U) = E(U) = \mathcal{C}(U, \mathbb{C}^n) = \mathcal{C}(U, \mathbb{C})^n \equiv \mathcal{C}(U)^n,$$

within isomorphisms of the $\mathbb{C}$-vector spaces concerned (see also (6.13)); in this regard, cf. also D. Husemoller [1: p. 12, Proposition 1.5, along with the subsequent relevant comments therein], as well as A. Mallios [5: p. 403; (1.5)] for the particular case considered.

Thus, by further commenting on (6.20), one also concludes, what amounts to the same thing as (6.20), that

> each local (continuous) *section* of E, or equivalently, of $\mathcal{E}$ (cf. (6.13)), can be construed as a (locally defined) $\mathbb{C}$-vector space-valued continuous map, its range being varied with the point of application of the map at issue. Indeed, one has here an important consequence of the notion of section, where one has (see (6.18′))

(6.21)

$$(6.21.1) \qquad s(x) \in \mathcal{E}_x = E_x (\cong \mathbb{C}^n), \qquad x \in U,$$

> with $s \in \mathcal{E}(U) = E(U)$, as in (6.20). (Cf. also A. Mallios [VS: Chapt. I, p. 40; (9.16)]). Accordingly, its significance for physics, let alone, since, these same ranges (hence, the corresponding values of s, as well) are "covariantly" varied(!) (cf., for instance, (7.17) in the sequel, in conjunction with (6.13) above), while, of course, are still, connected, through the topology of $\mathcal{E}$.

Still referring to (6.21), we further remark that it is actually our last comments that provide the possibility of getting relations (viz., in effect equations) pertaining to objects living on $\mathcal{E}$: The equations at issue are expressed by means of sections of $\mathcal{E}$, thus, finally, in terms of $\mathcal{E}$ itself and of the objects under consideration, that is, invariantly(!), in other words, independently of any local study by means of which we arrived at the equations concerned (principle of local gauge invariance; that is, physics does not depend on our descriptions, viz. on how we describe it). However, in point of fact, this also substantiates the so-called principle of general relativity, that is, technically speaking,

(6.22)

> physical equations (admittedly representing physical laws) do not depend on the way ("local calculations," viz. sections, alias "coordinates") we arrive at them.

In this connection, see also M. Nakahara [1: p. 28, along with p. 10, comments in the beginning of Section 1.2]).

Accordingly, it is certainly a consequence of the preceding that

(6.23)   the notion of a sheaf (in particular, that of a vector/principal (cf. Section 8 below) sheaf) fits well with the principle of local gauge invariance, hence too with the principle of general relativity.

6.1 Smooth (C^∞-) Case

Up to this point we have mainly considered in this section continuous $\mathbb{C}$-vector bundles on a given topological space X.

On the other hand, we have already seen in Section 5 that the C^∞-analogue of Serre–Swan theorem is also in force (see (5.10), along with Scholium 5.1). Thus, it is certainly clear that all the preceding have an analogous formulation in the case of smooth (C^∞-) $\mathbb{C}$-vector bundles on a (compact Hausdorff) smooth manifold X. So, based on what has been said in Section 5 and looking at the $\mathbb{C}$-algebraized space (6.2), one obtains, by analogy with (6.13), the following bijection of the sets concerned:

$$(6.24) \qquad {}^\infty \mathcal{V}ect^n_{\mathbb{C}}(X) = \Phi^n_{\mathcal{A}}(X), \quad n \in \mathbb{N},$$

where we still set (cf. (6.2'))

$$(6.25) \qquad \mathcal{A} \equiv C^\infty_X.$$

(See also (6.20), concerning (6.24).)

Remark 6.1 A similar interpretation to (6.13'), pertaining here to the notation employed in (6.24), where smooth is now replacing continuous, is certainly clear.

On the other hand, we still remark that

(6.26)   on a compact (Hausdorff) C^∞-manifold continuous and differentiable (smooth) $\mathbb{C}$-vector bundles are (categorically speaking) the same.

The assertion is a straightforward application of (5.7) and (5.9) in the preceding. ∎ In this connection, see also A. Mallios [2: p. 490, Theorem 6.2]. Thus, in more technical terms, (6.26) can be given via the following category equivalence (cf. also (5.6) above for the notation applied herewith):

$$(6.27) \qquad \mathcal{P}(\mathcal{C}(X)) = \mathcal{P}(\mathcal{C}^\infty(X)).$$

That is, every continuous (finite-dimensional) $\mathbb{C}$-vector bundle over X (with X a smooth manifold, as in (6.26)) carries a differentiable (smooth) structure, as well. [We notice that the same result as above can be obtained as a consequence of the standard Karoubi's density theorem (cf. M. Karoubi [1: p. 109; 6.15]) when appropriately generalized; cf. A. Mallios [6]. In this regard, cf. also, for instance, J.

Rosenberg [1: p. 40; 1.6.16. (2)], however, for order of differentiability C^1. See H. Inassaridze [1: p. 312, Theorem 2.12], still within the normed algebra theory context. On the other hand, by further referring to the order of differentiability as above, we might bear in mind the classical aspect that whatever we can do on a (topological) manifold with (class) C^∞, we can actually do it with (class) C^1, as well; cf. J.R. Munkres [1: p. viii, and p. 46, Corollary 4.9, along with p. 57, Theorem 5.11].

In this connection, we should also remark that the same (6.27), as above is still in accord with, or even justifies, recent tendencies in physics, where differentiability questions are tending to be excluded altogether(!) Being in agreement with our hypothesis for X in (6.27) (however, see Remark 6.2 in the sequel), we further note that

(6.28)   in many cases of physical interest a compact manifold arises,

due to vanishing type boundary conditions that we usually impose on the particular problems (fields); see also S.J. Avis–C.J. Isham [1: p. 353], as well as the following Scholium 6.1, along with the ensuing Remark 6.2.

Thus, based now on the preceding discussion, and in particular (5.11) and (6.24), as well as (3.15) and (4.1), we thus arrive at the desired

(∗)   classification of (bare) particle states in terms of (sections of) vector sheaves.

See also (6.14), along with (6.29) in the sequel, our starting point being the corresponding spin-classification of the (elementary) particles under consideration (cf. (2.4) above). So, still referring to the framework of (6.2) (however, see also Note 6.2 in the sequel), we conclude with the following Selesnick's correspondence principle:

(6.29)   Fields of (states of bare) bosons correspond to (sections of) line sheaves on X (cf. Definition 6.1 above). On the other hand, fields of (states of bare) fermions correspond to (sections of) vector sheaves over X (ibid.) of rank greater than 1.

Note 6.2 We have already remarked in the preceding that

(6.30)   the framework within which (6.29) is valid is that of (6.2), the smooth manifold X being compact (Hausdorff).

The above, formulated within the sheaf-theoretic set-up that has been advocated by the present treatise, is referred, though, to the $\mathbb{C}$-*algebra sheaf* (6.2′), viz. to the *structural sheaf of the standard differential geometry of smooth manifolds*.

However, as we shall see by the subsequent discussion, one can actually shift too far ahead from the aforementioned framework by employing, instead of (6.2′), the so-called *Rosinger's algebra sheaf* (cf. Vol. II; Chapt. IV, Section 5 in the sequel), whose sections contain the biggest so far possible number of singularities, in the standard sense of the term. This marks a situation that might be proved of paramount importance in physics, connected with potential applications, for instance in general relativity, concerning problems of quantum gravity (loc. cit.).

Scholium 6.1 The topological space X that appears in (6.29) above refers, in particular, to an empty finite universe, alias "vacuum" ("bare" particles have been considered, so far; cf. Section 2.2 in the preceding, or (6.29)). Thus, we assume that

(6.31) X, as in (6.29), is a (*Hausdorff*) *compact connected complete flat* 4-*dimensional Lorentz manifold.*

In this regard, we still note that for purposes connected with the preceding discussion, other types of manifolds could also be considered as well (cf., for instance, the following Remark 6.2). This, however, does not affect any local arguments as above related to the application of vector sheaves; in this connection, see also the relevant comments in S.A. Selesnick [1: p. 37].

Remark 6.2 The compactness of X, which we assumed in the foregoing (see, for instance, (6.31) above), is also connected with the application of the Serre–Swan theorem (cf. (5.2), or (5.7)). However, more general types of spaces can also be employed, provided the said theorem holds: Thus, to paraphrase slightly the terminology of S.E. Landsburg [1: p. 271, Remark], Swan spaces, or paracompact spaces of finite type (cf. R.G. Swan [1: p. 277, Remark] as well as L.V. Vaserstein [1]) can be applied instead, all these spaces having the desired property (viz. the Serre–Swan theorem is in force) as before.

More generally, any topological space with an appropriate definition of a space of finite type, in the sense of L.V. Vaserstein (loc. cit.), can still be applied in that connection; in other words, such a space is, in fact, a Swan space in the previous sense (see A. Mallios [5: p. 420]); yet, within the same vein of ideas, see A. Mallios [11: (5.12)] for a further extended generalization of the sort of topological algebras considered in the preceding see (5.7) above; thus, one can consider inductive limits.

On the other hand, the preceding are also in order with the fact that the standard machinery of classical differential geometry (of smooth manifolds), which we are going to consider throughout the sequel, will finally be employed within the abstract (axiomatic) approach that has been recently advocated by the present author (cf. A. Mallios [VS]). As a result, the relevant argument holds essentially true for any topological space, whatsoever (since, as already said in the preceding, within that context, no calculus, in the classical sense of the term, is involved at all!). This, modulo occasionally certain cohomological expediencies, compels one to consider, for instance, paracompact (Hausdorff) spaces.

7 Cohomological Classification of Elementary Particles

The classification alluded to in the title of this section is in fact an immediate consequence of the classification of elementary particles that has already been given by (6.29) above (see also Section 2.1) and the well-known cohomological classification of (finite-dimensional) $\mathbb{C}$-vector bundles (continuous/smooth), or equivalently (cf. (6.13) and/or (6.24)), of the corresponding vector sheaves (see (6.12), as well as, A. Mallios [VS: Chapt. V]). To keep the present exposition as self-contained as possible, we highlight, in brief, following mainly the previous quotation, the relevant situation as this concerns, first, the analogous classification of vector sheaves, according to the following subsection:

7.1 Vector Sheaves

Applying the terminology of Section 6, assume that we are given a $\mathbb{C}$-algebraized space

$$(7.1) \qquad (X, \mathcal{A})$$

on a topological space X, and let $\mathcal{E}$ be a given vector sheaf on X (see Definition 6.1) such that

$$(7.2) \qquad rk_A(\mathcal{E}) \equiv rk\mathcal{E} = n \in \mathbb{N}$$

(rank of $\mathcal{E}$, cf. (6.4) and/or A. Mallios [VS: Chapt. II; p. 125, (4.2)]. In particular, for $n = 1$, one considers line sheaves; see also Section 7.2).

On the other hand, let us further look at an open covering of X, say

$$(7.3) \qquad \mathcal{U} = (U_\alpha)_{\alpha \in I},$$

that also satisfies the relation

$$(7.4) \qquad \mathcal{E}\big|_{U_\alpha} = \mathcal{A}^n\big|_{U_\alpha},$$

for any $\alpha \in I$, within an $\mathcal{A}\big|_{U_\alpha}$-isomorphism of the $\mathcal{A}\big|_{U_\alpha}$-modules (in point of fact, free ones, hence vector sheaves too) concerned. We call (7.3) a local frame of $\mathcal{E}$, the given vector sheaf on X such that (7.2) holds. Thus, one obtains an important family of (matrices of) local (continuous) sections of $\mathcal{A}$, which, as we shall see, can actually describe $\mathcal{E}$. That is, we have

$$(7.5) \qquad (\phi_{\alpha\beta})$$

such that

$$(7.6) \qquad \phi_{\alpha\beta} \in GL(n, \mathcal{A}(U_{\alpha\beta})),$$

where we have set

$$(7.7) \qquad U_{\alpha\beta} \equiv U_\alpha \cap U_\beta \neq \emptyset,$$

for any α, β in I, as in (7.3), with (7.7) holding. Namely, we define

$$(7.8) \qquad \phi_{\alpha\beta} := \phi_\alpha \circ \phi_\beta^{-1} : \mathcal{A}^n\big|_{U_{\alpha\beta}} \longrightarrow \mathcal{A}^n\big|_{U_{\alpha\beta}},$$

where $\phi_\alpha, \alpha \in I$, stands for the $\mathcal{A}|_{U_\alpha}$-isomorphism as in (7.4). So the previous family (7.5) supplies, in effect, a 1-cocycle of the local frame $\mathcal{U}$ of $\mathcal{E}$ as above. Thus one has

$$(7.9) \qquad\qquad (\phi_{\alpha\beta}) \in Z^1(\mathcal{U}, \mathcal{GL}(n, \mathcal{A})).$$

Indeed, as we shall presently see (cf. (7.20) below), the 1-cocycle at issue determines $\mathcal{E}$ uniquely modulo an $\mathcal{A}$-isomorphism in the category of $\mathcal{A}$-modules over X,

$$(7.10) \qquad\qquad \mathcal{A} - \mathcal{Mod}_X,$$

with $(X, \mathcal{A})$ as in (7.1). We still note that

$$(7.11) \qquad\qquad \mathcal{GL}(n, \mathcal{A})$$

in (7.9) stands for the so-called *general linear group sheaf on X of order $n \in \mathbb{N}$* (cf. (7.2)), which can be associated with any pair $(X, \mathcal{A})$, as in (7.1). It is, by definition, a sheaf of groups (nonabelian, unless $n = 1$), or else a group sheaf on X, that is generated by the (complete) presheaf (of groups) on X,

$$(7.12) \qquad\qquad U \longmapsto GL(n, \mathcal{A}(U)),$$

where U runs over all the open subsets of X, while the target of (7.12) is the usual general linear group of order n that is associated with the (unital commutative) $\mathbb{C}$-algebra $\mathcal{A}(U)$. Of course, one has by definition

$$(7.13) \qquad\qquad GL(n, \mathcal{A}(U)) := M_n(\mathcal{A}(U))^{\boldsymbol{\cdot}},$$

where U varies as before, while the second member of (7.13) denotes the *group of units* (invertible elements) of the (full matrix) $\mathbb{C}$-algebra

$$(7.14) \qquad\qquad M_n(\mathcal{A}(U)) = M_n(\mathcal{A})(U);$$

that is, the $n \times n$ matrix with entries from $\mathcal{A}(U)$ (local (continuous) sections of $\mathcal{A}$ over the open set $U \subseteq X$; thus, this also explains our previous terminology, as applied in (7.5)).

On the other hand, the second member of (7.14) stands for the (unital noncommutative, unless $n = 1$) $\mathbb{C}$-algebra of local (continuous) sections over $U \subseteq X$, as before, of the (full matrix) $\mathbb{C}$-algebra sheaf over X

$$(7.15) \qquad\qquad M_n(\mathcal{A}).$$

The latter is thus defined as the sheaf over X generated by the (complete) presheaf of (full) matrix ($\mathbb{C}$-) algebras defined by the first member of (7.14) as U varies over the open subsets of X. Consequently, and based on (7.13), one still defines, equivalently, (7.11) according to the relation

$$(7.16) \qquad\qquad \mathcal{GL}(n, \mathcal{A}) := M_n(\mathcal{A})^{\boldsymbol{\cdot}},$$

that is, as the group sheaf of units of the $\mathbb{C}$-algebra sheaf (7.15) over X, this being, indeed, a sheaf of groups on X. (See A. Mallios [VS: Chapt. IV; p. 282, Lemma 1.1].)

Now, by coming back to the 1-cocycle of the given vector sheaf $\mathcal{E}$, as in (7.9), we also refer to it as a representative 1-cocycle of $\mathcal{E}$ that is associated with the local frame $\mathcal{U}$ of $\mathcal{E}$ (cf. (7.3), (7.4)), or a system of coordinate transformations (cf. (7.8)) provided by $\mathcal{U}$.

Thus, by further commenting on the $\mathcal{A}$-isomorphism in the category (7.10), as alluded to above in connection with (7.9), we remark that

> two given vector sheaves $\mathcal{E}$ and $\mathcal{F}$ on X, of the same rank $n \in \mathbb{N}$, are $\mathcal{A}$-isomorphic, that is, one has (cf. also (6.13), (6.13′))
>
> (7.17.1) $$[\mathcal{E}] = [\mathcal{F}] \in \Phi_{\mathcal{A}}^n(X),$$
>
> if and only if their respective coordinate 1-cocycles relative to a common local frame $\mathcal{U}$ of $\mathcal{E}$ and $\mathcal{F}$ yield similar (section-) matrices (see (7.6)). That is whenever there exists a 0-cochain of $\mathcal{U}$,

(7.17) (7.17.2) $$(\eta_\alpha) \in C^0(\mathcal{U}, \mathcal{GL}(n, \mathcal{A}))$$

> such that the following relation holds:
>
> (7.17.3) $$\psi_{\alpha\beta} = \eta_\alpha \circ \phi_{\alpha\beta} \circ \eta_\beta^{-1},$$
>
> for any α, β in I (cf. (7.3)) for which (7.7) is in force; of course $(\phi_{\alpha\beta})$ and $(\psi_{\alpha\beta})$ are the 1-cocycles of $\mathcal{E}$ and $\mathcal{F}$ corresponding to $\mathcal{U}$, respectively.
>
> One can still refer to (7.17.3) as a gauge transformation of coordinates between $\mathcal{E}$ and $\mathcal{F}$. (In this regard, cf. also Chapt. III; Note 1.2.)

In this connection, see also A. Mallios [VS: Chapt. V; p. 353, Lemma 2.1]. The above, in particular (7.17.3), is going to be employed often in the sequel as an equivalent form of (7.17.1); see, for instance, Chapter III.

Accordingly, one thus obtains, in view of (7.17), a one-to one (and onto) correspondence

(7.18) $$\mathcal{E} \longleftrightarrow (\phi_{\alpha\beta}),$$

modulo the isomorphisms as indicated by (7.17.1), or equivalently, (7.17.3). Thus, by passing to the respective sets of isomorphism classes, one gets the basic relation

(7.19) $$\Phi_{\mathcal{A}}^n(X) = H^1(X, \mathcal{GL}(n, \mathcal{A})),$$

valid within a bijection given by (7.18); the same relation (7.19) yields also a co-homological classification of vector sheaves on X of a given rank, say $n \in \mathbb{N}$, and therefore, according to (6.29), the desired one for the elementary particles as well (see the ensuing subsections). In fact, the bijection (7.19) is given by means of (7.18) through the respective equivalence classes, that is, via the bijection,

(7.20) $[\mathcal{E}] \longleftrightarrow [\phi_{\alpha\beta})]$.

In other words, a given vector sheaf $\mathcal{E}$ on X of rank $n \in \mathbb{N}$ is identified with the corresponding 1-dimensional cohomology class

(7.21) $[(\phi_{\alpha\beta})] \in H^1(X, \mathcal{GL}(n, \mathcal{A}))$,

as in (7.20), which is also uniquely defined through any given local frame $\mathcal{U}$ of $\mathcal{E}$ as above. In this connection, we still note that the 1st cohomology set appearing in the second member of (7.19) is given, by definition, via the relation

(7.22)
$$H^1(X, \mathcal{GL}(n, \mathcal{A})) := \varinjlim_{\mathcal{U}} H^1(\mathcal{U}, \mathcal{GL}(n, \mathcal{A}))$$
$$= \prod_{\mathcal{U}} H^1(\mathcal{U}, \mathcal{GL}(n, \mathcal{A})) = \sum_{\mathcal{U}} H^1(\mathcal{U}, \mathcal{GL}(n, \mathcal{A})).$$

Here $\mathcal{U}$ may be varied over the set of (proper) local frames of $\mathcal{E}$, this being in effect a cofinal subset of the set of all proper open coverings of X; see also A. Mallios [VS: Chapt. II, p. 127; (4.9), and Chapt. III, p. 275; (11.26)]. For a detailed proof of (7.19), we refer to loc. cit., Chapt. V; p. 358, Theorem 2.1.

7.2 Line Sheaves

In particular, for $n = 1$, one obtains, in view of (7.19), the following relation, yielding a cohomological classification of line sheaves on X (see also Definition 6.1 in the preceding). One has

(7.23) $\Phi^1_{\mathcal{A}}(X) = H^1(X, \mathcal{A}^{\cdot})$

within an isomorphism of abelian groups. Thus for $n = 1$, one obtains, by virtue of (7.16), the relation

(7.24) $\mathcal{GL}(1, \mathcal{A}) = \mathcal{A}^{\cdot}$,

that is (see the hypothesis of $\mathcal{A}$), the abelian group sheaf of units of $\mathcal{A}$ as in the second member of (7.24). On the other hand,

(7.25) $\Phi^1_{\mathcal{A}}(X)$,

the set of (isomorphism classes of) line sheaves on X, with $(X, \mathcal{A})$ as in (7.1), becomes an abelian group as well under the tensor product relative to $\mathcal{A}$; that is, one sets

(7.26) $[\mathcal{L}] \otimes_{\mathcal{A}} [\mathcal{L}'] := [\mathcal{L} \otimes_{\mathcal{A}} \mathcal{L}']$

(see also [VS: Chapt. II, p. 132; (5.27)]), the neutral (identity) element of the group being $\mathcal{A}$ itself (in fact, its isomorphism class, according to the definition of (7.25);

cf. also loc. cit., p. 130; (5.15)). The inverse of a given line sheaf $\mathcal{L}$ on X is defined as the dual of $\mathcal{L}$,

(7.27) $$\mathcal{L}^* := \mathcal{H}om_A(\mathcal{L}, A),$$

so that one sets

(7.28) $$[\mathcal{L}]^{-1} := [\mathcal{L}^*] \equiv [\mathcal{H}om_A(\mathcal{L}, A)].$$

Thus, based on (7.26) and (7.28), one now obtains

(7.29)
$$\mathcal{L} \otimes_A \mathcal{L}^{-1} \equiv \mathcal{L} \otimes_A \mathcal{L}^* \equiv \mathcal{L} \otimes_A \mathcal{H}om_A(\mathcal{L}, A)$$
$$= \mathcal{H}om_A(\mathcal{L}, \mathcal{L}) \equiv \mathcal{E}nd\mathcal{L} = A,$$

which also justifies the previous terminology. (In this connection, see also loc. cit., Chapt. V; p. 365: (4.3) and (4.4).)

The abelian group (7.25), as defined above (cf. (7.26)), is referred to as the *Picard group* of X with respect to the pair (X, A), as in (7.1), denoted by

(7.30) $$Pic_A(X) \equiv Pic(X).$$

Thus, one sets

(7.31) $$Pic(X) := (\Phi_A^1(X), \otimes_A) \equiv \Phi_A^1(X),$$

so that in accordance with (7.23), one obtains

(7.32) $$Pic(X) = H^1(X, A^{\cdot}),$$

up to an isomorphism of the (abelian) groups concerned. (See also ibid., Chapt. V; p. 367, Theorem 4.1.)

Note 7.1 By referring to the sheaf cohomology that is considered in (7.19), therefore, in (7.23) and (7.32) as well, we simply look at it as the Čech cohomology (set, or even groups, as the latter two cases, as before) of X with the pertinent coefficients, a fact that is a consequence of the definition of the 1st cohomology set (or group, as the case may be) of X. It is known that sheaf cohomology coincides with Čech cohomology in dimension 1, for any topological space X (loc. cit., p. 234, Lemma 8.1).

On the other hand, having in mind later applications, we also recall, in brief, the relevant useful notion of the determinant line sheaf

(7.33) $$\det \mathcal{E} \equiv [\det \mathcal{E}] \subset \Phi_A^1(X)$$

of a given vector sheaf $\mathcal{E}$ on X: Namely, by looking at the coordinate 1-cocycle of $\mathcal{E}$,

(7.34) $$(\phi_{\alpha\beta}) \in Z^1(\mathcal{U}, \mathcal{GL}(n, A)),$$

associated with a given local frame $\mathcal{U}$ of $\mathcal{E}$ (see, for example, (7.20), while we have set $n = rk\mathcal{E} \in \mathbb{N}$), one obtains a corresponding coordinate 1-cocycle of $\det \mathcal{E}$ by the relation

$$(7.35) \qquad (\det(\phi_{\alpha\beta})) \in Z^1(\mathcal{U}, \mathcal{A}^{\bullet})$$

(cf. also (7.6) and (7.24), along with A. Mallios [VS: Chapt. V; p. 368, Section 4.1, in particular, p. 369, (4.25)]). The last relation justifies our claim in (7.33), by virtue of (7.23). ∎

We come in the next subsection to the classification that we promised at the outset.

7.3 Elementary Particles

As already said at the introduction of this section, the sought-for classification is now a straightforward result of the preceding discussion, namely, of (7.19) and/or (7.23), in conjunction with the classification of (bare) elementary particles that we already obtained through (6.29) (Selesnick correspondence principle). Thus, to fix the terminology employed, we sum up:

(7.36)

> Bare elementary particles (in fact, fields of their states; see (6.29)) may be identified with appropriate cohomology classes; in particular, with
>
> $$(7.36.1) \qquad \mathcal{GL}(n, \mathcal{A})\text{-torsors}$$
>
> (cf. Yu.I. Manin [1: p. 117], for the terminology). Thus, with elements of the 1st cohomology set of X,
>
> $$(7.36.2) \qquad H^1(X, \mathcal{GL}(n, \mathcal{A})),$$
>
> pertaining for $n \geq 2$ to (bare) fermions. On the other hand, the set at issue is reduced to
>
> $$(7.36.3) \qquad \mathcal{A}^{\bullet}\text{-torsors},$$
>
> that is, to the 1st cohomology group of X,
>
> $$(7.36.4) \qquad H^1(X, \mathcal{A}^{\bullet}),$$
>
> in the case of (bare) bosons (cf. also (7.24) above). So here $\mathcal{A}$ stands for the $\mathbb{C}$-algebra sheaf on X, as given by (6.25), while X corresponds to the space-time manifold under consideration. (In this regard, see also Remark 6.2 in the preceding). Yet, $n \in \mathbb{N}$ in (7.36.1) indicates the particular rank of the vector sheaf (viz. (bare) fermion) at issue. More on this point of view will be supplied later on.

Thus, we are justified, in view of the preceding discussion, if we say that

(7.37)   the arithmetic of current elementary-particle physics is that of the characteristic classes.

The above is reminiscent of von Westenholz's remark that

(7.38)   *"the structure underlying an intrinsic approach to physics is essentially de Rham-cohomology."*

See C. von Westenholtz [1: p. 321, *Discussion*, beginning, along on p. 323, comments following (6.3)]. Another justification of (7.37) is the ever-increasing recent interest in the physics literature in the theory of characteristic classes, even of secondary ones, in particular, of the so-called *Chern–Simons* (*characteristic*) *classes*; see, for instance, M. Manoliu [1], [2], K.B. Marathe-G. Martucci [1: Chapt. 5, in particular, p. 124, Subsection 5.3.1]. On the other hand, the abstract theory of characteristic classes, that is, that formulated in terms of the abstract (axiomatic) differential-geometric point of view, which is also adopted by the present treatise as concerns its application in gauge theories, which is our concern here, has been already exhibited in A. Mallios [VS: Chapt. IX], to which we refer for the relevant technical details.

8 Elementary Particles as Principal Sheaves

Up to this point, we have looked at elementary particles, in effect, bare ones (viz. free, no interactions involved), through their (fields of) states, the latter being interpreted as sections of appropriate $\mathbb{C}$-vector bundles, or, equivalently, as those of (sheaves of germs of sections of) vector sheaves; see, for instance, (6.29) along with (6.12), as well as (6.14).

On the other hand, by considering a structured elementary particle, that is, an elementary particle with an internal structure, or a set of (internal) states (as, for instance, spin, charge, color and the like), these states are usually parametrized by the elements of a (compact matrix) Lie group (abelian, or not), as, for instance, $SU(1) \equiv U(1)$ or $SU(n)$, with $n \geq 2$. Thus, the mathematical framework that is best suited to describe the above situation is that of a principal fiber bundle, or of a (principal) G-bundle, whose structure group, viz. fiber, is the Lie group, say, G, that parametrizes the internal states of the (structure elementary) particle under consideration. The base space of the bundle at issue is a space-time (C^∞-)manifold X, as in the preceding (see Scholium 6.1, along with Remark 6.2), hence the (fiber) bundle involved is a smooth one too.

The fiber of the previous bundle, or its structure group, is still what the physicists call the internal symmetry group of the physical system (elementary particle) under discussion. In this connection, as already said in the preceding, we still remark that according to the so-called symmetry axiom (see, for example, D.J. Simms–N.M.J. Woodhouse [1: pp. 21, 150]), one further assumes that

(8.1)   the same group G, as above, is the symmetry group of the classical system at issue, as well as of the underlying quantum system.

Remark 8.1 In this connection, it is still good to remark at this point that an occasional violation of the previous symmetry axiom is also the reason for the occurrence of an anomaly in quantum field theory, in particular, if one employs a Hamiltonian point of view for the latter theory, so that the corresponding invariance group is changed. See, for instance, C. Nash [1: Chapt. X, in particular, pp. 269, 291]. On the other hand, one can still say that

(8.2)
> it is, in effect, the same group as before that characterizes the particular physical system under discussion, usually disguised throughout the sequel under various other names (as, for example, gauge group); the same is further construed in a twisted form, that is, as a principal fiber (G)-bundle, due to the particular form of our space-time manifold (curved, nonflat), already accepted as a base space of the previous bundle.

However, it is certainly preferable, from the point of view of physics, to think of the above fiber bundles in terms of their sections rather than the bundles themselves; this is still profitable, among other things, from the point of view of our calculations: To quote R.P. Feynman, *"physics is number"*; after all(!), our contribution being simply our own interpretation of the physical meaning of these numbers, which we gather through our experiments. In point of fact, what we actually are doing here is describing, always, not explaining nature! Cf., for instance, N. Bohr:

(∗)
> *"It is wrong to think that the task of physics is to find out how nature is. Physics concerns what we can say about nature."*

See S.Y. Auyang [1: p. 229]. Einstein too used always to talk about the *"description"* of reality, not of its *"explanation"*; cf., for example, A. Einstein [1: p. 166]. Admittedly, *the Creator only can explain it*! Thus, the previous principal fiber (G)-bundle, as in (8.2), appears, still, equivalently (see (6.13′), (6.14)), through the corresponding sheaf of germs of its *sections*, as a

(8.3)   principal sheaf

(see also Definition 8.1 in the sequel).

Thus, for the reader's convenience, we discuss below, in brief, the notion that was pointed out above, the sheaf-theoretic analogue of the classical concept of a principal G-bundle. On the other hand, several applications of the same notion in the case of differential geometry, along the lines of the standard theory of smooth ($\mathcal{C}^\infty$-) principal G-bundles, however within the abstract framework (no calculus, at all) of the geometry of vector sheaves (cf. A. Mallios [VS]) have been already started to be considered lately by E. Vassiliou, indeed, in a very revealing and unravelling way pertaining to the classical theory (ibid. [1]).

The same notion as above can be regarded in effect as a particularization of the classical concept of a fiber space with structure sheaf, in the sense of A. Grothendieck [1]. The latter account is still our guiding track in the sequel, in conjunction with our previous considerations, as cited above. We discus the aforementioned notion, as in (8.3), in the ensuing subsection.

8.1 Principal Sheaves

Suppose we are given a sheaf of groups (nonabelian, in general), or else a group sheaf

$$(8.4) \qquad\qquad\qquad\qquad \mathcal{G}$$

on a topological space X (see, for instance, A. Mallios [VS: Chapt. II, p. 86, Definition 1.1] for technical details). Now, on the analogy of the notion of a G-set S (a set S equipped with a group G of operators, N. Bourbaki [3: Chap. I: p. 29ff], or a set S on which a group G acts), one defines a

$$(8.5) \qquad\qquad\qquad \mathcal{G}\text{-sheaf (of sets)} \quad \mathcal{E}$$

on the given topological space X, as above, where $\mathcal{E}$ is by definition a sheaf (of sets) over X and $\mathcal{G}$ a given group sheaf on X (cf. (8.4)), in such a manner that

$$(8.6) \qquad\qquad \mathcal{G} \text{ operates on E(on the left)},$$

or $\mathcal{G}$ is a sheaf of operators (on the left) on (the sheaf of sets) $\mathcal{E}$; this means, by definition, that one has, first, a sheaf morphism

$$(8.7) \qquad\qquad\qquad \mathcal{G} \underset{X}{\times} \mathcal{E} \longrightarrow \mathcal{E}$$

(a morphism of sheaves (of sets), see [VS: Chapt. I; p. 5, Definition 1.2]), where the source of (8.7) stands for the fiber product over X of the sheaves of sets concerned, as above (ibid., p. 87), such that for every $x \in X$, the corresponding fiber $\mathcal{E}_x$ of $\mathcal{E}$ is a $\mathcal{G}_x$-set: that is, as already explained above, the group $\mathcal{G}_x$ acts (on the left) on the set $\mathcal{E}_x$. (See also A. Grothendieck [1: p. 31, Definition 3.4.1]). Of course, one has an analogous notion pertaining to a right action of $\mathcal{G}$ on $\mathcal{E}$.

In particular, we say that the group sheaf $\mathcal{G}$ operates faithfully on $\mathcal{E}$ whenever this happens fiberwise; viz. the corresponding sheaf morphism ("action"), as in (8.7), is such that, for every $x \in X$, the respective $\mathcal{G}_x$-set $\mathcal{E}_x$, as above, exhibits a faithful group action (of the group $\mathcal{G}_x$ on the set $\mathcal{E}_x$) in the standard sense of the term; see P. Tondeur [1: p. 23, "effective operators"].

Analogously, one defines a transitive action of $\mathcal{G}$ on $\mathcal{E}$, when we are given (8.7). (See also loc. cit., p. 24, Definition 1.4.5, as well as A. Grothendieck [1: p. 32, Definition 3.4.2]).

In this regard, it is certainly clear that

$$(8.8) \quad \begin{array}{l} \text{any group sheaf } \mathcal{G} \text{ on a topological space } X \text{ can be considered, by right} \\ \text{translations, as a fiber space over } X, \text{ with faithful sheaf } \mathcal{G} \text{ of right oper-} \\ \text{ators, which is also transitive.} \end{array}$$

See also A. Grothendieck [1: p. 41, §4.2]. More specifically, one concludes that

$$(8.9) \quad \begin{array}{l} \text{every group sheaf } \mathcal{G} \text{ on } X, \text{ as in (8.8), can be construed as a principal} \\ \text{fiber space over } \mathcal{G} \text{ (the latter group operating on itself on the right).} \end{array}$$

See loc. cit., p. 42. Thus, the preceding leads us now to the following general notion, as pointed out by (8.3). That is, one has the following definition.

Definition 8.1 Given a topological space X, a principal sheaf over X is a $\mathcal{G}$-*sheaf* $\mathcal{P}$ on X (see (8.5)), where $\mathcal{G}$ is a group sheaf on X operating on $\mathcal{P}$ on the right (cf. (8.7)) in a faithful and transitive manner. (We also speak of a simply transitive action of $\mathcal{G}$ on $\mathcal{P}$.)

The above definition thus agrees with the analogous notion of A. Grothendieck (loc. cit.) pertaining to a principal fiber space (with a group bundle of operators), given that both $\mathcal{G}$ and $\mathcal{P}$ are, in view of our hypothesis, sheaves on X (see also loc. cit., p. 32, Definition 3.4.2, along with the ensuing relevant comments therein).

Scholium 8.1 In this connection, we still remark, for later use as well, that

(8.10)
    a $\mathcal{G}$-sheaf $\mathcal{P}$ on X is equivalent to a morphism of group sheaves

$$(8.10.1) \qquad \rho : \mathcal{G} \longrightarrow \mathcal{A}ut(\mathcal{P}).$$

Here $\mathcal{A}ut(\mathcal{P})$ stands for the group sheaf of (germs of) automorphisms of $\mathcal{P}$ (self-isomorphisms of $\mathcal{P}$ onto itself). See A. Grothendieck [1: p. 35, Proposition 3.5.1].

(8.11)
    The group sheaf $\mathcal{G}$ on X operates faithfully on (the sheaf of sets on X) $\mathcal{P}$, cf. (8.7), if and only if the map ρ, as in (8.10.1), is one-to-one.

(Loc. cit.; see A. Mallios [VS: Chapt. I; p. 60, Lemma 12.1, cf. also Chapt. II, p. 135; (6.10)].) The preceding can still be viewed, of course, as the sheaf-theoretic version of the relevant part of the standard theory of transformation groups (see, for instance, P. Tondeur [1]).

On the other hand, by referring to the group sheaf-morphism ρ, as in (8.10.1), we also say that it defines a representation of $\mathcal{G}$ through the automorphisms of $\mathcal{P}$, or simply a representation of $\mathcal{G}$ on $\mathcal{P}$. (See Section 9, where the representation set $\mathcal{P}$ is further endowed, as usual, within the present sheaf-theoretic set-up, with an algebraic structure, precisely with that of a vector sheaf: representation vector sheaf of the group sheaf $\mathcal{G}$; see (9.2).)

We have already discussed at the beginning of this section the significance of the preceding in connection with its potential physical description (application). Now, we further explain, in Section 9, how the above are related, physically speaking, with particle representations, that is, with the so-called matter, or particle fields; see also D. Bleecker [1: p. 42ff], Y. Choquet-Bruhat et al. [1: p. 403f]. In this regard, it is worth noting the relevant comment of W. Heisenberg [1: p. 48], that

(8.12)
    " ... *the wave picture* [of matter] *has ... its limitations, which may be derived from the particle representation.*"

In other words, through our calculations (coordinates), hence the appearance of singularities!

9 Vector Sheaves Associated with Principal Sheaves and Physical Interpretation

Assume that we are given a principal sheaf $\mathcal{P}$ on a topological space X, that is (cf. Definition 8.1 in the preceding), a $\mathcal{G}$-sheaf $\mathcal{P}$ on X, where $\mathcal{G}$ is a group sheaf on X acting on (the sheaf of sets on X) $\mathcal{P}$ simply transitively.

Furthermore, suppose that $\mathcal{E}$ is a vector sheaf on X such that

$$(9.1) \qquad rk_{\mathcal{A}}(\mathcal{E}) \equiv rk\mathcal{E} = n \in \mathbb{N}$$

(see Definition 6.1 in the foregoing, along with (6.5)). By further employing the terminology of the previous Scholium 8.1, let us also consider a group sheaf morphism

$$(9.2) \qquad \rho : \mathcal{G} \longrightarrow \mathcal{A}ut(\mathcal{E}),$$

that is (ibid., see (8.10.1)), a representation of $\mathcal{G}$ through ($\mathcal{A}$-)automorphisms of $\mathcal{E}$. As already mentioned in Scholium 8.1, we have here, by (9.2), a fundamental example, where the representation set $\mathcal{E}$ of $\mathcal{G}$ is, in fact, a structured one, precisely, according to our assumption, a vector sheaf on X, which thus, in view of (9.2), might be called a *representation vector sheaf* of the given group sheaf $\mathcal{G}$. Concerning the range of the map ρ, as above, one has

$$(9.3) \qquad \mathcal{A}ut(\mathcal{E}) \equiv \mathcal{A}ut_{\mathcal{A}}(\mathcal{E}) = \mathcal{H}om_{\mathcal{A}}(\mathcal{E}, \mathcal{E})^{\boldsymbol{\cdot}} \equiv (\mathcal{E}nd\mathcal{E})^{\boldsymbol{\cdot}},$$

that is, the group sheaf on X of the germs of ($\mathcal{A}$-)automorphisms of (the $\mathcal{A}$-module, in effect, by hypothesis, of the vector sheaf) $\mathcal{E}$, being also, in view of (9.3), the group sheaf of units of the $\mathcal{A}$-algebra sheaf of germs of ($\mathcal{A}$-)endomorphisms of $\mathcal{E}$,

$$(9.4) \qquad \mathcal{E}nd\mathcal{E} \equiv \mathcal{E}nd_{\mathcal{A}}\mathcal{E} := \mathcal{H}om_{\mathcal{A}}(\mathcal{E}, \mathcal{E}).$$

Concerning the applied terminology, see [VS: Chapt. II; p. 138, Chapt. IV; p. 282, Chapt. VI; p. 87, as well as, Chapt. V, p. 357, (2.31)]. In this connection, see also (6.1) in the preceding, pertaining to the $\mathbb{C}$-algebra sheaf $\mathcal{A}$ on X, as above.

Thus, given a principal sheaf $\mathcal{P}$ on X (Definition 8.1), a vector sheaf $\mathcal{E}$ on X, for which there exists a group sheaf morphism ρ as in (9.2) is called a

(9.5)   *vector sheaf associated with the principal sheaf $\mathcal{P}$ on X via* (the representation) ρ.

On the other hand, we note for later use as well that according to the definitions,

every vector sheaf $\mathcal{E}$ on X with

(9.6)

$$(9.6.1) \qquad rk_{\mathcal{A}}(\mathcal{E}) \equiv rk\mathcal{E} = n \in \mathbb{N},$$

i.e., of rank $n \in \mathbb{N}$, is a *fiber space over X of structure type $\mathcal{A}^n$* and *with structure group $\mathcal{GL}(n, \mathcal{A})$*

in the sense of A. Grothendieck [1: p. 39, Definition 4.1.1; see also p. 42, Example 4.2.c]. See A. Mallios [VS: Chapt. V; p. 361, Section 2.1, in particular, p. 362, (2.63)].

Furthermore, one has *locally* the relation

$$(9.7) \qquad (Aut\mathcal{E})\big|_U = \mathcal{GL}(n, \mathcal{A})\big|_U,$$

within an isomorphism of the group sheaves concerned: That is, one proves, in view of our hypothesis for $\mathcal{E}$ (cf. (9.6.1)), that

(9.8)   for every point $x \in X$, there exists an open neighborhood U of x such that (9.7) holds. In fact, one can prove then that there exists a basis of the topology of X for the open sets of which (9.7) is in force.

Indeed, first we remark that one can take as an open set U for which (9.7) is valid, a local gauge of the given vector sheaf $\mathcal{E}$ (cf. (6.5), along with (6.7) in the preceding). So our assertion is, in effect, an immediate consequence of the following quite general result.

Lemma 9.1 Suppose we are given two $\mathcal{A}$-modules $\mathcal{E}, \mathcal{F}$ on X, and an open $U \subseteq X$, such that one has

$$(9.9) \qquad \mathcal{E}\big|_U = \mathcal{F}\big|_U,$$

within an $\mathcal{A}\big|_U$-isomorphism of the $\mathcal{A}\big|_U$-modules concerned. Then, one further obtains;

$$(9.10) \qquad (\mathcal{E}nd\mathcal{E})\big|_U = (\mathcal{E}nd\mathcal{F})\big|_U$$

as well as

$$(9.11) \qquad (Aut\mathcal{E})\big|_U = (Aut\mathcal{F})\big|_U,$$

within, of course, the pertinent isomorphisms.

Proof. This is indeed straightforward, according to the definitions. Thus, one has

$$
\begin{aligned}
(9.12) \qquad (\mathcal{E}nd\mathcal{E})|_U &\equiv \mathcal{H}om_{\mathcal{A}}(\mathcal{E}, \mathcal{E})|_U = \mathcal{H}om_{\mathcal{A}|_U}(\mathcal{E}|_U, \mathcal{E}|_U) \\
&\equiv \mathcal{E}nd_{\mathcal{A}|_U}(\mathcal{E}|_U) \equiv \mathcal{E}nd(\mathcal{E}|_U) = \mathcal{E}nd(\mathcal{F}|_U) \\
&= \mathcal{H}om_{\mathcal{A}|_U}(\mathcal{F}|_U, \mathcal{F}|_U) = \mathcal{H}om_{\mathcal{A}}(\mathcal{F}, \mathcal{F})|_U \equiv (\mathcal{E}nd\mathcal{F})|_U,
\end{aligned}
$$

modulo, of course, the pertinent $\mathcal{A}\big|_U$-isomorphisms, which thus proves (9.10), an immediate consequence of which is certainly, by restriction, (9.11). ∎

Concerning our argument in (9.12), see also A. Mallios [VS: Chapt. II; p. 138, (6.27)]. (Remark: For convenience, we have considered in Lemma 9.1 $\mathcal{A}$-modules on X; indeed, one might still take in general sheaves of sets on X. See also A.

Grothendieck, loc. cit., p. 24, Example 2.6.c.) On the other hand, by looking at the set of those open sets in X for which (9.7) is valid, this yields a basis of the topology of X, as claimed in the second part of (9.8) (see also (9.23) in the sequel), which thus proves completely our assertion. ■ Accordingly, one infers that

(9.13)   if a relation like (9.9) holds locally in the sense of the first part of (9.8) ("for every point ... "), then equivalently, the same is in force along a basis of the topology of X.

In this connection, we note for later use that as a consequence of (9.13), one also concludes that

whenever two given sheaves (of sets) on X, say $\mathcal{E}$ and $\mathcal{F}$, are locally isomorphic, in the previous sense (see the expression in the first part of (9.8)), then $\mathcal{E}$ and $\mathcal{F}$ have the same stalks (fibers) as well, viz. one has

$$(9.14.1) \qquad \mathcal{E}_x = \mathcal{F}_x, \quad x \in X,$$

within a bijection.

Indeed, the assertion is a straightforward outcome of the definitions; that is, one obtains

(9.15)
$$\mathcal{E}_x := \varinjlim_{U \in \mathcal{B}(x)} \mathcal{E}(U) = \varinjlim_{U \in \mathcal{B}(x)} (\mathcal{E}|_U)(U) = \varinjlim_{U}(\mathcal{F}|_U)(U) = \varinjlim_{U}\mathcal{F}(U) = \mathcal{F}_x,$$

for every $x \in X$, where $\mathcal{B}(x)$ stands for a fundamental system of open neighborhoods of x, which can be taken from a basis of the topology of X, for the elements of which an analogous relation to (9.9) holds (cf. (9.13)). See also A. Mallios, loc. cit.; Chapt. I, p. 55, (11.40). ■

$\mathcal{S}$ **Remark 9.1 (Warning!)** Equation (9.14.1) does not mean, of course, that $\mathcal{E}$ and $\mathcal{F}$ are in general isomorphic(!): See B.R. Tennison [1: p. 54, Example 7.5. C]. Hence, a conjunctive map (sheaf morphism) between $\mathcal{E}$ and $\mathcal{F}$, specializing to (9.14.1), should first be secured! (See also A. Mallios [VS: Chapt. I; p. 68, Theorem 12.1].)

Now, by taking in particular in the preceding (see (9.12)) $\mathcal{F} = \mathcal{A}^n$, one obtains, for the given vector sheaf $\mathcal{E}$ on X, the relations

(9.16)
$$(\mathcal{E}nd\mathcal{E})|_U = (\mathcal{E}nd\mathcal{A}^n)|_U = \mathcal{H}om_{\mathcal{A}}(\mathcal{A}^n, \mathcal{A}^n)|_U$$
$$= M_n(\mathcal{A})|_U = M_n(\mathcal{A}|_U),$$

modulo $\mathcal{A}|_U$-isomorphisms of the $\mathcal{A}|_U$-modules involved. Moreover (cf. (9.11)), one has

$$(9.17) \quad (\mathcal{A}ut\mathcal{E})|_U = (\mathcal{A}ut\mathcal{A}^n)|_U = M_n(\mathcal{A})^{\cdot}|_U \equiv \mathcal{GL}(n, \mathcal{A})|_U = \mathcal{GL}(n, \mathcal{A}|_U),$$

within group sheaf-isomorphisms, as indicated. See also [VS: Chapt. IV; p. 281, (1.8), as well as, p. 285, (1.26) and p. 294, (3.25)], concerning the above notation.

In fact, in conjunction with (9.17), one actually obtains the following general result, a particular case of which is our argument in (9.17).

Lemma 9.2 Let $\mathcal{E}$ be a given $\mathcal{A}$-algebra sheaf on a topological space X. Then, one obtains

$$(9.18) \qquad \mathcal{E}^{\cdot}\big|_U = (\mathcal{E}\big|_U)^{\cdot}$$

within a group sheaf-isomorphism for every open $U \subseteq X$, where $\mathcal{E}^{\cdot}$ stands for the group sheaf of units of $\mathcal{E}$.

See A. Mallios [VS: Chapt. II, p. 137; (6.24), proof of (6.23)]. ∎ In conjunction with (9.16) and (9.17), as above, cf. also loc. cit., p. 137; (6.24′).

By still referring to a local gauge of a given vector sheaf $\mathcal{E}$ on a topological space X, with $rk\mathcal{E} = n \in \mathbb{N}$ (see (6.5)), one means, by definition, a local isomorphism, say ϕ_U, of $\mathcal{E}$ onto $\mathcal{A}^n$ over an open set $U \subseteq X$; that is, one has

$$(9.19) \qquad \phi_U \in Isom_{\mathcal{A}|_U}(\mathcal{E}|_U, \mathcal{A}^n|_U) = \mathcal{I}som_{\mathcal{A}}(\mathcal{E}, \mathcal{A}^n)(U).$$

Thus, according to the definitions, one concludes that

(9.20) given a vector sheaf $\mathcal{E}$ on X of rank $n \in \mathbb{N}$, a local gauge of $\mathcal{E}$ over an open $U \subseteq X$ is a (continuous local) section over U of

$$(9.20.1) \qquad \mathcal{I}som_{\mathcal{A}}(\mathcal{E}, \mathcal{A}^n),$$

viz. of the sheaf of germs of ($\mathcal{A}$-)isomorphisms of $\mathcal{E}$ onto $\mathcal{A}^n$.

See also [VS: Chapt. II; p. 133, (6.1), and Chapt. V; p. 357, (2.31)]. In this connection, we further remark, with A. Grothendieck [1: p. 36, 3.6.a], by still taking into account (9.1) and (6.5), that

(9.21) $\mathcal{I}som_{\mathcal{A}}(\mathcal{E}, \mathcal{A}^n)$, as in (9.20.1), is a $\mathcal{GL}(n, \mathcal{A})$-principal sheaf over X having $\mathcal{GL}(n, \mathcal{A})$ as a group sheaf of (left) operators.

Cf. also Definition 8.1. Accordingly, one may further consider (cf. also (9.20))

(9.22) $\mathcal{I}som_{\mathcal{A}}(\mathcal{E}, \mathcal{A}^n)$ as the sheaf of germs of local gauges of the vector sheaf $\mathcal{E}$ on X, where $rk\mathcal{E} = n \in \mathbb{N}$. On the other hand, in view of (9.21), the same sheaf as in (9.20.1) may be considered as the principal $\mathcal{GL}(n, \mathcal{A})$-sheaf of the local gauges of $\mathcal{E}$, which is associated with the given vector sheaf $\mathcal{E}$ on X.

In this regard, see also A. Grothendieck, loc. cit. p. 41, Example 4.2.a, and p. 42, Example 4.2.c. Therefore, as a result of (9.22), we can say that one finds here too the abstract (sheaf-theoretic) analogue of the classical situation one gets in the case of a given vector bundle and the so-called (*principal*) *bundle of local frames* associated with it. In this connection, we finally remark that

(9.23)   given a vector sheaf $\mathcal{E}$ on a topological space X, one can always find a basis of the topology of X consisting of (domains of definition of) local gauges of $\mathcal{E}$ (cf. also (9.20)).

Indeed, by taking any local frame of $\mathcal{E}$ (by definition, an open covering of X consisting of local gauges of $\mathcal{E}$, cf. [VS: Chapt. II; p. 126, (4.8)]), one can further consider its intersection with any basis of the topology of X, getting thus a local frame of $\mathcal{E}$ (see also loc. cit., p. 127; (4.9)), that is, a basis of the topology of X. In this connection, cf. also our previous general argument in (9.13). ∎ On the other hand, as a consequence of (9.23), see also (9.14), one thus concludes, in particular, that

for any point $x \in X$, one gets at the relation

(9.24)   (9.24.1) $\mathcal{I}som_{\mathcal{A}}(\mathcal{E}, \mathcal{A}^n)_x = \mathcal{GL}(n, \mathcal{A})_x$

within a bijection of the fibers concerned.

Thus, (9.24.1) vindicates the previously applied terminology as in (9.21) above, by analogy with the classical case of a G-principal bundle. However, by also taking into account our previous warning in Remark 9.1, we further remark that

(9.25) $\mathcal{S}$ (9.24.1) does not necessarily mean, in general that $\mathcal{I}som_{\mathcal{A}}(\mathcal{E}, \mathcal{A}^n)$ is either a group sheaf on X or that it is isomorphic to $\mathcal{GL}(n, \mathcal{A})$!

Nevertheless, by virtue of our hypothesis that $\mathcal{E}$ is locally isomorphic to $\mathcal{A}^n$ (cf. (6.4) and (6.5) in the preceding), one actually proves that

(9.26)   (9.26.1) $$\mathcal{I}som_{\mathcal{A}}(\mathcal{E}, \mathcal{A}^n) = \sum_{x \in X} \mathcal{I}som_{\mathcal{A}}(\mathcal{E}, \mathcal{A}^n)_x$$

is in fact a group sheaf on X.

Indeed, in view of (9.24), it is also enough to prove (cf. [VS: Chapt. II; p. 106, (1.76)]) that

for any open $U \subseteq X$, the set

(9.27)   (9.27.1) $\mathcal{I}som_{\mathcal{A}}(\mathcal{E}, \mathcal{A}^n)(U)$

is in effect a group.

Since continuity is a local matter, it suffices to check the continuity of, say, $s + t$, with s, t elements of (9.27.1), by looking at any local gauge U of $\mathcal{E}$, still in view of our hypothesis for $\mathcal{E}$ (cf. (6.5)), so that one then actually obtains

$$\mathcal{I}som_{\mathcal{A}}(\mathcal{E}, \mathcal{A}^n)\big|_U = \mathcal{I}som_{\mathcal{A}|_U}(\mathcal{E}\big|_U, \mathcal{A}^n\big|_U) = \mathcal{I}som_{\mathcal{A}|_U}(\mathcal{A}^n\big|_U, \mathcal{A}^n\big|_U)$$

(9.28)
$$= \mathcal{I}som_{\mathcal{A}}(\mathcal{A}^n, \mathcal{A}^n)\big|_U \equiv (\mathcal{A}ut\,\mathcal{A}^n)\big|_U \equiv \mathcal{GL}(n, \mathcal{A})\big|_U$$

$$= \mathcal{GL}(n, \mathcal{A}\big|_U) = \mathcal{A}ut(\mathcal{E}\big|_U) = (\mathcal{A}ut\mathcal{E})\big|_U,$$

within group sheaf-isomorphisms. This, of course, proves the assertion. (In this connection, cf. also [VS: Chapt. I, p. 11: (2.15)].) ■

Scholium 9.1 Our previous conclusion for (9.27) is in fact true quite generally: That is, suppose that $\mathcal{E}$ and $\mathcal{F}$ are $\mathcal{A}$-modules on X that are also locally isomorphic; namely, we assume that for every $x \in X$, there exists an open $U \subseteq X$, with $x \in U$, such that

(9.29) $$\mathcal{E}\big|_U = \mathcal{F}\big|_U,$$

within an $\mathcal{A}\big|_U$-isomorphism of the $\mathcal{A}\big|_U$-modules involved. According to the previous definition, it is clear that (cf. also (9.13)

(9.30)   whenever two $\mathcal{A}$-modules $\mathcal{E}$, $\mathcal{F}$ on a topological space X are locally isomorphic, then there exists a basis of the topology of X with respect to which this is still in force.

Our claim, referring to the aforementioned extension of (9.27), is that

if two $\mathcal{A}$-modules $\mathcal{E}$, $\mathcal{F}$ on X are locally isomorphic, then

(9.31.1) $$\mathcal{I}som_{\mathcal{A}}(\mathcal{E}, \mathcal{F})$$

(9.31)   (see (9.20.1) for the notation) is locally a group sheaf on X (in general, not globally! Cf. Remark 9.1). That is, we claim that for every $x \in X$, there exists an open $U \subseteq X$, with $x \in U$, such that

(9.31.2) $$\mathcal{I}som_{\mathcal{A}}(\mathcal{E}, \mathcal{F})(U)$$

is a group.

In fact, based on a similar argument as in (9.30) or in (9.13), one easily concludes that

(9.32)   our previous assertion for (9.31.2) holds for every open $V \subseteq U$, hence in particular for a basis of the topology of X. Therefore, one finds a local basis at (fundamental system of (open) neighborhoods of) x for which the above holds.

To prove our assertion in (9.31), we remark that according to our hypothesis for $\mathcal{E}$ and $\mathcal{F}$ and the definitions (cf., for instance, A. Mallios [VS: Chapt. II; p. 134, Definition 6.1]), one has

$$
\begin{aligned}
\mathcal{I}som_{\mathcal{A}}(\mathcal{E}, \mathcal{F})(U) &= Isom_{\mathcal{A}|_U}(\mathcal{E}|_U, \mathcal{F}|_U) \\
&= Isom_{\mathcal{A}|_U}(\mathcal{E}|_U, \mathcal{E}|_U) = \mathcal{I}som_{\mathcal{A}}(\mathcal{E}, \mathcal{E})(U) \\
&\equiv (\mathcal{A}ut\mathcal{E})(U) = Aut(\mathcal{E}|_U) = Aut(\mathcal{F}|_U) = (\mathcal{A}ut\mathcal{F})(U),
\end{aligned}
$$
(9.33)

which proves our claim. ∎

Furthermore, within the same vein of ideas, one obtains (cf. also A. Mallios [VS: Chapt. II;p. 137, (6.23)])

$$
\begin{aligned}
\mathcal{I}som_{\mathcal{A}}(\mathcal{E}, \mathcal{F})|_U &= \mathcal{I}som_{\mathcal{A}|_U}(\mathcal{E}|_U, \mathcal{F}|_U) = \mathcal{I}som_{\mathcal{A}|_U}(\mathcal{E}|_U, \mathcal{E}|_U) \\
&= \mathcal{I}som_{\mathcal{A}}(\mathcal{E}, \mathcal{E})|_U \equiv (\mathcal{A}ut\mathcal{E})|_U = \mathcal{A}ut_{\mathcal{A}|_U}(\mathcal{E}|_U) \equiv \mathcal{A}ut(\mathcal{E}|_U),
\end{aligned}
$$
(9.34)

which also generalizes (9.28); similarly, of course, for $\mathcal{F}$, based on our hypothesis in (9.29). Here we have implicitly used the fact that for any $\mathcal{A}$-module $\mathcal{E}$ on X, one has

$$
(\mathcal{A}ut\mathcal{E})|_U = \mathcal{A}ut(\mathcal{E}|_U),
$$
(9.35)

within an isomorphism of the group sheaves concerned, for any open $U \subseteq X$.

On the other hand, by looking at the reverse direction and by analogy with the standard case of a principal bundle one concludes that

(9.36)
> a vector sheaf $\mathcal{E}$ of rank $n \in \mathbb{N}$ that is associated with a given principal $\mathcal{G}$-sheaf $\mathcal{P}$ via a faithful representation
>
> $$\rho : \mathcal{G} \longrightarrow \mathcal{A}ut\mathcal{E}$$
> (9.36.1)
>
> is given by the relation
>
> $$\mathcal{E} = (\mathcal{P} \underset{X}{\times} \mathcal{A}^n)/\mathcal{G},$$
> (9.36.2)
>
> within an $\mathcal{A}$-isomorphism.

Indeed, taking for convenience the case of the principal $\mathcal{GL}(n, \mathcal{A})$-sheaf $\mathcal{I}som_{\mathcal{A}}(\mathcal{E}, \mathcal{A}^n)$, or the sheaf of germs of local gauges of $\mathcal{E}$ (cf. (9.21), (9.22)), one obtains

$$
\mathcal{E} = (\mathcal{I}som_{\mathcal{A}}(\mathcal{E}, \mathcal{A}^n) \underset{X}{\times} \mathcal{A}^n)/\mathcal{GL}(n, \mathcal{A}),
$$
(9.37)

within an $\mathcal{A}$-isomorphism: Thus, by virtue of (9.21), the corresponding representation ρ of $\mathcal{GL}(n, \mathcal{A})$, through "operators" on $\mathcal{A}^n$, as in (9.34.1), is just the identity (group) sheaf morphism of $\mathcal{GL}(n, \mathcal{A})$ onto itself, hence, *a fortiori* faithful. On the other hand, in view of the commutativity, according to our assumption of the (structure) $\mathbb{C}$-algebra sheaf $\mathcal{A}$, the previous group sheaf can be viewed as acting on $\mathcal{A}^n$ on either side, this action being compatible with the quotient sheaf appearing in (9.37), so that the same quotient sheaf is, in effect, an $\mathcal{A}$-module. Furthermore, it is also locally free, of finite rank, hence by definition a vector sheaf on X too, its rank being in particular equal to that of the given vector sheaf $\mathcal{E}$ (cf. (9.20)). To see this, one can still argue by analogy with the classical case of a bundle of frames; see, for instance, S. Kobayashi–K. Nomizu [1: p. 54], or L. Conlon [1: p. 151f, and p. 154, Theorem 5.5.5]. In this connection, cf. also the relevant recent work on principal sheaves by E. Vassiliou [1] for further details on the adaptation of the classical case alluded to above to the abstract framework that is adopted by the present treatise.

9.1 Physical Applications

As already said in Section 8 (cf., for instance, the comments at the end of that section), one usually considers particle representations, that is mathematically speaking, representations of the principal sheaves, that correspond to the ("structured") elementary particles under consideration (cf. the beginning of Section 8), more precisely, representations of the structure group sheaf of the principal sheaf concerned (one thus gets here in effect a sheafification of the internal symmetry group of the physical system at issue) in the group sheaf (of germs) of automorphisms of a vector sheaf (cf. (9.36.1)). On the other hand, the (finite) rank of the vector sheaf involved corresponds to the standard case of a (finite-dimensional $\mathbb{C}$-)vector bundle, which, as already said (cf. Section 3, (3.15), as well as (6.29) concerning the corresponding sheaf-theoretic point of view), can be associated with a particular elementary particle field ("matter field"; see also S.Y. Auyang [1: p. 45]).

Now, the role here of the representation (vector) sheaf $\mathcal{E}$, as above (cf. (9.2) or (9.36.1)), consists in that we comprehend (or at least try to comprehend, see the relevant remarks of W. Heisenberg, as quoted at the end of Section 8) the internal structure of our particle by actually studying the action of $\mathcal{G}$ on $\mathcal{E}$, or by allowing an interaction of $\mathcal{G}$ (particle represented by it) with $\mathcal{E}$ (again cf. (6.29)), to apply physics terminology; thus, by looking at the above vector sheaf $\mathcal{E}$ as the vector sheaf of states of the (interacting) particle at issue, we still recall (cf. (6.14)) that the same vector sheaf can actually be perceived through its sections, whose values lie, in view of our hypothesis for $\mathcal{E}$ (see (6.5.1) along with (9.39)), in the stalks (of a finite power) of our structure sheaf ("generalized arithmetic") $\mathcal{A}$. Therefore,

(9.38)
> a (local) section of $\mathcal{E}$, as above, stands for a *state of the field* (particle) under consideration, which is affected by (or even, on which acts) the structure group sheaf $\mathcal{G}$ (or (group) sheaf of internal symmetries), again through its corresponding sections. So the states of the initial particle field ($\mathcal{G}$-principal sheaf, say, $\mathcal{P}$) can now be conceived, in view of the representation ρ, as in (9.36.1), in terms of the states of $\mathcal{E}$ (cf. also (9.36.2) along with (9.17), viz. the states of $\mathcal{P}$; one should also recall here Definition 8.1, act as operators on those of $\mathcal{E}$). Therefore, the states of $\mathcal{P}$ can now acquire, through their values, coordinates, as hinted at above; see (9.39) below.

As a byproduct of the preceding, we arrive, though through a different path, at a similar conclusion as in (3.8), reminiscent of Bohr's correspondence principle; cf.

(3.8′): Indeed, as already remarked, the values of the previous sections, as in (9.38), always lie in the fibers of (our representation vector sheaf) $\mathcal{E}$, so that by employing a local gauge U of $\mathcal{E}$, one obtains, for any point $x \in U$, the relation

$$(9.39) \qquad \mathcal{E}_x \cong \mathcal{A}_x^n \cong (\underbrace{\mathcal{A} \oplus \cdots \oplus \mathcal{A}}_{n\text{-times}})_x,$$

within an $\mathcal{A}_x$-isomorphism of the $\mathcal{A}_x$-modules concerned (cf. (6.5.1) along with (9.14)). Thus, one is led either to classical c-numbers or to generalized ones from our extended arithmetic, especially appertaining to second quantization (see (3.6) along with subsequent comments); that is, from the commutative $\mathbb{C}$-algebra (structure) sheaf $\mathcal{A}$, where one has (cf. also (3.5) as well as (6.6)),

$$(9.40) \qquad \mathbb{C} \underset{\longrightarrow \varepsilon}{\subset} \mathcal{A}.$$

9.2 Interacting Particles

In the case the particle (field), as above is acted on by some "external field" (that is, by another particle, so self-interactions are here not excluded), then any particular (internal) state of the initial particle will be changed, as this particle flows along the integral (alias, solution) curves, which are associated with a vector field (differential equation) that represents the external field; the presence here of an external field can be expressed, as we shall see later (see Chapt. V; (5.106)), by another vector sheaf, while the corresponding vector field-differential equation will be represented by an ($\mathcal{A}$-)connection (covariant derivative or differential operators in our case acting on the sections (states) of the vector sheaf involved; see also (3.42), (3.43)). Thus, the interaction under consideration is finally conceived through a "tensoring" of the respective ($\mathcal{A}$-)connections of the vector sheaves concerned. As we shall see in the subsequent discussion (see Chapts. III, IV, and Vol. II, Chapt. I),

(9.41)   *every (bare) elementary particle can be construed as carrying by itself an ($\mathcal{A}$-)connection,*

provided we dispose the appropriate background to detect it, that is, when appropriately picking out our $\mathbb{C}$-algebraized space $(X, \mathcal{A})$ as in (6.1). Of course, this extends to our abstract setting the situation one has, by considering X as the usual 4-dimensional $\mathcal{C}^\infty$-(space-time) manifold (see (6.2) as well as Chapt. V; (5.5)). However, on this material we shall also comment in several places throughout the ensuing discussion. We can depict the above through the following diagram.

(9.42)

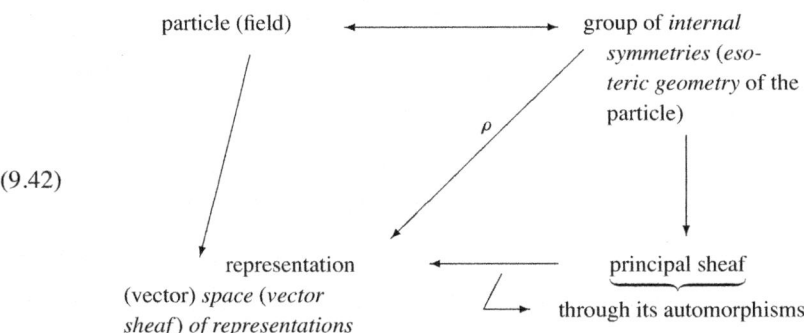

3

Electromagnetism

> "We shall regard the electromagnetic field as a connection ∇ on a one-dimensional vector bundle $\mathcal{E}$ over the space-time M"
>
> Yu. I. Manin, in *Gauge Field Theory and Complex Geometry* (Springer-Verlag, Berlin, 1988). p. 71.

> "For the rest of my life I will ponder on the question of what light is!"
>
> A. Einstein (quoted by W. Pauli in *Writings on Physics and Philosophy*, Springer-Verlag, Berlin, 1994), p. 121.

We discuss in the present chapter the standard theory of electromagnetism, in particular the situation that pertains to Maxwell's equations (the latter are finally considered in Chapter IV) on a classical C^∞-manifold X, however in the much more general context, when one takes instead an abstract differential triad

$$(0.1) \qquad\qquad (X, \mathcal{A}, \partial) \equiv (\mathcal{A}, \partial, \Omega^1),$$

along with the necessary differential equipment (again abstract) to the extent that each particular case may demand; here we refer the reader to Chapter I for details on the applied terminology. Thus, according to the standard model, the carrier of the electromagnetic field (hence, the space attributed to it) is the (quantum of light [A. Einsten], that is, the) photon, in other words, a boson (see Chapter II; (2.2′)), so that (Chapter II; (6.29)) line sheaves will mainly be employed in the sequel. Of course, we shall always argue within an empty space, namely, in the vacuum, so that our discussion will be in accord with what we have already considered (cf. Chapter II; Section 6, e.g., (6.29)).

As already noted (cf. the preface of the present volume), a justification for dealing with the above more general framework than the standard one comes from its potential application to particular instances, where the classical apparatus (viz. calculus) is no lower applicable within the standard differential geometric point of view, due, for instance, to occasional singularities that appear in those cases. These singularities, which thus plague the classical theory, become here ineffectual(!) in view of our more general arithmetic (i.e., of our generalized (abstract) differential machinery) employed. On the other hand, as already said, the same abstract (differential) machinery has always a very special bearing on the classical differential geometry of smooth (C^∞-)manifolds, hence, in particular, on the physical applications of the latter in the case considered.

1 The Electromagnetic Field. The Maxwell Category

The field at issue is, according to the present standard physics model, one of the four fundamental interaction fields, the other three being the *weak*, the *strong*, and the *gravitational* interaction fields. The same field can be viewed as a

(1.1)   *"connection on a 1-dimensional vector bundle over the space-time manifold X."*

See Yu.I. Manin [1: p. 71, §16].

Accordingly, by taking into account the equivalence between vector bundles and their respective sheaves (of germs) of sections (viz. vector sheaves, see Chapter II; Section 6, in particular, (6.13), (6.29) as well as Remark 6.2), we now set the following basic assumption.

Definition 1.1 The *electromagnetic field* on (a space-time manifold) X is a pair

(1.2) $(\mathcal{L}, D)$

consisting of a line sheaf $\mathcal{L}$ on X, along with an $\mathcal{A}$-connection D on $\mathcal{L}$.

Concerning the space X in Definition 1.1, cf. also the relevant comments in Note 1.1. In this connection, one still considers on X a differential triad $(\mathcal{A}, \partial, \Omega^1)$, as in (0.1), whose existence is, of course, always guaranteed in the particular case that X is a usual space-time manifold.

Note 1.1 By still commenting on Definition 1.1, we further remark that the only reason to refer therein to a space-time manifold X was simply to be in accordance with the classical theory, viz. with the so-called *classical model*. On the other hand, the point of view that is advocated by the present treatise is that our argument is still in force for an arbitrary, in general, topological space X, not necessarily a manifold (topological or otherwise) in the ordinary sense of the term, with the proviso that the corresponding (differential) framework we apply remains valid. Of course, there do exist important particular cases that meet the aforementioned situation; see A. Mallios [VS: Vol. II; Chapts X, XI], as well as Vol. II. Chapter IV; Sections 5, 6 of the present account.

Motivated by the above fundamental example (in physics) as presented by Definition 1.1, whenever one has a differential triad

(1.3) $(\mathcal{A}, \partial, \Omega^1)$,

as before (cf. Chapter I; (1.13)), a given pair

(1.4) $(\mathcal{L}, D)$

as in (1.2) is called a *Maxwell field* on the topological space X under consideration. Therefore, by extending Definition 1.1, henceforward,

(1.5) we shall argue in terms of a given Maxwell field $(\mathcal{L}, D)$ on a topological space X, in principle an arbitrary one, unless otherwise specified.

Equivalently, we argue henceforth within the framework of

(1.6) $$\mathcal{M}_X,$$

that is, within the *Maxwell category* of the topological space X considered, carrier of the given differential triad $(\mathcal{A}, \partial, \Omega^1)$, as in (1.3), whose objects are Maxwell fields $(\mathcal{L}, D)$ on X, as in (1.4). Consequently, as a consequence of the previous terminology, one concludes that

(1.7) the electromagnetic field (Definition 1.1) can be construed simply as a Maxwell field $(\mathcal{L}, D)$ (see (1.4)) on the particular space X under consideration, this being chosen in such a manner that the relevant theory of electromagnetism (especially that of Maxwell's equations) can be applied (cf. also Chapter IV; Section 6).

Continuing within the same vein of ideas, we call a given topological space X a *Maxwell space* whenever it can be used as the carrier (the underlying space) of electromagnetism in the previous sense, namely, as that space on which the (abstract) Maxwell's equations (cf. Chapter IV; Section 6 below) hold.

On the other hand, by considering the Maxwell category $\mathcal{M}_X$ as above (cf. (1.6)), we still have to specify its morphisms: Thus, given two objects of $\mathcal{M}_X$, that is, two Maxwell fields $(\mathcal{L}, D)$ and $(\mathcal{L}', D')$, a *morphism* between them, say

(1.8) $$\phi : (\mathcal{L}, D) \longrightarrow (\mathcal{L}', D'),$$

is a sheaf morphism (by an obvious abuse of notation we retain the same symbol)

(1.9) $$\phi : \mathcal{L} \longrightarrow \mathcal{L}'$$

of the respective line sheaves, as indicated (cf. [VS: Chapter I; p. 6]), which further relates the two $\mathcal{A}$-connections D and D', as in (1.8). Thus, by definition,

(1.10) the $\mathcal{A}$-connections D and D', as in (1.8), are ϕ-*related* (with ϕ given by (1.9)) if the following relation holds:

(1.10.1) $$D' \circ \phi = (\phi \otimes 1) \circ D.$$

In (1.10.1) we have set, for convenience,

(1.11) $$1 \equiv 1_{\Omega^1} \equiv id_{\Omega^1};$$

that is, 1 in (1.10.1) stands for the *identity $\mathcal{A}$-automorphism* (or *identity $\mathcal{A}$-isomorphism*) of (the $\mathcal{A}$-module) Ω^1 (onto itself). Equivalently, (1.10.1) is expressed through the following commutative diagram:

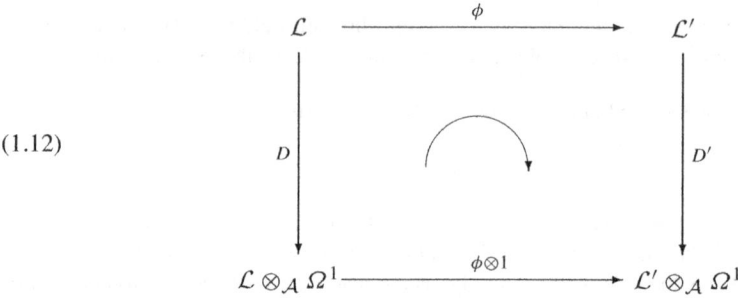

(1.12)

In particular, if ϕ, as above, is an $\mathcal{A}$-isomorphism of $\mathcal{L}$ onto $\mathcal{L}'$, that is, in the case that one has

(1.13) $$\phi \in Isom_A(\mathcal{L}, \mathcal{L}') = \mathcal{I}som_A(\mathcal{L}, \mathcal{L}')(X)$$

(see also A. Mallios [VS: Chapt. V; p. 357, (2.33)]), then the above (1.10.1) takes the form

(1.14) $$D' = (\phi \otimes 1) \circ D \circ \phi^{-1} \equiv \phi D \phi^{-1} \equiv Ad(\phi) \cdot D \equiv \phi_*(D).$$

In this regard, one still refers to (1.14) as a relation defining gauge equivalent $\mathcal{A}$-connections. Namely, by definition,

(1.15)
gauge equivalent $\mathcal{A}$-connections are characterized as (ϕ-) related connections (see (1.10.1) or (1.14)) modulo an ($\mathcal{A}$-) isomorphism, say ϕ, of the carriers (viz. of the vector sheaves, or, in particular, for the case in hand, line sheaves) of the ($\mathcal{A}$-) connections concerned.

Note 1.2 As one remarks in the previously applied terminology, the word "gauge" refers to something spatial, viz. for the case at issue, to the carriers of the fields (vector sheaves) involved, in particular to a transformation (in fact, (space) isomorphism). As we shall see, this in effect will always be the case in the sequel as well. However, it is still to be noticed here that this "spatial" concept is, at the end, expressed algebraically (cf. also the quotations given at the end of this note). Thus here too we *transcribe geometry into algebra* (F. Klein).

In this connection, we also remark that even in Chapter I, (6.5.1), when referring to an open $U \subseteq X$ as a local gauge of an $\mathcal{A}$-module $\mathcal{E}$, that was simply, by an abuse of language, the relevant gauge, being actually the (local) $\mathcal{A}|_U$-isomorphism of the $\mathcal{A}|_U$-modules concerned. (See also Chapt. II; (9.19) or (9.20), along with (9.22)).

On the other hand, by further referring to the case that ϕ, as in (1.9), is in particular an $\mathcal{A}$-isomorphism such that (1.10.1) holds, one gets in effect what we call

gauge equivalent Maxwell fields, denoted by

(1.16)   (1.16.1) $$(\mathcal{L}, D) \underset{\phi}{\sim} (\mathcal{L}', D'),$$

where (the sheaf morphism) ϕ satisfies (1.13) and (1.14).

In this regard, it is quite straightforward (cf. (1.14)) that (1.16.1) defines in effect an *equivalence relation*. Thus, we further look at the set

(1.17)
$$\text{(1.17.1)} \qquad\qquad \Phi^1_{\mathcal{A}}(X)^\nabla,$$

viz. the *set of isomorphism classes of (gauge equivalent) Maxwell fields on X*.

In fact, as we shall see (cf. Theorem 2.1), the previous set (1.17.1) is actually an abelian group. We thus call it, in anticipation, the *Maxwell group of X*.

Now, the physical meaning of the equivalence relation defined by (1.16.1), hence, also of the set (1.17.1), is that

(1.18)
equivalent (viz. gauge equivalent, see (1.16.1), or even isomorphic) Maxwell fields on X provide the same field strength. In other words, we do not distinguish, through (1.17.1), Maxwell fields on X, that have the same field strength (cf. also (3.11)).

In this connection, an equivalence class in (1.17.1), that is, an element

$$\text{(1.19)} \qquad\qquad [(\mathcal{L}, D)] \in \Phi^1_{\mathcal{A}}(X)^\nabla,$$

is called a *beam* (of electromagnetic fields, in fact *of photons*, that carry the fields in question) or a *light ray* (see also (3.11) below). A complete proof of our assertion in (1.18), as well as details concerning the unexplained terms employed therein, will be given in Section 3. Indeed,

(1.20)
the space X in (1.18) has to be further suitably specialized so that the applied terminology, for instance "field strength," is meaningful (see (3.2)).

Concerning our terminology applied above, as in (1.20), for the space X considered in (1.18), we still remark that

(1.21)
by an abuse of terminology, although we refer to an *occasional specification of the space X* at issue, this actually corresponds to the particular *structure elements* (structure sheaf, differential operators, etc.) that accompany it, while X remains in general an *arbitrary topological space*. In other words, we do not have to specialize the space X, but simply the objects (sheaves etc.) living on it, as well as eventual interrelations (sheaf morphism) between them, a fact that is characteristic of the abstract point of view, advocated by the present exposition.

Furthermore, by still commenting on the terminology applied in the preceding, in particular on that connected with (1.19), we also remark that by a light ray, as above, we actually refer to a particular *line sheaf* (in effect, to a certain *equivalence class* of such; see (1.16.1)), not necessarily to a trivial "straight" line sheaf on X. In this connection, it is also of a particular interest to mention at this point the respective apostrophe of A. Einstein [2: p. 14] that also fits in with the corresponding epigram of the present chapter:

(1.22)   "light rays ... might be ... involved in the origin of the concepts and laws of geometry."

On the other hand, within the same vein of ideas, one can still refer to the relevant "splitting principle" of A. Grothendieck in K-Theory (see, for instance, M. Karoubi [1: p. 193, Theorem 2.15]), while concerning the (compact) base space X considered therein, we also refer to Chapt. II; Remark 6.2 in the preceding. So this, in conjunction with (1.22), leads us to the eventual reflection that

(1.23)   the only geometry we have might be that determined by the light rays (in the previous sense of the term, cf. (1.19)).

2 Characterization of the Maxwell Group Through Local Data

We have already seen (cf. Chapter II; Section 7.(b)) that the set of isomorphism (or even equivalence) classes of line sheaves on X,

$$(2.1) \qquad \Phi^1_{\mathcal{A}}(X)$$

is *cohomologically characterized* through the following isomorphism of abelian groups (loc. cit.; (7.23)):

$$(2.2) \qquad \Phi^1_{\mathcal{A}}(X) = \check{H}^1(X, \mathcal{A}^{\bullet})$$

(cf. also Note 7.1 concerning the 1st Čech cohomology group of X, appearing in the second member of (2.2)). In this connection, we still recall that the group structure in the set (2.2) is defined through the factor $\otimes_{\mathcal{A}}$ (see also Chapter I), so that one has, by definition, concerning the *Picard group of X* (or else group of *invertible sheaves*, viz. line sheaves on X, in point of fact, their set of isomorphism classes, as in (2.1)), the relation

$$(2.3) \qquad Pic(X) := (\Phi^1_{\mathcal{A}}(X), \otimes_{\mathcal{A}}) \equiv \Phi^1_{\mathcal{A}}(X) = \check{H}^1(X, \mathcal{A}^{\bullet})$$

(see Chapt. II; (7.31), (7.32)).

Now, our first objective, by the ensuing discussion, is to show that the set (1.17.1), as above, can also similarly be made into an abelian group. The corresponding natural map

$$(2.4) \qquad \Phi^1_{\mathcal{A}}(X)^{\nabla} \longrightarrow \Phi^1_{\mathcal{A}}(X) \equiv Pic(X)$$

is an *abelian group morphism* (cf. (2.6) below).

Theorem 2.1 Suppose we are given a differential triad $(\mathcal{A}, \partial, \Omega^1)$ (cf. (1.3)) on a topological space X. Then, the set of isomorphism classes of Maxwell fields on X,

$$(2.5) \qquad \qquad \Phi^1_{\mathcal{A}}(X)^\nabla$$

(cf. (1.17.1)), can be endowed with an abelian group structure, through the functor $\otimes_{\mathcal{A}}$. We call then (2.5) the *Maxwell group* of X.

Proof. The asserted group structure in (2.5) is given by virtue of the following relations:

(i) group multiplication (see also (1.19)):

$$(2.6) \qquad \begin{aligned} [(\mathcal{L}, D)] \otimes_{\mathcal{A}} [\mathcal{L}', D')] &\equiv [(\mathcal{L}, D)] \otimes [(\mathcal{L}', D')] \\ &:= [(\mathcal{L} \otimes_{\mathcal{A}} \mathcal{L}', D \otimes D')]. \end{aligned}$$

(ii) inversion:

$$(2.7) \qquad \begin{aligned} [(\mathcal{L}, D)]^{-1} &:= [(\mathcal{L}, D)^{-1}] := [(\mathcal{L}^{-1}, D^{-1})] \\ &:= [(\mathcal{L}^*, D^*)] \end{aligned}$$

(iii) identity (neutral) element:

$$(2.8) \qquad \qquad [(\mathcal{A}, \partial)] \in \Phi^1_{\mathcal{A}}(X)^\nabla.$$

(See also Chapter I for the notation applied in the previous relations.) Concerning our assertion for (2.8), we still note that

the $\mathcal{A}$-connections

$$(2.9.1) \qquad \qquad D \otimes \partial := D \otimes 1 + 1 \otimes \partial$$

(2.9)   on $\mathcal{L} \otimes_{\mathcal{A}} \mathcal{A}(\cong \mathcal{L})$ and D on $\mathcal{L}$ are "gauge equivalent" (cf. (1.15)), with respect to the $\mathcal{A}$-isomorphism

$$(2.9.2) \qquad \qquad \mathcal{L} \otimes_{\mathcal{A}} \mathcal{A} = \mathcal{L};$$

viz. $s \otimes \alpha \longmapsto \alpha \cdot s = s \cdot \alpha$, for any $\alpha \in \mathcal{A}(U)$ and $s \in \mathcal{L}(U)$, with U open in X.

The above proves, of course, our claim for (2.8), by virtue of (2.6) and (2.9.2). On the other hand, we further remark that

(2.10)   the equivalence relation (1.16.1) between Maxwell fields is compatible with the group operations, as defined by (2.6) and (2.7) (hence the latter are well defined).

Indeed, our previous assertion in (2.10) is a straightforward consequence of the definitions, in conjunction with a local characterization of (1.16.1), as will be given by Lemma 2.2, and this completes the proof of the theorem. (See also Chapter II; (7.26)–(7.29).) ∎

We continue in the next section to explain the aforementioned local characterization of the equivalence relation (1.16.1) between Maxwell fields that was of use in anticipation for the proof of (2.10) along a similar result for a given Maxwell field on X (viz. a pair, as in (1.2) or in (1.4)). These both will be of particular use in the sequel. So we start with the following.

2.1 Local Characterization of Maxwell Fields

Suppose we are given a Maxwell field

(2.11) $(\mathcal{L}, D)$

(cf. Definition 1.1, along with (1.4) in the preceding) on a topological space X (see also Note 1.1, as well as, (1.5) above), base space of a given differential triad (cf. (1.3))

(2.12) $(\mathcal{A}, \partial, \Omega^1)$.

On the other hand, suppose that

(2.13) $\mathcal{U} = (U_\alpha)_{\alpha \in I}$

is a given local frame of the line sheaf $\mathcal{L}$ at issue as in (2.11), that is, a family of local realizations, or else, mathematically speaking (see Chapter I; (2.29)), of local gauges, of the given Maxwell field $(\mathcal{L}, D)$, as in (2.11) (cf. also Note 1.2).

Given (2.13), one concludes that (1) we can identify $\mathcal{L}$ with a 1-cocycle (viz. with the so-called "coordinate 1-cocycle" of $\mathcal{L}$, associated with $\mathcal{U}$)

(2.14) $(g_{\alpha\beta}) \in Z^1(\mathcal{U}, \mathcal{A}^{\cdot})$

(see Chapt. II; (7.18), (7.20), (7.21), or (7.23)). Furthermore, (2) the $\mathcal{A}$-connection D of $\mathcal{L}$, as in (2.11) (viz. the corresponding gauge potential of the given electromagnetic, or else Maxwell field) yields a 0-*cochain of 1-forms* on X (see Chapt. I; (2.54), for $n = 1$), say

(2.15) $(\theta_\alpha) \in C^0(\mathcal{U}, \Omega^1)$

such that the given $\mathcal{A}$-connection D as above is in effect identified with it in the sense of the following lemma.

Lemma 2.1 Suppose we have the framework that corresponds to (2.11)–(2.15). Then, *a given pair*

(2.16) $((g_{\alpha\beta}), (\theta_\alpha)) \in Z^1(\mathcal{U}, \mathcal{A}^\bullet) \times C^0(\mathcal{U}, \Omega^1)$

entails a Maxwell field on X if and only if one has

(2.17) $\delta(\theta_\alpha) = \tilde{\partial}(g_{\alpha\beta}).$

Before we come to the proof of the above lemma, we recall for convenience (see Chapter I) the relevant notation employed in (2.17). Thus,

(2.18) $\tilde{\partial} : \mathcal{A}^\bullet \to \Omega^1$

stands for the corresponding *logarithmic derivation* to the given (flat) $\mathcal{A}$-connection ∂ on $\mathcal{A}$ (cf. (2.12)), as given by the relation

(2.19) $\tilde{\partial}(\alpha) := \alpha^{-1} \cdot \partial(\alpha),$

for any local (continuous) section $\alpha \in \mathcal{A}^\bullet(U) = \mathcal{A}(U)^\bullet$ and any open $U \subseteq X$. On the other hand, the first member of (2.17) denotes the *difference connection 1-form* on the open set in X,

(2.20) $U_{\alpha\beta} \equiv U_\alpha \cap U_\beta \neq \emptyset$

(for any indices α, β in I, as in (2.13), for which (2.20) is in force), which is provided by applying the *0th coboundary operator* (cf. (2.14))

(2.21) $\delta \equiv \delta^0 : C^0(\mathcal{U}, \Omega^1) \to C^1(\mathcal{U}, \Omega^1)$

(see also [VS: Chapt. III; p. 178, (4.29)]).

Note 2.1 In this connection, we remark that the (local) 1-forms θ_α, $\alpha \in I$, as in (2.15), stand within our (abstract) framework for the classical gauge potential 1-forms,

(2.22) $A_\mu dx^\mu \quad (\mu = 0, 1, 2, 3),$

(see Yu. I. Manin [1: p. 71, §16]). Precisely speaking, one has (of course locally) that

(2.22′) $A_\mu dx^\mu \in \partial(\mathcal{A}) \equiv \operatorname{im} \partial \subseteq \Omega^1,$

in terms always of local sections, for example over a local gauge (chart) of the line sheaf (bundle) involved (see also Chapt. II; (6.25), as well as, Definition 1.1 of the present Chapter, along with chapter I).

After the above preliminary material on the terminology employed, we come next to the

Proof of Lemma 2.1. Equation (2.17) is in fact a special case of the so-called *transformation law of potentials* (see Chapter I or Vol. II: Chapter I; Section 9), valid for any vector sheaf $\mathcal{E}$ on X with $rk\mathcal{E} = n \in \mathbb{N}$, endowed with an $\mathcal{A}$-connection D, that is, for any Yang–Mills field $(\mathcal{E}, D)$ on X (cf. Chapter I, along with Vol. II: Chapter I; Section 4(a)). Thus, by referring to this general law, one has the relation

$$(2.23) \qquad \omega^{(\beta)} = Ad(g_{\alpha\beta}^{-1})\omega^{(\alpha)} + \tilde{\partial}(g_{\alpha\beta}).$$

As we know, the previous relation provides, in effect, a criterion for the existence of an $\mathcal{A}$-connection D on $\mathcal{E}$ when in general one is given a 0-cochain of $n \times n$ matrices of 1-forms,

$$(2.24) \qquad \omega \equiv (\omega^{(\alpha)}) \in C^0(\mathcal{U}, M_n(\Omega^1)),$$

along with a coordinate 1-cocycle of $\mathcal{E}$,

$$(2.25) \qquad g \equiv (g_{\alpha\beta}) \in Z^1(\mathcal{U}, \mathcal{GL}(n, \mathcal{A})),$$

i.e., a family of local gauge transformations (or even transformations of local realizations in terms of $\mathcal{A}$) of the given vector sheaf $\mathcal{E}$; cf. Chapt. II, Note 6.1, as well as, (7.8), along with (7.18)–(7.21)). In this connection, see also [VS: Chapt. VII, p. 116, Theorem 3.1] for a complete proof of (2.23). The same reference, p. 119, Theorem 3.2, pertaining to a reformulation of the same general law, as above, in the point of view of the present Lemma 2.1, under consideration.

Thus, as already said, (2.17) is a particular case of (2.23) when one considers a Maxwell field $(\mathcal{L}, D)$ on X, where $rk\mathcal{L} = 1$, that is, for $n = 1$, and this terminates the proof of Lemma 2.1. ∎

As a result of the preceding, we may represent the above bijective correspondence, as established by Lemma 2.1, through the diagram

$$(2.26) \qquad (\mathcal{L}, D) \xleftarrow[\delta(\theta_\alpha) = \tilde{\partial}(g_{\alpha\beta})]{\mathcal{U} = (U_\alpha)_{\alpha \in I}} ((g_{\alpha\beta}), (\theta_\alpha)),$$

in such a manner that the items on the right-hand satisfy (2.17), as indicated. On the other hand, based on the preceding correspondence, we can also give the

(2.27) group operations of the Maxwell group $\Phi^1_{\mathcal{A}}(X)^\nabla$ (cf. (2.5)), as defined by (2.6) and (2.7) through local data as well.

That is, by virtue of (2.6) and (2.7), one obtains, in the sense of (2.26),

$$(2.28) \quad (\mathcal{L}, D) \otimes (\mathcal{L}', D') := (\mathcal{L} \otimes_{\mathcal{A}} \mathcal{L}', D \otimes D') \longleftrightarrow ((g_{\alpha\beta} \cdot g'_{\alpha\beta}), (\theta_\alpha + \theta'_\alpha)),$$

in such a manner that one still has

$$(2.29) \quad \delta(\theta_\alpha + \theta'_\alpha) = \delta(\theta_\alpha) + \delta(\theta'_\alpha) = \tilde{\partial}(g_{\alpha\beta}) + \tilde{\partial}(g'_{\alpha\beta}) = \tilde{\partial}(g_{\alpha\beta} \cdot g'_{\alpha\beta}),$$

viz. (2.17), for the case at issue, which also justifies (2.28) (or equivalently, (2.6)).

Furthermore, one also obtains (cf. (2.7))

(2.30) $(\mathcal{L}, D)^{-1} := (\mathcal{L}^{-1}, D^{-1}) := (\mathcal{L}^*, D^*) \longleftrightarrow ((g_{\alpha\beta}^{-1}), (-\theta_\alpha)),$

such that one obtains

(2.31) $$\delta(-\theta_\alpha) = -\delta(\theta_\alpha) = -\tilde{\partial}(g_{\alpha\beta}) = \tilde{\partial}(g_{\alpha\beta}^{-1}),$$

justifying thus (2.7). Therefore, one also concludes that

(2.32)
$$\begin{aligned}
(\mathcal{L}, D) \otimes (\mathcal{L}, D)^{-1} &= (\mathcal{L}, D) \otimes (\mathcal{L}^*, D^*) \\
&= (\mathcal{L} \otimes_{\mathcal{A}} \mathcal{L}^*, D \otimes D^*) \\
&= (\mathcal{A}, D \otimes D^*) \longleftrightarrow ((1_{\alpha\beta} = 1), (0)) \\
&\equiv (1, 0) \longleftrightarrow (\mathcal{A}, \partial),
\end{aligned}$$

which further justifies (2.8) (cf. also (2.9)).

Accordingly, by further using physical language, we can say that

a given pair

(2.33.1) $(\mathcal{L}, D) \longleftrightarrow ((g_{\alpha\beta}), (\theta_\alpha)),$

as above (cf. (2.16)), satisfying the relation

(2.33.2) $\delta(\theta_\alpha) = \tilde{\partial}(g_{\alpha\beta})$

can be interpreted, according to the particular case at hand, as an electromagnetic field (Maxwell field, cf. Definition 1.1), that is,

(2.33)
the *carrier* (*photon*) represented by

$$(g_{\alpha\beta}) \in Z^1(\mathcal{U}, \mathcal{A}^\cdot),$$

(2.33.3)   along with the field ($\mathcal{A}$-connection) given by

$$(\theta_\alpha) \in C^0(\mathcal{U}, \Omega^1),$$

so that (2.33.2) be in force.

On the other hand, more generally,

a pair

(2.34.1) $((g_{\alpha\beta}), (\omega^{(\alpha)})) \in Z^1(\mathcal{U}, \mathcal{GL}(n, \mathcal{A})) \times C^0(\mathcal{U}, M_n(\Omega^1))$

(2.34)
can be construed as representing a *physical field* (*Yang–Mills field*) if and only if the transformation law of potentials holds, that is, one has

(2.34.2) $\omega^{(\beta)} = Ad(g_{\alpha\beta}^{-1})\omega^{(\alpha)} + \tilde{\partial}(g_{\alpha\beta}).$

Of course, (2.33) is a particular case of (2.34) for $n = 1$ (Maxwell field). We can also write (2.34.2) in the form of (2.33.2), since one has

(2.35)
$$\delta(\omega^{(\alpha)}) = \omega^{(\beta)} - Ad(g_{\alpha\beta}^{-1})\omega^{(\alpha)},$$

so that one finally obtains, in view of (2.34.2),

(2.36)
$$\delta(\omega^{(\alpha)}) = \tilde{\partial}(g_{\alpha\beta}),$$

as desired. (In this connection, see also [VS: Chapt. VII; p.119, Theorem 3.2].)

We give right away a physical meaning to (2.33.2) and/or (2.36), by arguing in particular in terms of a Maxwell field (viz. (2.33.2) or (2.36), for $n = 1$; the general case, Yang–Mills field, is of course quite similar):

(2.37)   The 0-cochain (θ_α) (cf. (2.15)) yields an $\mathcal{A}$-connection D of the line sheaf $\mathcal{L}$, the latter being determined by the 1-cocycle $(g_{\alpha\beta})$ (cf. (2.14)), if and only if the items of the 0-cochain as above, viz. the local 1-forms $\theta_\alpha, \alpha \in I$ (cf. (2.13)), are pairwise gauge equivalent through (cf. (2.20))

(2.37.1)
$$g_{\alpha\beta} \in \mathcal{A}^\bullet(U_{\alpha\beta}) = \mathcal{A}(U_{\alpha\beta})^\bullet,$$

the latter denoting a *change of a local gauge* (i.e., of *local coordinates*, cf. Chapt. II; (6.7) and (7.8)) *of the line sheaf* $\mathcal{L}$ concerned.

That is

a given 0-cochain (of (local) 1-forms),

(2.38)
$$(\theta_\alpha) \in C^0(\mathcal{A}, \Omega^1),$$

entails a (Maxwell) field, e.g., the electromagnetic field, if and only if the corresponding transformation law of potentials, viz. (2.33.2), is in force.

The above are susceptible of further physical parlance (transcription), since we can still remark that

(2.39)   the above transformation law of potentials as given by (2.36), or in particular for $n = 1$ by (2.33.2), manifests in effect the covariance of the field (as the latter is expressed via the gauge potential or the $\mathcal{A}$-connection D) with its corresponding carrier (viz. the vector/line sheaf). Precisely speaking, we can say that

(2.39.1)   the field changes at the same rate as its carrier.

In fact, (2.39.1) can be viewed as an equivalent formulation of the same transformation law of potentials (in either form of it, as in (2.36) or (2.33.2)). The picture resulting from (2.39.1) can explain an occasional identification of the two (physical) objects in question (field/carrier), although they are, of course, different in nature. In this connection, cf. also our remarks in (2.40) below. Thus, we can still say, as a result of the preceding, that

(2.40)   the field accompanies the carrier, or even, equivalently, the carrier follows the field, of course in the same manner, viz. within the same rate of variation (see (2.36)).

So, by adding to the motto of D. R. Finkelstein [1: p. 126], we can further say that

(2.41)   "flow follows fracture," in accordance with the transformation law of potentials, viz., in such a manner that the aforesaid law is still in force.

Finally, the same law can be construed as the mathematical expression of how things are varied(!), yet one thus obtains a quantitative account of the way things behave during our observation (gauging; furthermore, the "if" part of the same law entails, of course, a criterion for the existence of a field!).

Having now a local expression of a Maxwell field $(\mathcal{L}, D)$ in the form of a pair, as in (2.16), under the proviso, of course, that (2.17) is valid (see also e.g. (2.26)), we further examine in the same spirit, viz. locally, the equivalence relation between two given Maxwell fields (cf. (1.16)), in the next subsection. The same relation defines, in effect, the Maxwell group of X (see (1.17)); that is, one obtains a local characterization of the Maxwell group.

2.2 Local Characterization of (Gauge) Equivalent Maxwell Fields

In this connection, we start with our main result, in the form of the following:

Lemma 2.2 Assume that we are given a differential triad

$$(2.42) \qquad (\mathcal{A}, \partial, \Omega^1)$$

(cf. (1.3)), along with two Maxwell fields $(\mathcal{L}, D)$ and $(\mathcal{L}', D')$ on X. Then, one has

$$(2.43) \qquad (\mathcal{L}, D) \sim (\mathcal{L}', D')$$

("gauge equivalent" Maxwell fields, cf. (1.16.1)) if and only if there exists a 0-cochain

$$(2.44) \qquad (s_\alpha) \in C^0(\mathcal{U}, \mathcal{A}^\cdot)$$

such that (i)

$$(2.45) \qquad g'_{\alpha\beta} = \delta(s_\alpha^{-1}) \cdot g_{\alpha\beta},$$

and (ii)

$$(2.46) \qquad \theta'_\alpha = \theta_\alpha + \tilde{\partial}(s_\alpha^{-1}).$$

Before we come to the proof of the previous lemma, we comment, for convenience, a bit on the notation employed in its statement: Thus,

(2.47) $$\mathcal{U} = (U_\alpha)_{\alpha \in I}$$

in (2.44) stands for a *common local frame of $\mathcal{L}$ and $\mathcal{L}'$*; $(g_{\alpha\beta})$ and $(g'_{\alpha\beta})$ denote the coordinate 1-cocycles of $\mathcal{L}$ and $\mathcal{L}'$, respectively, with respect to (2.47), while (θ_α) and (θ'_α) the 0-cochains of 1-forms of D and D', respectively corresponding to $\mathcal{U}$. (It is easy to see, in view of the definitions, that our previous assumption for (2.47) is always true.) So we come next to the

Proof of Lemma 2.2. By looking at the definition of the equivalence relation in (2.43) (see (1.16)), one realizes that

(2.48)   the existence of an $\mathcal{A}$-isomorphism ϕ between $\mathcal{L}$ and $\mathcal{L}'$ (cf. (1.9)) is equivalent to the existence of a 0-cochain, as in (2.44), satisfying (2.45).

Indeed, cf. [VS: Chapt. V; p. 353, Lemma 2.1, for $n = 1$, along with p. 356, Scholium 2.1].

On the other hand, the same $\mathcal{A}$-isomorphism

(2.49) $$\phi \longleftrightarrow (s_\alpha)$$

between $\mathcal{L}$ and $\mathcal{L}'$, as above, can further be construed as a local gauge transformation of either one of the two fields at issue, hence the associated, by our hypothesis (cf. (2.43)), gauge equivalence of D and D'. Accordingly (cf. [VS: Chapt. VII; p. 109, (2.15), as well as, p. 105, (1.41)]), one concludes that

(2.50)   the *gauge equivalence of D and D'*, through $\phi \longleftrightarrow (s_\alpha)$ (see (2.49)), is (locally) expressed equivalently via the relation

(2.50.1) $$\theta'_\alpha = \theta_\alpha + \tilde{\partial}(s_\alpha^{-1}).$$

Thus (2.48) together with (2.50) supplies the proof of Lemma 2.2. ∎

Note 2.2 (Terminological) Concerning (2.50.1), the transformation

(2.51) $$\theta_\alpha \mapsto \theta_\alpha + \tilde{\partial}(s_\alpha^{-1}) \equiv \theta'_\alpha$$

is also what we usually call, classically, a *gauge transformation*.

A substantial application of Lemma 2.2, along with a nice cohomological description of both Lemmas 2.1 and 2.2, will be given in the sequel (cf. Chapt. IV; Theorem 5.1 pertaining to a cohomological classification of Maxwell fields; see also the ensuing remark (2.52)). By virtue of our terminology in Chapt. II; Section 7.(b), one concludes that

(2.45), that is, the relation

(2.52.1) $$g'_{\alpha\beta} = \delta(s_\alpha^{-1}) \cdot g_{\alpha\beta}$$

(2.52)

or equivalently the relation

(2.52.2) $$\delta(g_{\alpha\beta}) = \delta(s_\alpha^{-1}),$$

where we set $g'_{\alpha\beta} \cdot g_{\alpha\beta}^{-1} \equiv \delta(g_{\alpha\beta})$, provides in effect a local characterization of the Picard group of X (cf. also Chapt. II; (7.31), (7.32)).

On the other hand, the following result is an outcome of Lemma 2.2 in the particular case $\mathcal{L} = \mathcal{L}'$. The same result extends a relevant conclusion of B. Kostant in the classical case; see B. Kostant [1: p. 116. Corollary 1 to Lemma 1.10.1].

Corollary 2.1 Suppose we have the framework of Lemma 2.2, and let $(\mathcal{L}, D)$ be a Maxwell field on X. Then an *automorphism of* $\mathcal{L}$, i.e.,

$$(2.53) \qquad \phi \in \mathcal{I}som_A(\mathcal{L}, \mathcal{L}) \equiv Aut\mathcal{L},$$

is still an automorphism of $(\mathcal{L}, D)$, as well; that is,

$$(2.54) \qquad \phi \in Aut((\mathcal{L}, D))$$

if and only if one has (cf. (2.44), (2.49))

$$(2.55) \qquad (s_\alpha) \in C^0(\mathcal{U}, \ker \tilde{\partial}).$$

In particular (in point of fact, this is Kostant's result in our abstract setting), if

$$(2.56) \qquad \ker \tilde{\partial} = \mathbb{C}^{\bullet},$$

then (2.53) implies (2.54) too if and only if one has

$$(2.57) \qquad \phi = \lambda \in \mathbb{C}^{\bullet} = Z^0(\mathcal{U}, \mathbb{C}^{\bullet})$$

(that is, ϕ is virtually reduced to a multiplication by a non-zero (constant) $\lambda \in \mathbb{C}$).

Proof. By virtue of (2.53) and (2.48) (for $\mathcal{L} = \mathcal{L}'$), we know that ϕ is given by a 0-cochain (see also (2.49))

$$(2.58) \qquad (s_\alpha) \in C^0(\mathcal{U}, \mathcal{A}^{\bullet}).$$

Thus, in view of (2.50) (for $\theta_\alpha = \theta'_\alpha$), our assertion for (2.54) is valid if and only if one has

$$(2.59) \qquad \tilde{\partial}(s_\alpha^{-1}) = 0,$$

that is, equivalently,

$$(2.60) \qquad s_\alpha \in \ker \tilde{\partial}, \quad \alpha \in I;$$

see also (2.47); in this connection, we remark that

$$(2.61) \qquad \ker \tilde{\partial} \subseteq \mathcal{A}^{\bullet}$$

is a *group subsheaf of* $\mathcal{A}^{\bullet}$ according to the properties of $\tilde{\partial}$; cf. Chapter I. So the preceding establish the first part of our claim, the rest being clear under the proviso of (2.56), and this terminates the proof. ∎

Scholium 2.1 Concerning the last part of Corollary 2.1, we further remark that

(2.62)   (2.56) is not, in general, true in abstract differential geometry,

as happens in the classical theory. Thus (2.56) is in effect in our case part of the classical Poincaré lemma, which in general is not valid in the abstract theory! However, we have important particular instances in which the same lemma (in its abstract form) does hold; see, for example, [VS: Chapts X, XI], as well as the present account, Chapt. IV; Section 5.

3 A Natural Fibration

The fibration in the title of this section is defined by the concept of *field strength*: Thus, by referring in particular to the electromagnetic field, that is, by our definition (see Definition 1.1) to a Maxwell field $(\mathcal{L}, D)$, the notion at issue is defined as the *curvature* of the respective $\mathcal{A}$-connection D (gauge potential), denoted henceforward by

$$(3.1) \qquad\qquad R(D) \equiv R.$$

See also Chapter IV, where the same concept refers to any Yang–Mills field $(\mathcal{E}, D)$.

However, to formulate the notion of curvature, thus of that fundamental concept of classical differential geometry of smooth manifolds within our abstract framework that we advocate throughout this treatment, we have, as already explained in Chapter I, to enhance our differential setting employed hitherto through another item (differential operator), so that the notion under consideration acquires a meaning.

Thus, apart from our basic differential triad $(\mathcal{A}, \partial, \Omega^1)$ on a topological space X, as before (cf., for instance, (2.12)), assume now that we are further given a curvature datum

$$(3.2) \qquad\qquad (\mathcal{A}, \partial, d^1)$$

on X; we still refer to X as a curvature space (cf. Chapter I for the terminology applied).

In this connection, we already know (ibid; Section 7) that given a Maxwell field

$$(3.3) \qquad\qquad (\mathcal{L}, D)$$

on X, its field strength (curvature) entails a "2-form" on X, viz. one has

$$(3.4) \qquad\qquad R(D) \equiv R \in \Omega^2(X).$$

Indeed, $R(D)$ is an $\mathcal{A}$-morphism of $\mathcal{L}$ into

$$(3.5) \qquad\qquad \Omega^2(\mathcal{L}) \equiv \Omega^2 \otimes_{\mathcal{A}} \mathcal{L} = \mathcal{L} \otimes_{\mathcal{A}} \Omega^2$$

(a tensor, as we say classically), so that one has

(3.6)
$$R \in Hom_{\mathcal{A}}(\mathcal{L}, \Omega^2(\mathcal{L})) = \mathcal{H}om_{\mathcal{A}}(\mathcal{L}, \Omega^2(\mathcal{L}))(X)$$
$$= \Omega^2(\mathcal{E}nd\mathcal{L})(X) = \Omega^2(X),$$

which also explains (3.4), given that

(3.7)
$$\mathcal{E}nd\mathcal{L} = \mathcal{A}$$

(see Chapt. I; (5.10), along with [VS: Chapt. II; p. 139, Lemma 6.2, and Chapt. IV; p. 304, Corollary 6.1]).

On the other hand, by further looking at the curvature R, in terms of a local frame of $\mathcal{L}$, say

(3.8)
$$\mathcal{U} = (U_\alpha)_{\alpha \in I}$$

(cf. Chapt. II; (7.4), for $n = 1$), one gets the relation

(3.9)
$$R = (d\theta_\alpha) \in Z^0(\mathcal{U}, \Omega^2),$$

where we have set (see (3.2))

(3.9')
$$d\theta_\alpha \equiv d^1(\theta_\alpha), \quad \alpha \in I,$$

while

(3.10)
$$(\theta_\alpha) \in C^0(\mathcal{U}, \Omega^1)$$

stands for a 0-*cochain of* 1-*forms* determining locally (with respect to $\mathcal{U}$, as in (3.8)) the given $\mathcal{A}$-connection D of $\mathcal{L}$ (see also [VS: Chapt. VIII; p. 197, Corollary 3.1]).

Now, based on the preceding, we can further state our first remark, as in (3.11) below, which also entails in effect the desired fibration (map), as alluded to by the title of this section. The same remark also justifies our comments in the foregoing pertaining to the equivalence relation between Maxwell fields (cf. (1.16), along with (1.18)). Namely, one concludes that (see also (3.9), (2.46) and Chapt. I; (7.5))

(3.11)
> given a curvature space X (cf. (3.2)), two equivalent Maxwell fields on X always have the same field strength (= curvature, cf. (3.9)).
>
> Therefore, an equivalence class in (an element of) the Maxwell group of X,
>
> (3.11.1) $$[(\mathcal{L}, D)] \in \Phi^1_{\mathcal{A}}(X)^\nabla,$$
>
> in other words, a beam (of photons, or even a light ray, cf. (1.19)) consists of (a bunch of) Maxwell fields having the same field strength (same "color").

Indeed, our assertion in (3.11) is a straightforward consequence of (3.9) and (2.50.1), when also taking into account the flatness of ∂ (that $d^1 \circ \partial \equiv d \circ \partial = 0$; cf. Chapt. I; (7.5), or [VS: Chapt. VIII; p. 187f, (1.12)–(1.14)]). ∎

Thus, being always within the framework of a given curvature space X, as before, we are now in position to define the following *map (fibration)*:

$$(3.12) \qquad \tau : \Phi_{\mathcal{A}}^1(X)^{\nabla} \longrightarrow \Omega^2(X)$$

(cf. also (3.6) in the preceding), such that one sets

$$(3.13) \qquad \tau([(\mathcal{L}, D)]) := R(D) \equiv R$$

for any beam $[(\mathcal{L}, D)] \in \Phi_{\mathcal{A}}^1(X)^{\nabla}$. In this connection, we still note that by virtue of (3.11), the above map τ is of course well-defined, viz. independently of the particular Maxwell field,

$$(3.14) \qquad (\mathcal{L}, D) \in [(\mathcal{L}, D)]$$

involved, as in (3.13).

So one gets a fibration, via τ, of the Maxwell group of X over $\Omega^2(X)$, hence also at the following *fiber decomposition of the Maxwell group of X*; that is, one has

$$(3.15) \qquad \Phi_{\mathcal{A}}^1(X)^{\nabla} = \sum_{R \in \mathrm{im}\, \tau} \Phi_{\mathcal{A}}^1(X)_R^{\nabla},$$

where we set

$$(3.16) \qquad \Phi_{\mathcal{A}}^1(X)_R^{\nabla} := \tau^{-1}(R), \quad R \in \mathrm{im}\, \tau,$$

that is, the *fiber of the map τ over $R \in \mathrm{im}(\tau) \subseteq \Omega^2(X)$*, or the *set of beams in the Maxwell group of X, $\Phi_{\mathcal{A}}^1(X)^{\nabla}$ that have a given field strength* (= curvature), say $R \in \Omega^2(X)$.

The preceding makes it clear that it becomes a primary concern to us to identify the image of τ in $\Omega^2(X)$. Indeed, as we shall see, not all of $R \in \Omega^2(X)$ belongs to $\mathrm{im}\,\tau$; that is, not every 2-form on X, say $R \in \Omega^2(X)$, can be construed, as the field strength of a Maxwell field on X! In this connection, the relevant answer comes through a classical result, as the latter is formulated, within our abstract setting, viz. the so-called Weil's integrality theorem (cf. Theorem 3.1). So the above will be our main objective in the next few subsections.

3.1 The Image of (the Natural Fibration) τ

The determination of the image of the map τ, as in (3.12), forces us to further enlarge the differential setting we have already. Thus, what we really need is the differential framework of what we may call a *Bianchi space* (or a *Bianchi datum*), viz. of the tetrad

$$(3.17) \qquad (\mathcal{A}, \partial, d^1, d^2)$$

on a topological space X, where in addition to a curvature datum (cf. (3.2)) that one considers on X, we still assume the existence of one more differential operator, viz. of the 2nd exterior differential,

$$(3.18) \qquad\qquad d^2 : \Omega^2 \longrightarrow \Omega^3,$$

satisfying the appropriate conditions (motivated by the classical case of smooth manifolds) relative to $d^\circ \equiv \partial$ and d^1 (cf. Chapt. I or Vol. II: Chapt. I; Section 1). Thus, as a first outcome of the previous terminology and of (3.9), one now gets the following result, which further locates the range of τ; that is, one concludes that

(3.19)
> given a Bianchi space X (cf. (3.17)), the field strength (= curvature) of a Maxwell field $(\mathcal{L}, D)$ on X is a closed 2-form on X, in the sense that one has
>
> $$(3.19.1) \qquad\qquad d^2(R) \equiv dR = 0.$$

Therefore, one obtains the first more special location of the image of τ, apart from (3.12); namely, one has

$$(3.20) \qquad\qquad \tau : \Phi^1_{\mathcal{A}}(X)^\nabla \longrightarrow \Omega^2(X)_{\mathrm{cl}} \subseteq \Omega^2(X),$$

viz. the range of τ is just the set of closed 2-forms on X, the latter being, precisely speaking, a $\mathbb{C}$-*vector subspace* of $\Omega^2(X)$ (given that d^2, as in (3.18), is only by definition a $\mathbb{C}$-*linear morphism* of the respective $\mathbb{C}$-vector space sheaves concerned). Thus, we set

$$(3.21) \qquad\qquad \Omega^2(X)_{\mathrm{cl}} = \ker d^2_X \equiv \ker(d^2)$$

such that

$$(3.22) \qquad\qquad d^2_X \equiv d^2 \equiv d$$

(see also (3.19.1)) denotes the respective global component of (the differential operator) d^2, as above. (In this connection, cf. also [VS: Chapt. I; p. 70, (12.49)], concerning our previous notation in (3.22).) Indeed, as we shall see in the sequel (cf. Section 3.(e) below), the image of τ is in effect a certain (additive) subgroup of (3.21).

Note 3.1 By looking at (3.19.1), we further remark that according to the classical theory of electromagnetism on a space-time manifold X,

(3.23)
> the relation
>
> $$(3.23.1) \qquad\qquad dR = 0$$
>
> stands for (is equivalent to) *the first pair of Maxwell's equations in vacuo* (no interactions are present).

See, for instance, J. Baez–J. P. Muniain [1: p. 71]; cf. also the relevant comments therein on p. 72, where it is further noticed that

(3.24)   "We can take our spacetime to be *any manifold M* [the italics are ours] of any dimension, and define the electromagnetic field to be a 2-form F on M."

However, as already pointed out (see the introductory comments on the present chapter), even much more general spaces are allowed within the abstract set-up, which is advocated by the present treatise.

A complete account within our abstract setting of Maxwell's equations (in vacuo) will also be given later on as an outcome of a cohomological description of the Maxwell group of X (see thus the next Chapter IV; Section 6).

Finally, concerning the terminology applied in (3.19), referring to a Bianchi space X, this is actually motivated by the classical situation, given that within such a context one can obtain in our case the classical *Bianchi's identities* (cf. [VS: Chapt. VIII; Section 7, p. 219ff]).

To identify further the image of τ, which is our aim in this subsection, we have to deviate for a while from our program in order to comment on an important issue: Weil's integrality theorem, as formulated within the present abstract setting. (In this connection, see also A. Mallios [7: p. 194, Theorem 7.1], as well as [VS: Chapt. VIII; p. 238, Theorem 11.1, and p. 241, Theorem 11.1′].) The same result is, as we shall also see, still of fundamental importance in geometric (pre)quantization theory (see Chapter V). So we continue by discussing the aforesaid result.

3.2 Weil's Integrality Theorem (Again)

We start by fixing, as usual, the relevant set-up: we are given a $\mathbb{C}$-*algebraized space* (see Chapter I),

$$(3.25) \qquad\qquad (X, \mathcal{A}),$$

which we further specialize. Thus, we next suppose that we have

(3.26)   an exact curvature datum, along with an exponential sheaf diagram (cf. (3.27) below) on a paracompact (Hausdorff) space (cf. (3.25)), while we still assume that our structure ($\mathbb{C}$-algebra) sheaf $\mathcal{A}$ (ibid.) as fine. Henceforth, we shall refer to the above framework as a *Weil space X*.

Before we explain the above terminology, we further depict our previous data in (3.16) by the following diagram, which we call a *Weil scheme*.

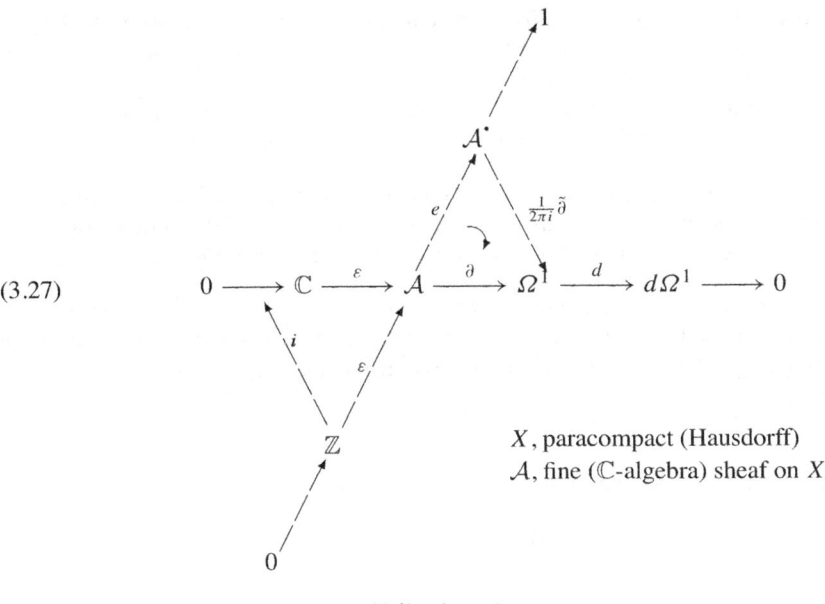

$$(3.27)$$

(Weil scheme)

The horizontal line in the preceding diagram represents the hypothesized (cf. (3.26)) *exact curvature datum*, that is, a curvature datum on X (cf. (3.2)), so that according to our terminology, X in (3.26) is, in particular, a *curvature space*, in such a manner that the (horizontal) sequence in (3.27) of $\mathbb{C}$-vector space sheaves is exact. In other words, we assume that

$$(3.28) \qquad \mathbb{C} \cong \operatorname{im} \varepsilon = \ker \partial, \quad \text{as well as} \quad \ker d = \operatorname{im} \partial$$

(of course, we set above $d^1 \equiv d$, cf. (3.2); see Chapt. II; (6.6) or (3.22)).

On the other hand, we further accept that we are given the exact sequence of (abelian) group sheaves, as indicated in (3.27) by the dashed arrows, while we also assume that the corresponding triangle therein is commutative; that is, one has by definition

$$(3.29) \qquad \tilde{\partial} \circ e = 2\pi i \cdot \partial.$$

(In this connection, cf. also (2.18) along with Chapter I; (1.5), concerning the exponential sheaf diagram involved in (3.27).) Our particular assumptions for X and $\mathcal{A}$ are due, in effect, to cohomological expediencies that will often be of use below, otherwise occuring in certain important particular cases, apart, of course, from the classical one (see Vol. II: Chapter IV; Section 5).

We are now in position to state the following fundamental result.

Theorem 3.1 (Weil's Integrality Theorem) Suppose that we are given a Weil space X. Then, the only 2-dimensional integral cohomology classes of X are just the field strengths of Maxwell fields on X.

Accordingly, what amounts to the same thing, whenever we have the framework of a Weil space X, in the sense of (3.26), one then concludes that

(3.30)   a (complex) 2-dimensional cohomology class of X is integral (cf. (3.33) below) if and only if it is the curvature (= field strength) of an $\mathcal{A}$-connection of a line sheaf on X (viz. of a Maxwell field).

Before we proceed to the proof of the above theorem, we comment a bit for convenience on the term *integral cohomology class* of X that is applied in our previous terminology. That is, by looking at the canonical embedding

$$(3.31) \qquad \mathbb{Z} \underset{\longrightarrow_i}{\subset} \mathbb{C},$$

one gets at the respective canonical map in cohomology,

$$(3.32) \qquad i^* : H^p(X, \mathbb{Z}) \longrightarrow H^p(X, \mathbb{C}),$$

for any $p \in \mathbb{Z}_+$ (nonnegative integers).

Thus, by a *p-dimensional integral cohomology class of X*, one means any class z in the image of the map i^*, as above viz. one has, by definition, that

$$(3.33) \qquad z \in \mathrm{im}(i^*) \equiv \mathrm{im}(H^p(X, \mathbb{Z}) \longrightarrow H^p(X, \mathbb{C})).$$

In this connection, we still remark that due to our hypothesis as in (3.26) that our topological space X is in particular paracompact (Hausdorff), one concludes (cf. [VS: Chapt. III; p. 234, Theorem 8.1]) that sheaf cohomology of X coincides (up to an isomorphism) with, for instance, Čech cohomoloy of X.

Now, concerning the previous Theorem 3.1, the relevant claim therein is that if $(\mathcal{L}, D)$ is a Maxwell field on X (a given Weil space), then the curvature of D provides a 2-dimensional integral cohomology class of X; viz., according to (1.33), for $p = 2$, one has

$$(3.34) \qquad R(D) \equiv R \in \mathrm{im}(H^2(X, \mathbb{Z}) \longrightarrow H^2(X, \mathbb{C}))$$

(and conversely, see below). Thus, we are now ready to come to the

Proof of Theorem 3.1. Based on (3.9) and (3.10), one concludes that the curvature $R(D) \equiv R$ of a given Maxwell field $(\mathcal{L}, D)$ on X (a Weil space, by hypothesis) satisfies the relation

$$(3.35) \qquad R = (d\theta_\alpha) \in Z^0(\mathcal{U}, d\Omega^1) \underset{\longrightarrow}{\subset} Z^0(\mathcal{U}, \Omega^2) = \Omega^2(X).$$

On the other hand, by virtue of Lemma 2.1 (cf. (2.17)), one has

$$(3.36) \qquad \delta(\theta_\alpha) = \tilde{\partial}(g_{\alpha\beta}),$$

where

(3.37)
$$(g_{\alpha\beta}) \in Z^1(\mathcal{U}, \mathcal{A}^{\bullet})$$

stands for a *coordinate* 1-*cocycle, determining* $\mathcal{L}$ (cf. (2.14) or (2.26)). Hence, due to the exactness of the exponential sequence in (3.27) and the paracompactness of X (cf. (3.26)), one obtains

(3.38)
$$(g_{\alpha\beta}) = e(f_{\alpha\beta}),$$

such that

(3.39)
$$(f_{\alpha\beta}) \in C^1(\mathcal{U}, \mathcal{A})$$

(see also [VS: Chapt. III; p. 196, (5.52), along with Lemma 4.2 therein]). Furthermore, concerning (3.39), one also gets (loc. cit., p. 190, Lemma 5.1)

(3.40)
$$\delta(f_{\alpha\beta}) \equiv (\lambda_{\alpha\beta\gamma}) \in Z^2(\mathcal{U}, \mathbb{Z}),$$

where we make use of the fact that

(3.41)
$$\mathbb{Z} \cong \mathrm{im}\,\varepsilon = \ker e,$$

due to the exactness of the exponential sequence in (3.27). Yet (3.36) together with (3.3) and (3.29) entails that

(3.42)
$$\delta(\theta_\alpha) = \tilde{\partial}(g_{\alpha\beta}) = 2\pi i \cdot \partial(f_{\alpha\beta}).$$

Thus, based on (3.40), we now define the (integral) cohomology class,

(3.43)
$$\frac{1}{2\pi i} R := [(\lambda_{\alpha\beta\gamma})] \in H^2(X, \mathbb{Z}),$$

which also establishes the first part (the "only (if)" one) of our assertion. (In this connection, cf. also [VS: Chapt. VIII; p. 512, (5.56) and p. 213, (5.60), (5.61)].)

Remark A more complete explanation of the way one is led to (definition) (3.43) will be given in Section 3.(d) below, under the supplementary hypothesis that X is also a Bianchi space. So (3.43) will become then a theorem! In this regard, we still note that a Bianchi datum has been already used to get (3.19.1).

Now, conversely, assuming as before that X is a Weil space (cf. (3.26)), suppose that we are given an element

(3.44)
$$z \in H^2(X, \mathbb{Z})$$

(see also (3.33), or even (3.34), for $p = 2$). Thus, in view of our hypothesis for X and $\mathcal{A}$ (viz. X paracompact, $\mathcal{A}$ fine, cf. (3.26)), one concludes that

$$(3.45) \qquad\qquad H^1(X, \mathcal{A}^{\bullet}) = H^2(X, \mathbb{Z}),$$

up to an isomorphism of the (abelian) groups concerned (viz. the so-called *Chern isomorphism*; cf. also [VS: Chapt. VIII; p. 2.39, (11.8)]. Yet, due to the same hypothesis for X and $\mathcal{A}$ as above, we further remark that relations (11.2) therein, loc. cit. p. 239, are fulfilled. In this connection, see also the subsequent Scholium 3.1, pertaining to the above Chern isomorphism, as in (3.45)).

Accordingly, in view of (3.44) and (3.45), one thus obtains a line sheaf $\mathcal{L}$ (in fact, an equivalence class of such, cf. (2.2)) through a coordinate 1-cocycle, say

$$(3.46) \qquad\qquad \mathcal{L} \longleftrightarrow (g_{\alpha\beta}) \in Z^1(\mathcal{U}, \mathcal{A}^{\bullet})$$

(cf., for instance, Chapt. II; (7.20), (7.21), in the particular case $n = 1$). Moreover, by virtue of the same hypothesis for X and $\mathcal{A}$ as before, one further concludes that $\mathcal{L}$ admits an $\mathcal{A}$-connection D (cf. [VS: Chapt. VI; p. 85, Theorem 16.1, along with Chapt. III; p. 207, (8.56)]), so that one finally gets a Maxwell field, say

$$(3.47) \qquad\qquad (\mathcal{L}, D) \longleftrightarrow ((g_{\alpha\beta}), (\theta_{\alpha})),$$

such that

$$(3.48) \qquad\qquad \delta(\theta_{\alpha}) = \tilde{\partial}(g_{\alpha\beta})$$

(see Lemma 2.1 or (2.26)). Yet, the curvature of D (the field strength of $(\mathcal{L}, D)$) is given by the relation

$$(3.49) \qquad\qquad R(D) \equiv R = (d\theta_{\alpha})$$

(*Cartan's structural equation*; cf. [VS: Chapt. VIII; p. 187, Corollary 3.1]).

On the other hand, in view of the Weil scheme, as in (3.37), cf. also (3.29), and based further on (3.46) and (3.48), one still obtains (3.42) and (3.40). Thus, by virtue of (3.44) and taking also into account the cohomology class of X determined by R, the latter given by (3.49) (cf. also (3.35)), a determination that is further rooted on the exactness of the two sequences in (3.27), along with the commutativity of the respective triangle therein (viz. (3.49), cf. also [VS: Chapt. III; p. 190, Lemma 5.1], while in this connection we refer to Section 3.(d) in the sequel), one finally concludes the coincidence of z, as given by (3.47), and R, which terminates the proof of Theorem 3.1. ■

Scholium 3.1 (Physical significance of the Chern isomorphism). We discuss below the physical meaning that might be associated with the previously employed Chern isomorphism, as in (3.45), as well as give a further delineation of the same relation according to the abstract setting we advocate in this study: Thus, based on the exactness of the exponential sheaf sequence, as appeared in (3.27), viz. on the short exact sequence

$$(3.50) \qquad\qquad 0 \longrightarrow \mathbb{Z} \xrightarrow{\varepsilon} \mathcal{A} \xrightarrow{e} \mathcal{A}^{\bullet} \longrightarrow 1,$$

one gets the corresponding long exact sequence in cohomology

(3.51)
$$\cdots \longrightarrow H^1(X, \mathcal{A}) \longrightarrow H^1(X, \mathcal{A}^{\cdot}) \overset{\delta}{\longrightarrow} H^2(X, \mathbb{Z})$$
$$\longrightarrow H^2(X, \mathcal{A}) \longrightarrow \cdots ,$$

so that by virtue of our hypothesis for X and $\mathcal{A}$ (see, for instance, (3.27)), one has

(3.52)
$$H^1(X, \mathcal{A}) = H^2(X, \mathcal{A}) = 0.$$

Therefore, in view of (3.51), one obtains the exact sequence

(3.53)
$$0 \longrightarrow H^2(X, \mathcal{A}^{\cdot}) \longrightarrow H^2(X, \mathbb{Z}) \longrightarrow 0,$$

hence finally the aforesaid Chern isomorphism

(3.54)
$$H^1(X, \mathcal{A}^{\cdot}) = H^2(X, \mathbb{Z}),$$

indeed an isomorphism of the abelian groups concerned. (In this connection, see [VS: Chapt. III; p. 207, Theorem 5.3, and p. 234, Theorem 8.1; see p. 238, (8.24)].) Thus, based on (3.54), one can now say that

(3.55)
the carrier of the electromagnetic field, viz. the photon, or even the corresponding line sheaf $\mathcal{A}$ (cf. Definition 1.1), as given by the respective cohomology class, viz. via the (bijective) correspondence (cf., for instance, (2.2)),

(3.55.1)
$$\Phi^1_{\mathcal{A}}(X) \ni [\mathcal{L}] \longleftrightarrow [(g_{\alpha\beta})] \in H^3(X, \mathcal{A}^{\cdot}),$$

is thus identified, through the Chern isomorphism (cf. (3.54)), with the effect of the field at issue, that is, with the corresponding field strength (curvature), yet via the cohomology class that can be associated with the latter (cf. (3.43)). That is, one has

(3.55.2)
$$[(g_{\alpha\beta})] = \frac{1}{2\pi i}[R] = [(\lambda_{\alpha\beta\gamma})] \in H^2(X, \mathbb{Z}).$$

(The first equality in (3.55.2) is based on the Chern isomorphism, cf. the "if" part in the above proof of Theorem 3.1.)

Accordingly, by further employing physical language, one realizes that (cf. (3.55.2))

(3.56)
in terms of cohomology, *matter* (*carrier*, or *space*) coincides with *curvature* (field strength), something certainly reminiscent of the well-known maxim that "matter tells space how to curve" (see [VS: Chapt. VIII; p. 185, epigrams]) or the relevant dictum of A. Einstein, concerning the potential significance of light rays for the geometry of space (see (1.23)).

In other words, when speaking in terms of (sheaf) cohomology theory, then as an outcome of the Chern isomorphism (cf. (3.54)), we may identify the carrier of the electromagnetic field (the photon) and the field strength; viz. one has

(3.57)

$$(3.57.1) \qquad [(g_{\alpha\beta})] = \frac{1}{2\pi i}[R].$$

In this connection, we still note that by an obvious abuse of language, we occasionally refer to

$$(3.58) \qquad [R] = 2\pi i \cdot [(\lambda_{\alpha\beta\gamma})]$$

as the (2-dimensional) *cohomology class of R*; indeed, one has here that

$$(3.59) \qquad [R] = 2\pi i \cdot [(\lambda_{\alpha\beta\gamma})] \in H^2(X, \mathbb{C}),$$

so that in particular one obtains

$$(3.60) \qquad \frac{1}{2\pi i}[R] \equiv [(\lambda_{\alpha\beta\gamma})] \in H^2(X, \mathbb{Z}),$$

cf., for instance, (3.55.5), along with (3.36).

Note 3.2 The (2-dimensional) integral cohomology class appearing in (3.55.2) is still called the *Chern characteristic class of the* given *line sheaf* $\mathcal{L}$ (cf. (3.55.1)), denoted by

$$(3.61) \qquad c_1(\mathcal{L}),$$

that is, precisely speaking, one sets

$$(3.62) \qquad c_1(\mathcal{L}) = -\frac{1}{2\pi i}[R] \equiv -\frac{1}{2\pi i}R \in H^2(R, \mathbb{Z}) \underset{\longrightarrow}{\subset} H^2(X, \mathbb{C}).$$

In this connection, see also [VS: Chapt. IT; p. 268, (5.14)].

3.3 The Image of the Map τ (Continued)

We continue in this subsection to identify exactly the image of the map τ: First, by referring to (3.20), we recall that the corresponding set-up to that relation was a Bianchi datum, as in (3.17).

Thus, within that context, we already know (cf. (3.20)) that the range of τ is

$$(3.63) \qquad \Omega^2(X)_{\mathrm{cl}},$$

that is, the $\mathbb{C}$-vector subspace of $\Omega^2(X)$ consisting of closed 2-forms on X. On the other hand, let us now denote by

$$(3.64) \qquad \Omega^2(X)_{\mathrm{cl}}^{\mathrm{int}} (\subseteq \Omega^2(X))$$

the *set* (subset of (3.63)) *of integral closed 2-forms on* X (cf. (3.33) or (3.34); see also Section 4, concerning the particular structure of the set (3.64).

In particular, suppose that we are given a

(3.65)   Bianchi–Weil space X, that is, a Bianchi space (cf. (3.17)) that is also a Weil space in the sense of (3.26) (cf. also (3.27)).

Thus, as a consequence of Theorem 3.1, we have the following description of the image of τ;

$$(3.66) \qquad \operatorname{im} \tau = \tau(\Phi_{\mathcal{A}}^1(X)^{\nabla}) \subseteq \Omega^2(X)_{\mathrm{cl}}^{\mathrm{int}},$$

which thus is valid for any Bianchi–Weil space X: Indeed, (3.66) is a straightforward outcome of Weil's integrality theorem as above, under the supplementary hypothesis that X is also a Bianchi space, so that (3.19.1), hence also (3.20), is in force. ∎ In point of fact, what we really have in (3.66) is an equality in the last term of that relation. However, to prove this, one needs to know that

(3.67)   every closed 2-form on X yields a 2-dimensional (complex) cohomology class of X.

Indeed, although this holds true in the classical case (even more generally, for any (closed) n-form; see also Section 3.(d)) in our abstract setting within which we work, one has in general to accept it. On the other hand, we do have important special instances where (3.67) is valid, even the general case of any closed n-form, $n \in \mathbb{N}$, as classically happens (see Vol. II: Chapt. IV; Section 5).

Accordingly, as a result of the preceding, we are thus in position to state the following assertion:

(3.68)   given a Bianchi–Weil space X (cf. (3.65)) for which (3.67) holds, one obtains

$$(3.68.1) \qquad \operatorname{im} \tau \equiv \tau(\Phi_{\mathcal{A}}^1(X)^{\nabla}) = \Omega^2(X)_{\mathrm{cl}}^{\mathrm{int}}.$$

Our previous claim follows immediately from (3.66) and the "if" part of Theorem 3.1 by also taking (3.67) into account. ∎

Thus, in this connection, we can further remark that

(3.69)   (3.68.1) may also be construed as another equivalent formulation of Weil's integrality theorem.

In point of fact, the "only if" part of Theorem 3.1 (cf. also (3.30)) is stronger than (3.68.1) in the sense that (3.67) is not needed for that theorem. Therefore, we can conclude the following:

Suppose that X is a Bianchi–Weil space, for which (4.67) is in force. Then a given closed 2-form on X, say

(3.70.1) $R \in \Omega^2(X)_{\mathrm{cl}} \subseteq \Omega^2(X)$

(3.70)   (cf. (3.63)), is the field strength (curvature form) of a Maxwell field $(\mathcal{L}, D)$, viz. one concludes that (see (3.68.1))

(3.70.2) $R \in \mathrm{im}\,\tau$

if and only if one has (cf. also (3.32))

(3.70.3) $[R] \in \mathrm{im}(i^*) \equiv \mathrm{im}(H^2(X, \mathbb{Z}) \overset{i^*}{\longrightarrow} H^2(X, \mathbb{C})),$

that is, if and only if R defines an integral (2-dimensional) cohomology class of X.

Now, by employing our previous terminology concerning (1.19), see also (3.11) we call the fiber of τ over $R \in \Omega^2(X)$ (cf. (3.12)) a light bundle over X. Namely, we set

(3.71) $\tau^{-1}(R) \equiv \Phi^1_{\mathcal{A}}(S)^{\nabla}_R, \quad$ light bundle,

for any $R \in \Omega^2(X)_{\mathrm{cl}}$ (cf. also (3.64)), where we assume that

(3.72) $\Phi^1_{\mathcal{A}}(X)^{\nabla}_R \neq \emptyset,$

so that, in other words, the above definition (3.71) is actually valid for any R belonging to the image of τ as in (3.68.1). Accordingly, one can still assert the following:

Assume that we have the framework of (3.70). Then a closed 2-form R on X (cf. (3.70.1)) corresponds to a light bundle over X if and only if one has (cf. (3.70.3))

(3.73)   (3.73.1) $[R] \in \mathrm{im}(i^*).$

Equivalently,

(3.73.2)   the only light bundles over X correspond to integral closed 2-forms on X.

In this connection, we further remark in anticipation (see Section 5), that

(3.74)   two light rays (cf. (1.19)) differ by just a polarized light ray, i.e., by an element, say

(3.74.1) $z \in H^1(X, \mathbb{C}^{\bullet}),$

or in particular (cf. Section 6; (3.68)) by an element

(3.74.2) $z \in H^1(X, S^1).$

Before we end the present subsection, we comment on our previous remarks in (3.11); see also (2.39) and (2.40): Sssuming the set-up of Lemma 2.1 (consider, for instance, a curvature space X, cf. (3.2), which we shall need anyway presently), we already know that (cf. (2.17))

$$(3.75) \qquad \delta(\theta_\alpha) = \tilde{\partial}(g_{\alpha\beta}).$$

Thus, one concludes that

(3.76)   the gauge potential (cf. (3.10)) varies together with the local gauge involved (cf. also (2.14), along with (2.39)).

However, in contradistinction to the above, one remarks that

(3.77)   the field strength (= curvature) of a Maxwell field $(\mathcal{L}, D)$ is in effect gauge (or space) invariant.

Indeed, one has

$$(3.78) \qquad R' = g_{\alpha\beta}^{-1} R g_{\alpha\beta} \equiv Ad(g_{\alpha\beta}^{-1}) R = R$$

(transformation law of field strengths), given that

$$(3.79) \qquad (g_{\alpha\beta}) \in Z^1(\mathcal{U}, \mathcal{A}^{\cdot})$$

(see (3.37) along with (2.37.1)), where according to our hypothesis, $\mathcal{A}$ is a $\mathbb{C}$-algebra sheaf on X whose stalks are (unital) commutative $\mathbb{C}$-algebras (cf. Chapt. II; (6.1), as well as, concerning (3.78), [VS: Chapt. VIII; p. 202, (4.27), and p. 203 (4.31), (4.32)]).

Now, by further looking at (3.78), we also remark that

(3.80)   in the case of a Maxwell field $(\mathcal{L}, D)$, the notion of curvature (= field strength) appears to be something inherent pertaining to the substance of the field itself (in this connection, see also our previous comments in (3.56) and (3.57)), independently of any local gauge (this is reminiscent of course, the *principle of general relativity* or the so-called *gauge principle*, cf., for instance, M. Nakahara [1: pp. 28 and 10, respectively]), in contrast to what happens with the corresponding gauge potential (= $\mathcal{A}$-connection); indeed, as we know, the latter is realized (detected) only through the respective transformation law of potentials. (See (3.75) or, for the general case, (2.34.2); cf. also Lemma 2.1).

The same remarks as above might further be combined with those in (1.22) and (1.23).

The above supplements the picture we already had for a given Maxwell field $(\mathcal{L}, D)$, as provided by our previous comments in (2.39)–(2.41), along with those in (3.55).

3.4 Cohomology Class Associated with the Field Strength of a Maxwell Field (Continued)

As already said, we discuss in this subsection the precise (direct) definition within the appropriate context (see thus (3.94) below) of the cohomology class that can be associated with the curvature (cf. (3.4))

$$(3.81) \qquad\qquad R \in \Omega^2(X)$$

of a given Maxwell field

$$(3.82) \qquad\qquad (\mathcal{L}, D).$$

In fact, our aim in the following discussion is to turn our previous definition, as in (3.43), into a theorem, under the proviso that one has the appropriate set-up.

Thus, assuming that we are given a curvature space X (cf. (3.2)), so that (3.81) is valid, we already know that R yields in effect, a closed 2-form on X (see (3.19.1)), provided X is, in particular, a Bianchi space (cf. (3.17), (3.19)). Now, to proceed further, we still assume that we have the following exact sequence of $\mathbb{C}$-vector space sheaves:

$$(3.83) \qquad 0 \longrightarrow \mathbb{C} \overset{\varepsilon}{\longrightarrow} \mathcal{A} \overset{\partial}{\longrightarrow} \Omega^1 \overset{d^1 \equiv d}{\longrightarrow} \Omega^2 \overset{d^2 \equiv d}{\longrightarrow} d\Omega^2 \longrightarrow 0.$$

To fix the terminology, we call X an *exact Bianchi space* (a Bianchi space for which (3.83) is exact). Thus, one now obtains the following basic result:

(3.84) Suppose we are given an exact Bianchi space X, being also a paracompact (Hausdorff) space. Then, every closed 2-form on X entails a 2-dimensional complex (Čech) cohomology class of X. So, by looking, for instance, at (3.81) (cf. also (3.19.1)), one thus obtains

$$(3.84.1) \qquad\qquad [R] \in H^2(X, \mathbb{C}).$$

The preceding assertion is in effect a particular case of a relevant general result, holding for any $n \in \mathbb{N}$, not only for $n = 2$, as above, under a corresponding extension, of course, of the pertinent exact sequence by analogy with (3.83). See [VS: Chapt. IX; p. 256, Lemma 3.1]. However, for the sake of completeness, we present here the proof of (3.84).

First, by virtue of (3.9), (3.19.1), and (3.21), together with the exactness of (3.83), one obtains

$$(3.85) \qquad \begin{aligned} R = (d\theta_\alpha) \in Z^0(\mathcal{U}, \ker d^2) &= Z^0(\mathcal{U}, \operatorname{im} d^1) \equiv Z^0(\mathcal{U}, d\Omega^1) \\ &\underset{\longrightarrow}{\subseteq} Z^0(\mathcal{U}, \Omega^2) \cong \Omega^2(X). \end{aligned}$$

Now consider the following short exact sequence (which is always true):

$$(3.86) \qquad 0 \longrightarrow \ker d \longrightarrow \Omega^1 \longrightarrow d\Omega^1 \longrightarrow 0.$$

Therefore, as a consequence of (3.85) and [VS: Chapt. III; p. 190, Lemma 5.1], one concludes the existence of a 0-cochain

$$(3.87) \qquad (\theta'_\alpha) \in C^0(\mathcal{U}, \Omega^1),$$

such that

$$(3.88) \qquad R = (d\theta'_\alpha),$$

while one also obtains

$$(3.89) \qquad (\delta(\theta'_\alpha)) \in Z^1(\mathcal{U}, \ker d = \operatorname{im} \partial) \equiv Z^1(\mathcal{U}, \partial \mathcal{A}).$$

We continue, by further looking at the short exact sequence (still always in force),

$$(3.90) \qquad 0 \longrightarrow \ker \partial \longrightarrow \mathcal{A} \longrightarrow \partial \mathcal{A} \longrightarrow 0.$$

Thus, based now on (3.89) and by applying a similar argument as before, one gets a 1-cochain

$$(3.91) \qquad (f'_{\alpha\beta}) \in C^1(\mathcal{U}, \mathcal{A}),$$

in such a manner that one has

$$(3.92) \qquad \delta(\theta'_\alpha) = \partial(f'_{\alpha\beta}),$$

as well as

$$(3.93) \qquad \delta(f'_{\alpha\beta}) \equiv (\lambda'_{\alpha\beta\gamma}) \in Z^2(\mathcal{U}, \ker \partial) = Z^2(\mathcal{U}, \operatorname{im} \varepsilon \cong \mathbb{C})$$

(see also (3.83)). So one now defines

$$(3.94) \qquad [R] := [(\lambda'_{\alpha\beta\gamma})] \in H^2(X, \mathbb{C})$$

(cf. also [VS: Chapt. III; p. 191, (5.21)]). Thus, this is by definition the 2-*dimensional* (*complex*) *Čech cohomology class of X*, which can be associated with R, the given closed 2-form on X, as in (3.85). This also terminates the proof of (3.84). ∎

On the other hand, as already promised, we proceed now to prove the coincidence, in effect (modulo, namely, the factor $2\pi i$, cf. (3.123) in the sequel), of (3.94) with (3.43): So, based on (3.85) and (3.88), one first obtains

$$(3.95) \qquad (d\theta_\alpha) = (d\theta'_\alpha) = R,$$

or

$$(3.96) \qquad d(\theta'_\alpha - \theta_\alpha) = 0,$$

so that one has (cf. also (3.83))

$$(3.97) \qquad (\theta'_\alpha - \theta_\alpha) \in C^0(\mathcal{U}, \ker d) = C^0(\mathcal{U}, \partial \mathcal{A}).$$

Hence, by still assuming X paracompact (Hausdorff) and in view of [VS: Chapt. III; p. 196, Lemma 5.2], one obtains a 0-cochain

$$(3.98) \qquad\qquad (s_\alpha) \in C^0(\mathcal{U}, \mathcal{A})$$

in such a manner that one gets (cf. (3.97))

$$(3.99) \qquad\qquad \theta'_\alpha - \theta_\alpha = \partial(s_\alpha)$$

for any $\alpha \in I$ (cf. (3.8)).

Now, to proceed further, in order to prove that (3.94) and (3.43) are virtually the same, as already claimed above, we make full use of the short exact exponential sheaf sequence as depicted by the dashed arrow in (3.27) (cf. also (3.50)). This in conjunction with the exact sequence in (3.83) leads us to the definition of what we briefly call an

(3.100) *exact Bianchi–Weil space* X, viz. a Bianchi–Weil space in the sense of (3.65) (in point of fact, we do not need, for what follows, $\mathcal{A}$ to be a fine sheaf on X) for which also (3.83) is exact.

Thus, assuming (3.100), we continue to prove the asserted coincidence of (3.94) with (3.43) (modulo $2\pi i$): First, based on (3.99) and (3.29), one obtains

$$(3.101) \qquad\qquad \theta'_\alpha = \theta_\alpha + \tilde{\partial}(t_\alpha^{-1}),$$

where we set

$$(3.102) \qquad\qquad (t_\alpha) := \left(e\left(\frac{1}{2\pi i} s_\alpha\right)^{-1} \right) \in C^0(\mathcal{U}, \mathcal{A}^\cdot).$$

On the other hand, we also set

$$(3.103) \qquad\qquad g'_{\alpha\beta} := \delta(t_\alpha^{-1}) \cdot g_{\alpha\beta}.$$

Hence, by virtue of (3.101), (3.103), and (3.75), one has

$$(3.104) \quad \delta(\theta'_\alpha) = \delta(\theta_\alpha + \tilde{\partial}(t_\alpha^{-1})) = \delta(\theta_\alpha) + \delta(\tilde{\partial}(t_\alpha^{-1}))$$
$$= \tilde{\partial}(g_{\alpha\beta}) + \tilde{\partial}(\delta(t_\alpha^{-1})) = \tilde{\partial}(\delta(t_\alpha^{-1}) \cdot g_{\alpha\beta}) = \tilde{\partial}(g'_{\alpha\beta})$$

(cf. also Chapt. I; (1.28)), so that in view of (3.87), (3.103), and Lemma 2.1, one actually gets a new Maxwell field on X (cf. (2.26)),

$$(3.105) \qquad\qquad (\mathcal{L}', D') \longleftrightarrow ((g'_{\alpha\beta}), (\theta'_\alpha)),$$

which is equivalent to the initial one $(\mathcal{L}, D)$. That is, one has, by virtue of (3.101), (3.103), and Lemma 2.2, that

$$(3.106) \qquad\qquad (\mathcal{L}, D) \sim (\mathcal{L}', D').$$

Thus, one finally obtains that;

(3.107) $$[(\mathcal{L}, D)] = [(\mathcal{L}', D')] \in \Phi^1_{\mathcal{A}}(X)^\nabla_R,$$

given that, in view of (3.95), one has

(3.108) $$(d\theta_\alpha) = (d\theta'_\alpha) = R,$$

pertaining to the field strength (= curvature) of the Maxwell fields involved. On the other hand, (3.108) can still be obtained in general from (3.106) by just differentiating (3.101) (cf. also Lemma 2.2, along with Chapt. I; (7.5)). In other words (see also (1.18) in the preceding), one concludes that

(3.109)   gauge equivalent $\mathcal{A}$-connections of Maxwell fields (see (1.14)) yield the same field strength (= curvature). That is,
$$D \sim D' \Rightarrow R(D) = R(D').$$

In this connection, we further note that (3.109) is in force within the framework of a curvature space X (cf. also [VS: Chapt. VIII; p. 197, (3.24)]).

We assume, in anticipation (cf. (5.116)), the fact that (see also (3.12, (3.71)))

(3.110) $$\Phi^1_{\mathcal{A}}(X)^\nabla_R \text{ is a } H^1(X, \mathbb{C}^\bullet)\text{-affine space.}$$

That is, one concludes that

(3.111)   the corresponding coordinate 1-cocycles of the line sheaves $\mathcal{L}$ and $\mathcal{L}'$, as in (3.106), differ by a constant 1-cocycle; viz. one has (cf. also (3.103))

(3.111.1) $$g'_{\alpha\beta} = \delta(t^{-1}_\alpha) \cdot g_{\alpha\beta},$$

such that

(3.111.2) $$\delta(t^{-1}_\alpha) \in Z^1(\mathcal{U}, \mathbb{C}^\bullet).$$

In this regard, see also Remarks 3.1 below. We further note, as an outcome of the preceding, the relation

(3.112) $$\delta(\theta_\alpha) = \delta(\theta'_\alpha).$$

Indeed, based on (3.101) and (3.111.2) (cf. also Chapt. I; (1.16)), one obtains

(3.113) $$\delta(\theta'_\alpha) = \delta(\theta_\alpha + \tilde{\partial}(t^{-1}_\alpha)) = \delta(\theta_\alpha) + \delta(\tilde{\partial}(t^{-1}_\alpha))$$
$$= \delta(\theta_\alpha) + \tilde{\partial}(\delta(t^{-1}_\alpha)) = \delta(\theta_\alpha),$$

which thus proves (3.112). Furthermore, in view of Lemma 2.1, one gets

(3.114) $$\delta(\theta'_\alpha) = \delta(\theta_\alpha) = \tilde{\partial}(g_{\alpha\beta}).$$

On the other hand, based on the exactness of (3.50) and on (3.29) (see also (3.100)), along with [VS: Chapt. III; p. 190, Lemma 5.1], one further obtains, by virtue of (3.37) and (3.114), that

$$(3.115) \qquad \delta(\theta_\alpha) = \tilde{\partial}(g_{\alpha\beta}) = \tilde{\partial}(e(f_{\alpha\beta})) = 2\pi i \cdot \partial(f_{\alpha\beta}) = \partial(2\pi i \cdot f_{\alpha\beta}),$$

such that

$$(3.116) \qquad (f_{\alpha\beta}) \in C^1(\mathcal{U}, A),$$

while one also has

$$(3.117) \qquad \delta(f_{\alpha\beta}) \equiv (\lambda_{\alpha\beta\gamma}) \in Z^2(\mathcal{U}, \ker(e) \cong \mathbb{Z}).$$

Furthermore, in view of (3.92), (3.114), and (3.115), one has

$$(3.118) \qquad \delta(\theta'_\alpha) = \partial(f_{\alpha\beta}) = \delta(\theta_\alpha) = \partial(2\pi i \cdot f_{\alpha\beta}),$$

so that one gets

$$(3.119) \qquad \partial(f'_{\alpha\beta}) = \partial(2\pi i \cdot f_{\alpha\beta});$$

therefore one has (cf. also (3.83))

$$(3.120) \qquad (f'_{\alpha\beta} - 2\pi i \cdot f_{\alpha\beta}) \in C^1(\mathcal{U}, \ker \partial \cong \mathbb{C}),$$

that is, one finally obtains

$$(3.121) \qquad f'_{\alpha\beta} = 2\pi i \cdot f_{\alpha\beta} + k_{\alpha\beta},$$

such that

$$(3.122) \qquad (k_{\alpha\beta}) \in C^1(\mathcal{U}, \mathbb{C}).$$

Thus, by virtue of (3.121) and (3.122), one has

$$(3.123) \qquad \delta(f'_{\alpha\beta}) = 2\pi i \cdot \delta(f_{\alpha\beta}),$$

or even, in view of (3.93) and (3.117), one has

$$(3.124) \qquad (\lambda'_{\alpha\beta\gamma}) = 2\pi i \cdot (\lambda_{\alpha\beta\gamma}).$$

Consequently (cf. also (3.94)), one now obtains

$$(3.125) \qquad \begin{aligned} \frac{1}{2\pi i}[R] &:= \frac{1}{2\pi i}[(\lambda'_{\alpha\beta\gamma})] = [(\lambda_{\alpha\beta\gamma})] \\ &= [\delta(f_{\alpha\beta})] \in H^2(X, \mathbb{Z}) \underset{\longrightarrow}{\subset} H^2(X, \mathbb{C}), \end{aligned}$$

which thus establishes the relation we were looking for between (3.43) and (3.94), or equivalently, among (3.94) and (3.117). ∎

Thus, (3.125) renders (3.43) into a theorem, which was our goal from the beginning of this subsection.

Remarks 3.1 (i) Throughout the preceding discussion, we retained, for convenience, the same local frame $\mathcal{U}$ of $\mathcal{L}$ (cf., for instance, (3.8)). Indeed, although this is not in general the case, we may employ the previous practice. Namely, since the various steps are finitely many, one can finally choose, by successive refinements, a common local frame as assumed. (In this connection, see also Chapter I; (2.53).)

(ii) Concerning our assertion in (3.111), being also an outcome of (3.110), we can further assume, in view of (3.106), that

(3.126)   the 1-cocycle that appears in (3.111.2) is just that one referring to the local data of the Maxwell fields involved (see (2.45) and (2.46)).

Of course, (3.126) holds within an occasional change of coordinate 1-cocycles of the Maxwell fields concerned. However, within the same vein of ideas, we still remark that

(3.127)   the cohomology class of R, as in (3.108) or (3.43), with R being the common field strength of $\mathcal{L}$ and $\mathcal{L}'$ (cf. (3.95)), remains the same.

The assertion is a consequence of the classical Chern–Weil theorem, as formulated in our abstract setting; cf. [VS: Chapt. IX; p. 258, Theorem 3.1].

We come now, in Section 4 and some of the subsequent ones, to examine further the inner structure of

$$(3.128) \qquad \Phi^1_{\mathcal{A}}(X)^\nabla,$$

the Maxwell group of X (cf. Theorem 2.1), mainly, by means of the natural action on it of the 1-dimensional (Čech) cohomology group of X,

$$(3.129) \qquad \check{H}^1(X, \mathbb{C}^\bullet)$$

("flat $\mathbb{C}$-line bundles" on X), or under suitable additional conditions on X, of that of

$$(3.130) \qquad \check{H}^1(X, S^1)$$

(see Section 6).

4 The Fibration τ as a Group Morphism

Our aim in the ensuing discussion is to prove the claim in the title of this section. Thus, by considering the corresponding set-up of the fibration at issue, viz. a curvature space X, we first recall the definition of the map considered; that is, one has

$$(4.1) \qquad \tau : \Phi^1_{\mathcal{A}}(X)^\nabla \to \Omega^2(X),$$

such that

(4.2) $$\tau([(\mathcal{L}, D)]) := R(D) \equiv R$$

(see (3.12), (3.13) in the preceding, as well as, (3.11)). Now, by looking at (4.1), we first remark (cf. Theorem 2.1) that the domain of definition of τ is an abelian group, with respect to the group structure induced on it by the *Picard group of X* (loc. cit., in particular, (2.6)–(2.8)), while the range of τ, as in (4.1) viz. $\Omega^2(X)$, being by the definition of Ω^2 an $\mathcal{A}(X)$-module, is in particular a $\mathbb{C}$-vector space (cf. also Chapt. II; (6.6)). Thus, what we are going to prove is that

(4.3)   the map τ as given by (4.2) is a morphism of the (abelian) groups involved, as in (4.1).

Indeed, based on (2.26) and (3.9), as well as on [VS: Chapt. VII; p. 233, (9.8), or even (9.10), along with p. 234; (9.16), (9.17)], one is led to the relations

$$
\begin{aligned}
(4.4) \quad R(D \otimes D') &= (d(\theta_\alpha + \theta'_\alpha)) = (d\theta_\alpha + d\theta'_\alpha) \\
&= (d\theta_\alpha) + (d\theta'_\alpha) = R(D) + R(D'),
\end{aligned}
$$

as well as

(4.5) $$R(D^{-1}) \equiv R(D^*) = (d(-\theta_\alpha)) = (-d\theta_\alpha) = -(d\theta_\alpha) = -R(D),$$

for any two given Maxwell fields $(\mathcal{L}, D)$ and $(\mathcal{L}', D')$ on X.

Therefore, by virtue of (4.2) and (2.6), one first obtains

$$
\begin{aligned}
(4.6) \quad \tau([(\mathcal{L}, D)] \cdot [(\mathcal{L}', D')]) &\equiv \tau([(\mathcal{L}, D)] \otimes_{\mathcal{A}} [(\mathcal{L}', D')]) \\
&= \tau([(\mathcal{L} \otimes_{\mathcal{A}} \mathcal{L}', D \otimes D')]) = R(D) + R(D') \\
&= \tau([(\mathcal{L}, D)]) + \tau([(\mathcal{L}', D')]),
\end{aligned}
$$

while in view of (4.5), (4.2), and (2.7), one also has

$$
\begin{aligned}
(4.7) \quad \tau([(\mathcal{L}, D)]^{-1}) &= \tau([(\mathcal{L}, D)^{-1}]) = \tau([(\mathcal{L}^{-1}, D^{-1})]) \\
&= \tau([(\mathcal{L}^*, D^*)]) = R(D^*) = -R(D).
\end{aligned}
$$

Thus, (4.6) and (4.7) prove our assertion in (4.3). ∎

On the other hand, *the set of closed 2-forms on X*,

(4.8) $$\Omega^2(X)_{\mathrm{cl}} \subseteq \Omega^2(X)$$

(cf. (3.21)), is of course (in view of the $\mathbb{C}$-linearity of $d^2 \equiv d$, see (3.19.1)) a $\mathbb{C}$-vector subspace of $\Omega^2(X)$. However, to formulate the latter notions, we assumed in the preceding the appropriate (abstract) set-up, thus in particular a Bianchi space X (cf. (3.17), as well as (3.19)). Hence, based now on our previous conclusion in (4.3), we can further state the following specification of it:

Suppose we are given a Bianchi space X (cf. (3.17)). Then, the map

(4.9)   (4.9.1) $$\tau : \Phi^1_{\mathcal{A}}(X)^{\nabla} \to \Omega^2(X)_{cl} \subseteq \Omega^2(X)$$

(see also (3.63)) is a *morphism of the (abelian) groups* concerned.

More particularly, we have already seen in Section 3 that the range of the same map τ, as in (4.1) or (4.9.1), is actually within

(4.10) $$\Omega^2(X)^{int}_{cl} \subseteq \Omega^2(X)_{cl} \subseteq \Omega^2(X),$$

that is, into the set of integral closed 2-forms on X (cf. (3.68.1)), provided, of course, one has again the appropriate general setting to formulate the relevant conclusion, thus a Bianchi–Weil space X (ibid. and (3.65)), the aforementioned result as in (3.68.1), being a consequence of Weil's integrality theorem (cf. Theorem 3.1). On the other hand, the set

(4.11) $$\Omega^2(X)^{int}_{cl},$$

as above, *is an additive (abelian) subgroup of* $\Omega^2(X)_{cl}$, as directly follows from the definition of an integral cohomology class, see (3.33), through the relevant map in cohomology, cf. (3.32). Accordingly, we are led to the following further specialization of (4.9), hence of (4.3) too. That is, we finally conclude that

given a Bianchi–Weil space X (cf. (3.65)), the map

(4.12)   (4.12.1) $$\tau : \Phi^1_{\mathcal{A}}(X)^{\nabla} \to \Omega^2(X)^{int}_{cl} \subseteq \Omega^2(X),$$

as given by (4.2), is a *morphism of* the respective *(abelian) groups*. (See also (3.68.1), along with Theorem 2.1.)

Now, by still looking at definition (4.2), we further note that

the morphism τ, as in (4.1), or in particular in (4.12.1), is in fact the curvature map $R(\cdot)$, viz. the map

(4.13.1) $$D \mapsto R(D),$$

(4.13)

which can be defined, for any given $\mathcal{A}$-connection D of a given $\mathcal{A}$-module $\mathcal{E}$, in general over a curvature (hence, in particular, over a Bianchi) space X. (See Chapt. I; (7.19), along with (8.8), or [VS: Chapt. VIII; p. 191, Definition 2.1, along with Note 2.1 therein].)

On the other hand, in view of (1.18), the same map (4.13.1) can be extended to the quotient set modulo the equivalence relation (1.16.1) defining the Maxwell group of X as in (1.17.1), yielding finally the definition (4.2) as above, X being always a curvature space. Furthermore, (4.4) and (4.5) are but properties of the same curvature map, as in (4.13.1), that are further extended, still in view of (1.18), as corresponding properties of the map τ.

Scholium 4.1 Now, within the same vein of ideas, and in anticipation of our detailed study on the matter in the sequel, we already remark here that by considering the so-called *moduli space* of a given line sheaf $\mathcal{L}$ on X, that is, the quotient set

$$(4.14) \qquad\qquad Conn_A(\mathcal{L})/\mathcal{A}^{\bullet},$$

one still gets an extension of the above curvature map (cf. (4.13.1)) to the previous space; thus, one has

$$(4.15) \qquad\qquad R(\cdot) : Conn_A(\mathcal{L})/A^{\bullet} \to \Omega^2(X),$$

such that one sets

$$(4.16) \qquad\qquad R(\cdot)[D] \equiv R([D]) := R(D),$$

where one has

$$(4.17) \qquad [D] := \{D' \in Conn_A(\mathcal{L}) : D' \sim D\} \in Conn_A(\mathcal{L})/A^{\bullet},$$

that is, the *orbit of D*, an $\mathcal{A}$-connection of $\mathcal{L}$, in the quotient set (4.14), as above (see also (1.14)). Thus, concerning (4.17), one has

$$(4.18) \qquad\qquad D' = Ad(\phi) \cdot D$$

for any $\phi \in \mathcal{A}ut(\mathcal{L})$.

Furthermore, (4.18) is locally expressed through the relation

$$(4.19) \qquad\qquad \theta'_{\alpha} = \theta_{\alpha} + \tilde{\partial}(s_{\alpha}^{-1})$$

in such a manner thet one has (cf. (2.26))

$$(4.20) \qquad\qquad D \longleftrightarrow (\theta_{\alpha}) \quad \text{and} \quad D' \longleftrightarrow (\theta'_{\alpha}),$$

while one still obtains, concerning (4.19), that

$$(4.21) \qquad\qquad (s_{\alpha}) \in C^0(\mathcal{U}, \mathcal{A}^{\bullet})$$

with respect to a local frame

$$(4.22) \qquad\qquad \mathcal{U} = (U_{\alpha})_{\alpha \in I}$$

of $\mathcal{L}$.

On the other hand, concerning the set (4.14), the moduli space of the line sheaf $\mathcal{L}$, the same space is actually given by the definitions (cf. Chapt. VII; (2.9)) by the quotient set

$$(4.23) \qquad\qquad Conn_A(\mathcal{L})/\mathcal{A}ut(\mathcal{L}),$$

where the group sheaf $\mathcal{A}ut(\mathcal{L})$ acts on the set (affine space) of $\mathcal{A}$-connections of $\mathcal{L}$,

(4.24) $Conn_A(\mathcal{L})$

((2.13), (2.14) and (2.15)). Yet in this connection, one obtains by the definitions (cf. [VS: Chapt. II; p. 139, Lemma 6.2] or Chapt. II; (9.3) in the preceding, for $\mathcal{E} = \mathcal{L}$) that

(4.25) $Aut(\mathcal{L}) \equiv Aut_A(\mathcal{L}) = (\mathcal{E}nd\mathcal{L})^{\cdot} = \mathcal{A}^{\cdot},$

given that

(4.26) $\mathcal{E}nd\mathcal{L} = \mathcal{A},$

within an $\mathcal{A}$-isomorphism of the line sheaves involved (loc. cit.), which thus, in view also of (4.23), completely justifies (4.14) as another version equivalent to the standard one (cf. (4.23)) of the moduli space of $\mathcal{L}$. ∎

Now, based on (4.19), as well as on (3.9), one further obtains

(4.27) $R(D') = (d\theta_\alpha') = (d\theta_\alpha) = R(D)$

for any two $\mathcal{A}$-connections D and D' of $\mathcal{L}$ that also are gauge equivalent, that is, (locally, viz. with respect to $\mathcal{U}$, as in (4.22)) they satisfy (4.19) or are elements of (the equivalence class) (4.17). This, of course, completely vindicates our previous definition in (4.16), hence the map in (4.15), extension of (4.13.1), that is, of the map (curvature map of $\mathcal{L}$)

(4.28) $R(\cdot) : Conn_A(\mathcal{L}) \to \Omega^2(X)$

to the moduli space (4.14), which was our initial claim in this scholium. ∎

Scholium 4.2 By looking at the 0-cochain (s_α), as in (4.21), that appears in (4.19), realizing the particular gauge transformation (viz. $\mathcal{A}$-automorphism ϕ) of $\mathcal{L}$ considered in (4.18), we further note that (4.19) is actually a consequence of [VS: Chapt. VII; p. 109, (2.15) or (2.23); cf. also p. 111, (2.29)], in the case of a line sheaf; that is, one has another version of the respective transformation law of potentials (see also loc. cit., p. 92; (17.30)): Namely, since by assumption (cf. (4.18)),

(4.29) $\phi \in Aut\mathcal{L} = (\mathcal{A}ut\mathcal{L})(X),$

the 0-cochain in (4.21) is a 0-cocycle, viz. one has

(4.30) $(s_\alpha) \in Z^0(\mathcal{U}, \mathcal{A}^{\cdot}) \subseteq C^0(\mathcal{U}, \mathcal{A}^{\cdot})$

(see also [VS: Chapt. V; p. 355, (2.22), (2.24)], for $\mathcal{E} = \mathcal{F} = \mathcal{L}$). Thus, for the case in hand, one actually concludes that the relevant criterion of Lemma 2.2 is fulfilled, so that finally one obtains that

(4.31) $(\mathcal{L}, D) \underset{\phi}{\sim} (\mathcal{L}, D'),$

for any D' in the orbit of D (cf. (4.17)). In this connection, we also refer to Chapt. V; Section 2, for a direct proof of (4.19).

Further properties of the same fibration (map) τ, as before, will also be considered in the subsequent discussion.

5 Action of $H^1(X, \mathbb{C}^\cdot)$ on the Maxwell Group $\Phi^1_{\mathcal{A}}(X)^\nabla$

For convenience we start, by first considering the natural (free) action

$$(5.1) \qquad\qquad \mathbb{C}^\cdot \times \mathbb{A}^\cdot \to \mathbb{A}^\cdot,$$

defined through (scalar) multiplication, viz.

$$(5.2) \qquad\qquad (\lambda, x) \mapsto \lambda \cdot x,$$

where the notation "$\cdot$", indicates the corresponding group of units (viz. of the invertible elements) of the unital $\mathbb{C}$-algebras $\mathbb{C}$, and $\mathbb{A}$. It is also clear, based on the definition (5.2), that the action considered is free. On the other hand,

(5.3)

> (5.3.1) the above action, as given by (5.2), can further be extended to a free action of the constant presheaf of groups $\mathbb{C}^\cdot$ (on a given topological space X, see below) on $\Gamma(\mathcal{A}^\cdot)$, the (complete) presheaf of sections of $\mathcal{A}^\cdot$; here $\mathcal{A}^\cdot$ stands for the (group) sheaf of units of (our structure sheaf) $\mathcal{A}$, where
>
> $$(5.3.1) \qquad\qquad (X, \mathcal{A})$$
>
> denotes a $\mathbb{C}$-*algebraized space* (cf., for instance, Chapter I). Thus, according to (5.2) one gets the following presheaf group action (see also [VS: Chapt. I; p. 43]):
>
> $$(5.3.2) \qquad\qquad \mathbb{C}^\cdot \times \Gamma(\mathcal{A}^\cdot) \to \Gamma(\mathcal{A}^\cdot),$$
>
> viz., by analogy with (5.2), one sets
>
> $$(5.3.3) \qquad\qquad (\lambda, \alpha) \mapsto \lambda \cdot \alpha \in \mathcal{A}^\cdot(U) = A(U)^\cdot,$$
>
> for any $\lambda \in \mathbb{C}^\cdot \equiv \mathbb{C} \setminus \{0\}$ and $\alpha \in \mathcal{A}^\cdot(U)$, with U open in X.

In this connection, we further note that the above extended action of (5.2) as given by (5.3.3) is defined by virtue of the (canonical) sheaf injections

$$(5.4) \qquad\qquad \mathbb{C}^\cdot \subseteq \mathbb{C} \underset{\varepsilon}{\subseteq} \mathcal{A}$$

that one obtains according to our hypothesis for the structure sheaf $\mathcal{A}$ (see, for example, Chapt. II; (3.5), (6.6)). Hence, in view of (5.4), one finally obtains

$$(5.5) \qquad\qquad \mathbb{C}^\cdot \underset{\varepsilon}{\subseteq} \mathcal{A}^\cdot,$$

which also vindicates our notation in (5.3.3). On the other hand, based on (5.3.3), one can further define a free action pertaining to the corresponding groups of 1-cocycles of the (abelian) group sheaves involved in (5.3.2). Namely, one has the following (abelian group) action:

(5.6)

$$\text{(5.6.1)} \qquad Z^1(\mathcal{U}, \mathbb{C}^{\cdot}) \times Z^1(\mathcal{U}, \mathcal{A}^{\cdot}) \longrightarrow Z^1(\mathcal{U}, \mathcal{A}^{\cdot})$$

such that one sets

$$\text{(5.6.2)} \qquad ((\lambda_{\alpha\beta}), (s_{\alpha\beta})) \longmapsto (\lambda_{\alpha\beta} \cdot s_{\alpha\beta}),$$

for any $(\lambda_{\alpha\beta}) \in Z^1(\mathcal{U}, \mathbb{C}^{\cdot})$ and $(s_{\alpha\beta}) \in Z^1(\mathcal{U}, \mathcal{A}^{\cdot})$.

Concerning the map (5.6.2) it is easy to prove that this is well defined, based on the definitions; indeed, one has

$$\text{(5.7)} \qquad (\lambda_{\alpha\beta} \cdot s_{\alpha\beta}) \cdot (\lambda_{\beta\gamma} \cdot s_{\beta\gamma}) = (\lambda_{\alpha\beta} \cdot \lambda_{\beta\gamma})(s_{\alpha\beta} \cdot s_{\beta\gamma}) = \lambda_{\alpha\gamma} \cdot s_{\alpha\gamma},$$

which proves our claim about (5.6.2). ∎

Finally, by virtue of (5.6.2), one gets the following conclusion:

the action (5.6.1) as given by (5.6.2) is finally extended to the respective 1-dimensional cohomology groups. That is, one obtains the following (abelian) group action:

$$\text{(5.8.1)} \qquad H^1(X, \mathbb{C}^{\cdot}) \times H^1(X, \mathcal{A}^{\cdot}) \to H^1(X, \mathcal{A}^{\cdot}),$$

such that one sets

(5.8)   (5.8.2)

$$\begin{aligned}([(\lambda_{\alpha\beta})], [(g_{\alpha\beta})]) &\mapsto [(\lambda_{\alpha\beta})] \cdot [(g_{\alpha\beta})] \\ &:= [(\lambda_{\alpha\beta}) \cdot (g_{\alpha\beta})] = [(\lambda_{\alpha\beta} \cdot g_{\alpha\beta})],\end{aligned}$$

(cf. also (5.6.2)) for any

$$\text{(5.8.3)} \qquad \lambda \equiv [(\lambda_{\alpha\beta})] \in H^1(X, \mathbb{C}^{\cdot}) \text{ and } \mathcal{L} \equiv [(g_{\alpha\beta})] \in H^1(X, \mathcal{A}^{\cdot}).$$

Thus, one can further write (5.8.2) in the form

$$\text{(5.8.4)} \qquad (\lambda, \mathcal{L}) \longmapsto \lambda \cdot \mathcal{L} \equiv \lambda \cdot [\mathcal{L}] := [\lambda \cdot \mathcal{L}].$$

In this connection, see also [VS: Chapt. III; p. 234, Lemma 8.1, and p. 182, Definition 4.2], as well as N. Bourbaki [1: Chapt. II; p. 117, no 5]. Concerning the previous notation, as in (5.8), cf. also the preceding Chapter II; (7.16), (7.18) and (7.19), for $n = 1$, along with (7.20) and (7.21) therein.

On the other hand, by looking at the Maxwell group of X,

$$\text{(5.9)} \qquad \Phi_{\mathcal{A}}^1(X)^{\nabla} \longrightarrow \Phi_{\mathcal{A}}^1(X) \cong H^1(X, \mathcal{A}^{\cdot})$$

(see also (2.3) and (2.4)), one can further consider the restriction of (5.8.1) to (5.9). In other words, one obtains a group action

$$(5.10.1) \qquad H^1(X, \mathbb{C}^{\cdot}) \times \Phi^1_{\mathcal{A}}(X)^{\nabla} \to \Phi^1_{\mathcal{A}}(X)^{\nabla}$$

in such a manner that one sets (cf. also the above notation in (5.8.4))

(5.10)

$$(5.10.2) \qquad \lambda \cdot [(\mathcal{L}, D)] := [(\lambda \cdot \mathcal{L}, D)],$$

for any

$$(5.10.3) \qquad \lambda \equiv [(\lambda_{\alpha\beta})] \in H^1(X, \mathbb{C}^{\cdot}) \text{ and } [(\mathcal{L}, D)] \in \Phi^1_{\mathcal{A}}(X)^{\nabla}.$$

Indeed, we have first to check that (5.10.2) is well defined: Thus, by virtue of (2.26) and (5.8.2), we can write (5.10.1) in the form

$$(5.11) \qquad [(\lambda_{\alpha\beta})] \cdot [((g_{\alpha\beta}), (\theta_{\alpha}))] := [((\lambda_{\alpha\beta} \cdot g_{\alpha\beta}), (\theta_{\alpha}))],$$

so that one has to verify whether the second member of (5.11) defines an element of $\Phi^1_{\mathcal{A}}(X)^{\nabla}$, viz. whether (2.17) is fulfilled: So one has (see also (5.5), along with Chapt. I; (1.30))

$$(5.12) \qquad \tilde{\partial}(\lambda_{\alpha\beta} \cdot g_{\alpha\beta}) = \tilde{\partial}(\lambda_{\alpha\beta}) + \tilde{\partial}(g_{\alpha\beta}) = \tilde{\partial}(g_{\alpha\beta}) = \delta(\theta_{\alpha}),$$

which establishes (5.10.2), hence, (5.10.1). ■

5.1 Freeness of the Action of $H^1(X, \mathbb{C}^{\cdot})$ on the Maxwell Group

The above group action (5.10.1) can be made into a free one under a supplementary cohomological restriction on X. One has the following lemma.

Lemma 5.1 Suppose we have a differential triad

$$(5.13) \qquad\qquad (\mathcal{A}, \partial, \Omega^1)$$

on a topological space X such that the following condition holds:

$$(5.14) \qquad\qquad \ker \tilde{\partial} = \mathbb{C}^{\cdot}.$$

Then the group action (5.10.1) is free.

Proof. Based on the definition (5.10.2), suppose that we have

$$(5.15) \qquad \lambda \cdot [(\mathcal{L}, D)] = [(\lambda \cdot \mathcal{L}, D)] = [(\mathcal{L}, D)],$$

so that (see (1.16.1), (1.17.1)), one obtains

$$(5.16) \qquad\qquad (\lambda \cdot \mathcal{L}, D) \sim (\mathcal{L}, D).$$

Accordingly, by looking at the last relation locally, viz. in terms of a local frame of $\mathcal{L}$,

(5.17) $$\mathcal{U} = (U_\alpha)_{\alpha \in I},$$

one gets (5.16), equivalently in the form of the two following relations (cf. Lemma 2.2):

(5.18) $$\lambda_{\alpha\beta} \cdot g_{\alpha\beta} = \delta(s_\alpha^{-1}) \cdot g_{\alpha\beta}$$

and

(5.19) $$\theta_\alpha = \theta_\alpha + \tilde{\partial}(s_\alpha^{-1}),$$

in such a manner that

(5.20) $$(s_\alpha) \in C^0(\mathcal{U}, \mathcal{A}^{\cdot})$$

(cf. also (5.8.2) and (5.11)). Therefore, by virtue of (5.18), one obtains

(5.21) $$(\lambda_{\alpha\beta}) = \delta(s_\alpha^{-1}) \in B^1(\mathcal{U}, \mathcal{A}^{\cdot}),$$

viz. a 1-*coboundary* of $\mathcal{U}$, *with coefficients in* $\mathcal{A}^{\cdot}$, so that one finally has

(5.22) $$[(\lambda_{\alpha\beta})] = 1 \in H^1(X, \mathcal{A}^{\cdot}).$$

Furthermore, one concludes from (5.19) that

(5.23) $$\tilde{\partial}(s_\alpha^{-1}) = 0,$$

that is,

(5.24) $$(s_\alpha) \in \ker \tilde{\partial} = \mathbb{C}^{\cdot},$$

in view also of (5.14); hence, based on (5.21) as well, one has

(5.25) $$(\lambda_{\alpha\beta}) = \delta(s_\alpha^{-1}) \in B^1(\mathcal{U}, \mathbb{C}^{\cdot}) \underset{\longrightarrow}{\subseteq} B^1(\mathcal{U}, \mathcal{A}^{\cdot}),$$

that is,

(5.26) $$[(\lambda_{\alpha\beta})] = 1 \in H^1(X, \mathbb{C}^{\cdot}),$$

and this terminates the proof. ∎

Note 5.1 By looking at the latter part of the previous proof, we realize that our cohomological condition (5.14) was a crucial point in that proof. Of course, in the classical case space-time manifolds for instance, the said condition is automatically fulfilled, by virtue of the Poincaré lemma. On the other hand, as we shall see later on, there exist important special cases apart from the classical ones for which the aforementioned assumption is in force (see Vol. II: Chapter IV, cf. [VS: Chapts. X, XI]).

Scholium 5.1 By considering the constant 1-cocycle

$$(5.27) \qquad\qquad \lambda \equiv (\lambda_{\alpha\beta}) \in Z^1(\mathcal{U}, \mathbb{C}^{\cdot})$$

(see, for instance, (5.10.3)), we further remark that this can still be considered as the carrier of a Maxwell field (see (1.5). In this connection, cf. also [VS: Chapt. I; p. 51, Theorem 11.1, along with Chapt. III; p. 275, (11.26.1), and p. 234, Lemma 8.1]). Namely, we conclude that

(5.28)

> any (constant) complex 1-cocycle as in (5.27) entails a Maxwell field $(\mathcal{L}, D)$; that is, one sets (see (2.26))
>
> $$(5.28.1) \qquad\qquad (\mathcal{L}, D) \longleftrightarrow ((\lambda_{\alpha\beta}), (\theta_\alpha)),$$
>
> where (cf. Lemma 2.1) we also assume that
>
> $$(5.28.2) \qquad\qquad \delta(\theta_\alpha) = \tilde{\partial}(\lambda_{\alpha\beta}).$$
>
> Thus, in view of (5.27) (see also Chapt. I; (1.16) or (1.34)), one has
>
> $$(5.28.3) \qquad\qquad \delta(\theta_\alpha) = \tilde{\partial}(\lambda_{\alpha\beta}) = 0.$$
>
> Therefore, by virtue of (5.28.3), one concludes that
>
> $$(5.28.4) \qquad (\theta_\alpha) \in \Omega^1(X) \cong Z^0(\mathcal{U}, \Omega^1) \subseteq C^0(\mathcal{U}, \Omega^1)$$
>
> (cf. also [VS: Chapt. III; p. 178, (4.28) and p. 183, (4.55)]).

Accordingly, one thus gets the following general result.

Lemma 5.2 Suppose we are given a differential triad

$$(5.29) \qquad\qquad\qquad (\mathcal{A}, \partial, \Omega^1)$$

on a topological space X. Then any pair (cf. also (4.22))

$$(5.30) \qquad ((\lambda_{\alpha\beta}), (\theta_\alpha)) \in Z^1(\mathcal{U}, \mathbb{C}^{\cdot}) \times Z^0(\mathcal{U}, \Omega^1) = Z^1(\mathcal{U}, \mathbb{C}^{\cdot}) \times \Omega^1(X)$$

yields a Maxwell field $(\mathcal{L}, D)$ on X, viz. (cf. (2.26)), one has

$$(5.31) \qquad\qquad (\mathcal{L}, D) \longleftrightarrow ((\lambda_{\alpha\beta}), (\theta_\alpha)).$$

Proof. The assertion follows from Lemma 2.1 and (5.28.3) by virtue of (5.30). ∎
   On the other hand, still referring to our previous argument concerning (5.22) and (5.26), we further remark that

$$(5.32) \qquad \mathcal{L} \equiv [(\lambda_{\alpha\beta})] \in \mathrm{im}(H^1(X, \mathbb{C}^{\cdot}) \to H^1(X, \mathcal{A}^{\cdot})) \equiv \mathrm{im}(\varepsilon^*),$$

which thus clarifies that argument (see also (5.5), (5.27), along with [VS: Chapt. III; p. 234, Lemma 8.1 and p. 233, (11.26.1)]).

We further note that

(5.33)

> the Maxwell field $(\mathcal{L}, D)$, as defined by (5.28.1) or (5.31), is at first sight far from being flat, namely, from having $R(D) \equiv R = 0$ (white light). One assumes here that our space X (cf. (5.29)) is in particular a curvature space (cf. (3.2)). On the other hand, the aforementioned flatness of $\mathcal{L}$, viz. $R = 0$, occurs under further suitable assumptions for X: See Theorem 5.2 in [VS: Chapt. VIII; p. 208]; in this connection, we note here the following further equivalent condition to those mentioned in that theorem, when, in particular, (5.14) is in force. Namely,

> $\mathcal{L}$ is a flat $\mathbb{C}$-line sheaf, viz. one has

(5.33.1)
$$\mathcal{L} \longleftrightarrow (\lambda_{\alpha\beta}) \in Z^1(\mathcal{U}, \mathbb{C}^{\cdot}) \underset{\longrightarrow}{\subseteq} Z^1(\mathcal{U}, \mathcal{A}^{\cdot}),$$

> if and only if for example $\mathcal{L}$ has a flat $\mathcal{A}$-connection.

However, in this regard, cf. also the relevant discussion in Section 7 below; in particular, see (7.3) as well as (7.6).

Note 5.2 By further referring to our previous comments pertaining to (5.32), we remark that

(5.34)
$$\mathrm{im}(\varepsilon^*) \equiv \varepsilon^*(H^1(X, \mathbb{C}^{\cdot})) \lhd H^1(X, \mathcal{A}^{\cdot}),$$

being a subgroup of $H^1(X, \mathcal{A}^{\cdot})$, acts freely o $H^1(X, \mathcal{A}^{\cdot})$, when looking at the respective action (5.8.1); cf. also (5.8.2) as well as (5.22).

We have treated so far the freeness of the action (5.10.1) (see Lemma 5.1). We continue in the next subsection to consider those conditions on X that guarantee the transitivity of the same action. This line of work is in agreement with our general philosophy throughout this treatise: that is, to have at each stage of our discussion the most precise as well as the most general set-up possible, within which one can work.

5.2 Transitivity of the Action of $H^1(X, \mathbb{C}^{\cdot})$ on the Maxwell Group

To fix the terminology employed in the sequel, we assume henceforward that we are given

(5.35)

> a curvature space X (cf. (3.2)) in such a manner that the following cohomological condition holds:

> (5.35.1) $\ker d = \mathrm{im}\, \tilde{\partial}.$

> We still express (5.35.1) by also saying, roughly speaking, that

> (5.35.2)   every closed 1-form is logarithmically exact.

> See also (5.39).

By still referring to (5.35.1), we further remark that according to our hypothesis for X (cf. (5.35), (3.2) and Chapt. I; (1.25)), we already conclude, concerning (5.35.2), that

$$(5.36) \qquad \operatorname{im} \tilde{\partial} = \tilde{\partial}(\mathcal{A}^{\cdot}) \equiv \tilde{\partial}\mathcal{A}^{\cdot} \subseteq \ker d,$$

or even, equivalently, that

$$(5.37) \qquad d \circ \tilde{\partial} = 0.$$

Thus (5.35.1) guarantees the reverse relation to (5.36).

On the other hand, the precise meaning of (5.35.1) is that

(5.38)

    for every 0-cochain

$$(5.38.1) \qquad \begin{aligned} (\theta_\alpha) \in C^0(\mathcal{U}, \ker d) &= C^0(\mathcal{U}, \operatorname{im} \tilde{\partial}) \\ &= C^0(\mathcal{U}, \tilde{\partial}\mathcal{A}^{\cdot}) = \tilde{\partial}(C^0(\mathcal{U}, \mathcal{A}^{\cdot})), \end{aligned}$$

    there exists a 0-cochain in $\mathcal{A}^{\cdot}$,

$$(5.38.2) \qquad (s_\alpha) \in C^0(\mathcal{U}, \mathcal{A}^{\cdot})$$

    (see also Remark 3.1, (i)), such that one has;

$$(5.38.3) \qquad \theta_\alpha = \tilde{\partial}(s_\alpha^{-1}), \qquad \alpha \in I$$

    (cf. also (5.17)).

We note here that an analogous relation to (5.38.3) has been already employed in the foregoing (cf. (3.101)).

Now, within the same vein of ideas, we further remark that

(5.39)

    (5.38.1) can still be constructed, in view of (5.38.3), as defining a trivial $\mathcal{A}$-connection of a line sheaf $\mathcal{L}$ on X (cf. (5.35)), a coordinate 1-cocycle of which is given by (cf. (5.38.2))

$$(5.39.1) \qquad (g_{\alpha\beta}) := \delta(s_\alpha^{-1}) = (s_\beta^{-1} s_\alpha) \in Z^1(\mathcal{U}, \mathcal{A}^{\cdot}).$$

Concerning the terminology applied in (5.39), cf. [VS: Chapt. VIII; p. 214f.]. On the other hand, one further verifies here the corresponding transformation law of potentials (see (2.39)) relating (5.38.3) with (5.39.1). Indeed, one gets

$$(5.40) \qquad \delta(\theta_\alpha) = \tilde{\partial}(g_{\alpha\beta})$$

when applying the above notation. That is,

$$(5.41) \qquad \delta(\theta_\alpha) = \theta_\beta - \theta_\alpha = \tilde{\partial}(s_\beta^{-1}) - \tilde{\partial}(s_\alpha^{-1}) = \tilde{\partial}(s_\beta^{-1} s_\alpha) = \tilde{\partial}(g_{\alpha\beta}),$$

in view of (5.38.3) and (5.39.1) (see also Chapt. I; (1.29)), which proves (5.40). ■ In other words, one concludes (see also Lemma 2.1) that

(5.35.1) entails the existence of a Maxwell field

(5.42)   (5.42.1) $(\mathcal{L}, D) \longleftrightarrow ((g_{\alpha\beta}), (\theta_\alpha)),$

as given by (5.39.1) and (5.38.3).

It is an immediate consequence of the definitions that

(5.43)
every trivial $\mathcal{A}$-connection D on a line sheaf $\mathcal{L}$, as above, is flat. That is, one has $R(D) = 0$. Hence, the Maxwell field $(\mathcal{L}, D)$, as defined by (5.42.1), corresponds to white light (see also (3.11), referring to the latter terminology).

On the other hand, the converse of the above conclusion concerning the triviality of a given flat $\mathcal{A}$-connection is still in force under a supplementary condition on our curvature space X, apart from (5.35.1), and in a sense that we exhibit presently in connection with our main issue of this subsection: the transitivity of the action (5.10.1). Indeed, one has the following lemma.

Lemma 5.3 Suppose we have a curvature space X (cf. (3.2)) for which the following sequence of (abelian) group sheaves is exact:

(5.44) $1 \longrightarrow \mathbb{C}^{\cdot} \xrightarrow{\;\varepsilon\;} \mathcal{A}^{\cdot} \xrightarrow{\;\tilde{\partial}\;} \Omega^1 \xrightarrow{\;d\;} d\Omega^1 \longrightarrow 0.$

That is, we assume that

(5.45) $\ker d = \operatorname{im} \tilde{\partial} \quad \text{and} \quad \ker \tilde{\partial} = \mathbb{C}^{\cdot}.$

Moreover, let $(\mathcal{L}, D)$ be a given Maxwell field on X with a flat $\mathcal{A}$-connection, viz. we assume that

(5.46) $R(D) = 0.$

Then there exists a Maxwell field $(\mathcal{L}', D)$ that belongs to the orbit of $(\mathcal{L}, D)$, viz. we have

(5.47) $(\mathcal{L}', D) \in \mathcal{O}([(\mathcal{L}, D)]) \subseteq \Phi^1_{\mathcal{A}}(X)^{\nabla},$

that is

(5.48) $\mathcal{L}' = \lambda \cdot \mathcal{L},$

such that

(5.49) $\lambda \equiv (\lambda_{\alpha\beta}) \in Z^1(\mathcal{U}, \mathbb{C}^{\cdot}),$

while $(\mathcal{L}', D)$ has D as a trivial $\mathcal{A}$-connection as well.

Proof. By looking at the given Maxwell field $(\mathcal{L}, D)$ in terms of local data (cf. (2.26)), viz. setting

$$(5.50) \qquad (\mathcal{L}, D) \longleftrightarrow ((g_{\alpha\beta}), (\theta_\alpha)),$$

and in view of (5.46), one has (cf. also (3.9))

$$(5.51) \qquad R(D) = (d\theta_\alpha) = 0,$$

that is (cf. (5.35.1)),

$$(5.52) \qquad (\theta_\alpha) \in C^0(\mathcal{U}, \ker d) = C^0(\mathcal{U}, \operatorname{im} \tilde{\partial}) = C^0(\mathcal{U}, \tilde{\partial}\mathcal{A}^{\cdot}) = \tilde{\partial}(C^0(\mathcal{U}, \mathcal{A}^{\cdot}))$$

(see also (5.38.1)), so that one obtains

$$(5.53) \qquad \theta_\alpha = \tilde{\partial}(s_\alpha^{-1}),$$

such that

$$(5.54) \qquad (s_\alpha) \in C^0(\mathcal{U}, \mathcal{A}^{\cdot}).$$

Therefore, by virtue of (5.39), or even (5.42), one gets a Maxwell field

$$(5.55) \qquad (\mathcal{L}', D) \longleftrightarrow ((g'_{\alpha\beta}), (\theta_\alpha)),$$

such that

$$(5.56) \qquad g'_{\alpha\beta} = \delta(s_\alpha^{-1}) \ and \ \theta_\alpha = \tilde{\partial}(s_\alpha^{-1}).$$

Thus, based on (5.56), along with (5.39), one concludes that

$$(5.57) \qquad \begin{array}{l} \text{the given (flat) } \mathcal{A}\text{-connection } D \text{ of } \mathcal{L} \text{ is by definition (ibid.) a trivial } \mathcal{A}\text{-} \\ \text{connection, however not of } \mathcal{L} \text{ as such but of } \mathcal{L}', \text{ an eventual, in general,} \\ \text{translation of } \mathcal{L}, \text{ viz. of an element of the orbit of } (\mathcal{L}, D), \text{ with respect to} \\ \text{the action of } H^1(X, \mathbb{C}^{\cdot}) \text{ on the Maxwell group of } X, \text{ as in (5.10.1).} \end{array}$$

The above also clarifies our previous conclusion in [VS: Chapt. VIII; p. 216, Theorem 6.1]. Now, for the sake of completeness, we come to the proof of (5.57):

By (5.50) and (2.17), one first obtains

$$(5.58) \qquad \delta(\theta_\alpha) = \tilde{\partial}(g_{\alpha\beta}),$$

so that in view of (5.56), one obtains (see also e.g. (5.41))

$$(5.59) \qquad \tilde{\partial}(s_\beta^{-1} s_\alpha) = \tilde{\partial}(g_{\alpha\beta}),$$

that is, the relation

$$(5.60) \qquad \tilde{\partial}(s_\beta^{-1} s_\alpha \cdot g_{\alpha\beta}^{-1}) = 0.$$

Consequently (cf. also (5.14)), one obtains

(5.61) $\qquad\qquad s_\beta^{-1} s_\alpha \cdot g_{\alpha\beta}^{-1} \in (\ker \tilde{\partial} = \mathbb{C}^{\cdot})(U_{\alpha\beta} \equiv U_\alpha \cap U_\beta)$,

so that one further gets the relation (cf. also (5.56))

(5.62) $\qquad\qquad g'_{\alpha\beta} = s_\beta^{-1} s_\alpha = \lambda_{\alpha\beta} \cdot g_{\alpha\beta}$,

where we set

(5.63) $\qquad\qquad \lambda \equiv (\lambda_{\alpha\beta}) \in Z^1(\mathcal{U}, \mathbb{C}^{\cdot})$.

Therefore, by virtue of (5.50), (5.55), and (5.62), one finally obtains

(5.64) $\qquad\qquad \mathcal{L}' = \lambda \cdot \mathcal{L}$,

which was our claim in (5.48), proving (5.57) as well, and this also completes the proof of Lemma 5.3. ∎

   In this connection, it is still to be noticed that Lemma 5.3 is a particular case for flat line sheaves (Maxwell fields with curvature zero) of a more general result, valid for any two Maxwell fields of the same field strength (see Theorem 5.1), or in other words, of the transitivity of the action of $H^1(X, \mathbb{C}^{\cdot})$ on the set of the previous type of Maxwell fields, viz. on light rays of the same color.

Note 5.3 The notion of a trivial $\mathcal{A}$-connection for line sheaves, thus for Maxwell fields (cf. (5.39)), can further be extended, analogously, to any vector sheaf $\mathcal{E}$, that is, to a Yang–Mills field $(\mathcal{E}, D)$; see [VS: Chapt. VIII; p. 214f.]. The analogous result to Lemma 5.3, for the general case, viz. for $rk\mathcal{E} = n \neq 1$, is still in force as an application of an appropriate formulation in the abstract case of the classical Frobenius integrability condition (loc. cit., Chapt. XI; p. 355, Theorem 9.1, and p. 357, (9.26), as well as, Chapt. VIII; p. 205, Section 5.(a)).

   Before we come to our main result of this subsection (see Theorem 5.1 below), we first give the following item, being also at the basis of the subsequent discussion. In this connection, we still observe that henceforth we are interested in Maxwell fields having the same field strength; in other words, we consider light bundles (see (3.71)), the structure of which is examined by the ensuing discussion. Thus, we come to the following basic result.

Lemma 5.4 Suppose we are given a light bundle

(5.65) $\qquad\qquad\qquad \Phi^1_{\mathcal{A}}(X)^{\nabla}_R$

on a curvature space X, the latter being also paracompact (Hausdorff); viz. consider the set of those (equivalence classes of) Maxwell fields on X whose field strength (curvature) equals R (see (3.71)). Thus, one has

(5.66)
$$R(D) = R$$

for any $(\mathcal{L}, D)$, or $[(\mathcal{L}, D)]$ (light ray) in (5.65) (cf. also (3.11)). We further assume that

(5.67)
$$\ker d = \operatorname{im} \tilde{\partial}.$$

Then for any pair of elements in (5.65), say $(\mathcal{L}, D)$ and $(\mathcal{L}', D')$, so that according to (5.66) one has

(5.68)
$$R(D) = R(D') = R,$$

one can find a common gauge potential ($\mathcal{A}$-connection) of the two fields at issue as well.

Proof. Based on (5.68), one first obtains (cf. also (3.9))

(5.69)
$$R = (d\theta_\alpha) = (d\theta'_\alpha),$$

where we have employed our previous notation in (2.26) for the Maxwell fields considered as in the statement of the lemma. Hence, by (5.69), one has

(5.70)
$$d(\theta'_\alpha - \theta_\alpha) = 0,$$

that is, in view of (5.67),

(5.71)
$$(\theta'_\alpha - \theta_\alpha) \in C^0(\mathcal{U}, \ker d) = C^0(\mathcal{U}, \operatorname{im} \tilde{\partial}) = C^0(\mathcal{U}, \tilde{\partial}(\mathcal{A}^\cdot)).$$

Thus, we have the following (group sheaf) epimorphism, or, equivalently, of the exact sequence (of abelian group sheaves):

(5.72)
$$\mathcal{A}^\cdot \xrightarrow{\tilde{\partial}} \underbrace{\operatorname{im} \tilde{\partial} = \tilde{\partial}(\mathcal{A}^\cdot) \equiv \tilde{\partial}\mathcal{A}^\cdot}_{\substack{\| \\ \ker d \ (\text{cf. (5.67)})}} \longrightarrow 0.$$

Consequently, in view of our hypothesis that the underlying space X is (Hausdorff) paracompact, one concludes from Lemma 5.2 in [VS: Chapt. III; p. 196] that (cf. (5.71))

(5.73)
$$C^0(\mathcal{U}, \tilde{\partial}(\mathcal{A}^\cdot)) = \tilde{\partial}(C^0(\mathcal{U}, \mathcal{A}^\cdot)),$$

so that one finally obtains from (5.71) that

(5.74)
$$(\theta'_\alpha - \theta_\alpha) = \tilde{\partial}(s_\alpha^{-1})$$

for some 0-cochain

(5.75)
$$(s_\alpha) \in C^0(\mathcal{U}, \mathcal{A}^\cdot).$$

Therefore, by virtue of (5.74), one now obtains

$$(5.76) \qquad\qquad \theta_\alpha = \theta'_\alpha + \tilde{\partial}(s_\alpha^{-1}).$$

Thus, one can now define the following Maxwell field:

$$(5.77) \qquad\qquad (\mathcal{L}'', D'') \longleftrightarrow ((g''_{\alpha\beta})(\theta''_\alpha)),$$

(cf. also (2.26)), where we set

$$(5.78) \qquad\qquad g''_{\alpha\beta} := \delta(s_\alpha^{-1}) \cdot g'_{\alpha\beta}$$

as well as (cf. (5.76))

$$(5.79) \qquad\qquad \theta''_\alpha := \theta_\alpha = \theta'_\alpha + \tilde{\partial}(s_\alpha^{-1}).$$

On the other hand, one has by the preceding

$$
\begin{aligned}
(5.80) \qquad \tilde{\partial}(g''_{\alpha\beta}) &= \tilde{\partial}(\delta(s_\alpha^{-1}) \cdot g'_{\alpha\beta}) = \tilde{\partial}(\delta(s_\alpha^{-1})) + \tilde{\partial}(g'_{\alpha\beta}) \\
&= \tilde{\partial}(\delta(s_\alpha^{-1})) + \delta(\theta'_\alpha) = \delta(\theta'_\alpha) + \delta(\tilde{\partial}(s_\alpha^{-1})) \\
&= \delta(\theta'_\alpha + \tilde{\partial}(s_\alpha^{-1})) = \delta(\theta''_\alpha),
\end{aligned}
$$

which thus justifies our definition in (5.77) (cf. Lemma 2.1). In this connection, see also (2.17), Chapt. I; (1.28) in the preceding, along with [VS: Chapt. III; p. 189, (5.8′)].

Accordingly, one obtains

a Maxwell field

$$(5.81) \qquad\qquad (5.81.1) \qquad\qquad\qquad (\mathcal{L}'', D)$$

that has the same $\mathcal{A}$-connection D as the given one $(\mathcal{L}, D)$ (cf. 5.77), (5.79)), being also equivalent to the second given one $(\mathcal{L}', D')$.

Indeed, the last part of our previous assertion in (5.81) follows from Lemma 2.2, in conjunction with the definition of $(\mathcal{L}'', D)$, by (5.78) and (5.79). That is,

$$(5.82) \qquad\qquad (\mathcal{L}', D') \sim (\mathcal{L}'', D).$$

This was our assertion in the statement of the lemma modulo the above equivalence (5.82), and this also completes the proof of Lemma 5.4. ■

It is our objective to give a precise description of $\mathcal{L}''$, as in (5.81.1). Indeed, we shall prove that

$$(5.83) \qquad\qquad \mathcal{L}'' = \lambda \cdot \mathcal{L},$$

for a uniquely defined

(5.84) $$\lambda \equiv [(\lambda_{\alpha\beta})] \in H^1(X, \mathbb{C}^{\cdot})$$

(see Theorem 5.1).

On the other hand, by using the physical terminology already employed for the set (5.65), we can express our previous result in Lemma 5.4 by saying that

(5.85)   two light rays (cf. also (1.19)) have the same color if and only if they stem from the same gauge potential.

Thus, the lesson so far from Lemma 5.4 (and its proof, see in particular (5.81) and (5.82)) is that

(5.86)   any time we are given two Maxwell fields of the same field strength (curvature), we can replace any one of them by an equivalent one having the same gauge potential (A-connection) as the other. Thus, finally, we get two new Maxwell fields (in point of fact, one of the given ones and another new one), which now have the same field strength, stemming from the same gauge potential.
Accordingly, and, what amounts to the same thing (modulo an eventual replacement, as above, cf. (5.82)),

(5.86.1)   two given Maxwell fields have the same field strength if and only if they have the same gauge potential as well.

Note 5.4 Of course, the previous remark (5.86.1) is still in force for any finitely many Maxwell fields, having the same field strength. Indeed, one can obviously apply a successive exploitation of our previous argument in the proof of Lemma 5.4.

5.3 $\Phi^1_{\mathcal{A}}(X)^{\nabla}_R$, as a Principal Homogeneous Space

We come now to our main result of this section, viz. Theorem 5.1, referring to the transitivity and freeness, as well as the action of $H^1(X, \mathbb{C}^{\cdot})$ when restricted to any given light bundle (cf. (3.71)). Indeed, the said result is just an application of the preceding lemmas. Thus, one obtains the following.

Theorem 5.1 Suppose we are given a (Hausdorff) paracompact curvature space X, such that the following (cohomological) conditions are in force:

(5.87) $$\ker \tilde{\partial} = \mathbb{C}^{\cdot},$$

as well as (cf. also (5.38))

(5.88) $$\ker d = \operatorname{im} \tilde{\partial}.$$

Equivalently, the following sequence of (abelian) group sheaves is exact:

(5.89) $$1 \longrightarrow \mathbb{C}^{\cdot} \overset{\varepsilon}{\longrightarrow} \mathcal{A}^{\cdot} \overset{\tilde{\partial}}{\longrightarrow} \Omega^1 \overset{d}{\longrightarrow} d\Omega^1 \longrightarrow 0.$$

Then the (group) action (5.10.1), when restricted to the light bundle (cf. (3.71))

$$(5.90) \qquad \qquad \Phi^1_{\mathcal{A}}(X)^\nabla_R \subseteq \Phi^1_{\mathcal{A}}(X)^\nabla$$

(cf. (3.73)) is free and transitive. Therefore, in other words, one concludes that

$$(5.91) \qquad \Phi^1_{\mathcal{A}}(X)^\nabla_R \text{ is a principal homogeneous } H^1(X, \mathbb{C}^\bullet)\text{-space.}$$

Proof. We already know that $H^1(X, \mathbb{C}^\bullet)$ acts freely on the Maxwell group of X, $\Phi^1_{\mathcal{A}}(X)$, under the assumption of (5.87), indeed, for any differential triad, thus without any extra hypothesis on X (cf. Lemma 5.1). Hence, the induced action on any light bundle, as in (5.90), is still free, and this terminates the first part of the proof.

Supposing further the rest of our assumptions for X, as well as in the statement of the theorem, we prove now the transitivity of the same action as above:

Assume that we are given two elements of the light bundle $\Phi^1_{\mathcal{A}}(X)^\nabla_R$ (cf. (3.71)), namely (by still employing, for simplicity, an obvious abuse of notation; see, for instance, (3.11.1) or (3.107)) two Maxwell fields

$$(5.92) \qquad \qquad (\mathcal{L}, D) \quad \text{and} \quad (\mathcal{L}', D')$$

such that

$$(5.93) \qquad \qquad R(D) = R(D') \equiv R.$$

Therefore, in view of (5.82), one obtains

$$(5.94) \qquad \qquad [(\mathcal{L}', D')] = [(\mathcal{L}'', D)].$$

On the other hand, by further employing a local description of the items appearing in (5.94) (cf., for instance, (2.26)), one gets, by virtue of (5.79) and (5.80), the relation

$$(5.95) \qquad \qquad \tilde{\partial}(g''_{\alpha\beta}) = \delta(\theta''_\alpha) = \delta(\theta_\alpha) = \tilde{\partial}(g_{\alpha\beta})$$

(however, see also (2.17)). Accordingly (cf. Chapt. I; (1.29)), one obtains

$$(5.96) \qquad \qquad \tilde{\partial}(g''_{\alpha\beta} \cdot g^{-1}_{\alpha\beta}) = 0,$$

or in other words (see also (5.87)),

$$(5.97) \qquad \qquad (g''_{\alpha\beta} \cdot g^{-1}_{\alpha\beta}) \in Z^1(\mathcal{U}, \ker \tilde{\partial}) = Z^1(\mathcal{U}, \mathbb{C}^\bullet),$$

so that we set

$$(5.98) \qquad \qquad (g''_{\alpha\beta} \cdot g^{-1}_{\alpha\beta}) \equiv (\lambda_{\alpha\beta}) \in Z^1(\mathcal{U}, \mathbb{C}^\bullet).$$

Therefore, we have

$$(5.99) \qquad \qquad g''_{\alpha\beta} = \lambda_{\alpha\beta} \cdot g_{\alpha\beta},$$

that is, equivalently (see, for example, (5.8)), one obtains

(5.100) $[\mathcal{L}''] = \mathcal{L}'' = [\lambda \cdot \mathcal{L}] = \lambda \cdot [\mathcal{L}] \equiv \lambda \cdot \mathcal{L},$

where one has (cf. (5.98))

(5.101) $\lambda \equiv [(\lambda_{\alpha\beta})] \in H^1(X, \mathbb{C}^{\cdot}).$

Consequently, based now on (5.94), (5.100), as well as on (5.10.2), one finally obtains

(5.102) $[(\mathcal{L}', D')] = [(\mathcal{L}'', D)] = [(\lambda \cdot \mathcal{L}, D)] = \lambda \cdot [(\mathcal{L}, D)],$

where λ is given by (5.101). This proves our assertion concerning the transitivity of the action (5.10.1) when restricted on (5.90) (cf. (5.93)), and also terminates the proof of the theorem. ∎

Note 5.5 (Terminological) By referring to the parlance that has been employed in (5.91), we remark that according to standard terminology, a *principal homogeneous G-space* (or a *principal homogeneous G-set*) is a set on which a group G acts freely and transitively (see, for instance, N. Bourbaki [3: Chapt. I; p. 58, Définition 7]).

On the other hand, by extending the classical terminology, applied in particular to (the additive group of) a vector space, we also call a set, as above, a *G-affine space* (cf. also loc. cit. 59, Example 4). See (5.116) in the sequel.

Scholium 5.2 Concerning our assumptions in Theorem 5.1, we remark that

(5.103)   the exactness of the sequence (5.89) can be obtained from the exactness of the horizontal sequence in (3.27), along with the commutativity of the exponential (sheaf) triangle of the same scheme (cf. (3.29)).

Indeed, by the definitions (see also Chapt. I; (1.26)), one first concludes that

(5.104) $\ker \tilde{\partial} = \ker \partial \cap \mathcal{A}^{\cdot},$

which proves that the first of the relations in (3.28), that is, the exactness at $\mathcal{A}$ of the horizontal sequence in (3.27), implies (5.87). Furthermore, assuming also the second relation in (3.28), viz. the exactness of the previous sequence at Ω^1 (hence exactness of the sequence at issue), in conjunction with (3.29), one easily concludes (5.88), taking also, however, into account that X is supposed to be paracompact (Hausdorff); thus, supposing that

(5.105) $\omega \in \Omega^1(U), \qquad \text{with } d\omega = 0,$

then, by the second of (3.28), along with [VS: Chapt. III, p. 196, Lemma 5.2] (cf. also, for instance, (5.73) in the preceding), one concludes that

(5.106) $\omega = \partial(\alpha),$

for some $\alpha \in \mathcal{A}(U)$. So by virtue of (3.29), one finally obtains that

$$(5.107) \qquad\qquad \omega = \tilde{\partial}(t^{-1}),$$

where we set

$$(5.108) \qquad\qquad t = e\left(\frac{1}{2\pi i}\alpha\right)^{-1} \in \mathcal{A}(U)^\cdot = \mathcal{A}^\cdot(U),$$

U being throughout our previous argument an open subset of X. This proves our last assertion (see also (5.35) and (5.38)), and with it the initial claim in (5.103). ∎

For convenience, we recapitulate our last conclusion in the form of the following.

Lemma 5.5 Suppose we are given a (Hausdorff) paracompact curvature space X for which the following logarithmic diagram is in force:

(5.109)

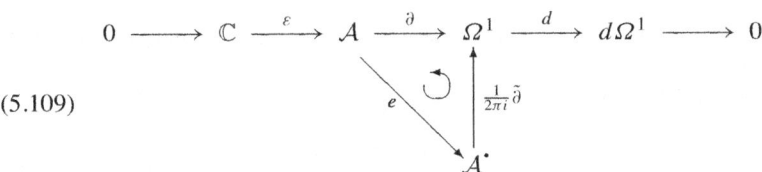

Here the horizontal sequence is exact, while the depicted triangle is also commutative.

Then every closed 1-form is logarithmically exact. (Cf. also, concerning the later terminology, (5.35.2), along with (5.38).) ∎

In this connection, we still note that we shall see in the sequel important particular cases for which (5.109) holds; indeed, the stronger Weil scheme (cf. (3.27)) is valid in those cases as well. See Vol. II: Chapt. IV; Section 5.

On the other hand, by applying physical language, we could express our conclusion in (5.91) by saying that

(5.110)　　two light rays (cf. (1.19) or (3.11)) have the same color if and only if they differ only by a phase factor, viz. by an element of $H^1(X, \mathbb{C}^\cdot)$. At the same time, they can also acquire the same gauge potential (see (5.85), (5.86)).

As we shall see in the sequel (cf. Section 6), one can consider the above phase factor as a normalized one, namely, as an element of $H^1(X, S^1)$ (cf. (6.).

Furthermore, by applying for the preceding a formal language, one obtains the following relation, an equivalent formulation, in effect, of (5.91), hence of Theorem 5.1:

$$(5.111) \qquad\qquad \Phi_{\mathcal{A}}^1(X)_R^\nabla = H^1(X, \mathbb{C}^\cdot) \cdot [(\mathcal{L}, D)],$$

within a bijection of the sets involved for any Maxwell field $[(\mathcal{L}, D)] \in \Phi^1_{\mathcal{A}}(X)^\nabla$, with $R(D) = R$.

Consequently, one gets the following equivalent formulation of (5.110):

(5.112)   two given Maxwell fields have the same field strength, say R, if and only if they belong to the same orbit with respect to the action of $H^1(X, \mathbb{C}^{\bullet})$ on the set $\Phi^1_{\mathcal{A}}(X)^\nabla_R$.

In fact, there actually exists just one orbit: Namely, one has

(5.113) $$\Phi^1_{\mathcal{A}}(X)^\nabla_R = \mathcal{O}([(\mathcal{L}, D)]) = \{\lambda \cdot [(\mathcal{L}, D)] : \lambda \in H^1(X, \mathbb{C}^{\bullet})\}$$

for any Maxwell field $[(\mathcal{L}, D)]$ with $R(D) = R$.

The previous relations are an application of the so-called *orbit map* (cf. also (5.8))

(5.114) $$\lambda \mapsto \lambda \cdot [(\mathcal{L}, D)] : H^1(X, \mathbb{C}^{\bullet}) \to \Phi^1_{\mathcal{A}}(X)^\nabla_R,$$

which is a bijection (Theorem 5.1) with respect to any Maxwell field $[(\mathcal{L}, D)]$ with $R(D) = R$.

Thus, we can still say that

(5.115) $$\Phi^1_{\mathcal{A}}(X)^\nabla_R = H^1(X, \mathbb{C}^{\bullet}),$$

within a bijection, established by any Maxwell field $[(\mathcal{L}, D)]$ with curvature R. Finally, by employing another familiar terminology (see also Note 5.5), we can say that

(5.116)   any light bundle $\Phi^1_{\mathcal{A}}(X)^\nabla_R$ is an $H^1(X, \mathbb{C}^{\bullet})$-affine space.

We come now, in Section 6, to look at the preceding material through a Hermitian metric, so that one can thus consider on a light bundle the action of the group $H^1(X, S^1)$, in place of $H^1(X, \mathbb{C}^{\bullet})$, as was the case in the foregoing.

6 The Hermitian Counterpart

To put the preceding into the perspective of an action of the group $H^1(X, S^1)$, in place of $H^1(X, \mathbb{C}^{\bullet})$, as was the case hitherto, we need to have at our disposal a Hermitian metric "on" X (cf. [VS: Chapt. IV; Section 9]), so that one can speak then of Hermitian $\mathcal{A}$-connections on the vector sheaves, in fact, on the line sheaves involved herewith. See also loc. cit. Chapt. VII; Section 10, for the relevant terminology.

Now, for convenience, we first fix the abstract framework within which we are going to argue in the sequel. On the other hand, for simplicity's sake, we do not consider at each stage the most general context possible, leaving it instead implicit, viz. the task of isolating, at each time, the particular hypotheses that are employed; indeed, this is not hard for those readers who are already familiar with the preceding discussion along with the fundamentals of the abstract differential geometry we apply [VS]. Anyhow, we will give hints throughout the sequence. To start with, we assume in the sequel that

(6.1)   we are given a Hausdorff paracompact curvature space X, base space of our $\mathbb{C}$-algebraized space $(X, \mathcal{A})$, with $(\mathcal{A}, \rho)$, being in particular a Hermitian $\mathcal{A}$-module and $\mathcal{A}$ a fine ($\mathbb{C}$-algebra) sheaf on X. Furthermore, we accept that X is an enriched ordered strictly involutive and strictly sheaf exponential space.

Concerning the terminology applied in the preceding, we also refer to Chapter I, while further explanations will be given throughout the ensuing discussion.

By considering a curvature space X, as in (6.1), we adopted a differential triad on X,

$$(6.2) \qquad (\mathcal{A}, \partial, \Omega^1)$$

(see, for instance, (0.1)), which is further endowed with one more differential operator $d^1 \equiv d$ (cf. (3.2), along with Chapter I), the 1st prolongation of ∂, so that one has the curvature datum;

$$(6.3) \qquad (\mathcal{A}, \partial, \Omega^1, d, \Omega^2)$$

(cf. also [VS: Chapt. VIII; p. 188, Definition 1.1, and the ensuing comments therein], or Vol. II: Chapter I; Section 1 in the sequel of the present discussion). In this connection, we still assume that

(6.4)   the $\mathcal{A}$-module Ω^1, as in (6.2), is in particular a vector sheaf on X as well.

Finally, we suppose that X satisfies the following (cohomological) condition:

$$(6.5) \qquad \ker \partial = \mathbb{C}$$

(cf. also the horizontal sequence in (5.109)). The last condition implies, of course, according to the definitions, (5.14) (see Chapter I, or (5.104) above). Thus, in sum,

(6.6)   we assume henceforward that (6.1), (6.4), and (6.5) are in force.

As a first consequence of the preceding, one gets the following specialization of (7.19) of Chapter II; that is,

$$(6.7) \qquad \Phi_{\mathcal{A}}^n(X) = H^1(X, \mathcal{SU}(n, \mathcal{A})),$$

for any $n \in \mathbb{N}$, within a bijection of the sets involved (see also [VS: Chapt. III; Section 11]. Here

$$(6.8) \qquad \mathcal{SU}(n, \mathcal{A})$$

stands for the special unitary group sheaf of $\mathcal{A}$ of order $n \in \mathbb{N}$ (loc. cit. Chapt. V; p. 402, (9.35)), the definition of which depends on the existence of a Hermitian metric on X (viz. in effect, on $\mathcal{A}$; see also [VS: Chapt. IV; p. 333, Theorem 9.1], along with our hypothesis in (6.1), as above). In this connection, we still express (6.7), by saying that

(6.9)   every vector sheaf $\mathcal{E}$ on X admits an $\mathcal{SU}(n, \mathcal{A})$-structure, where $n = rk\mathcal{E}$.

In particular, for $n = 1$, and setting

(6.10) $$\mathcal{SU}(1, \mathcal{A}) \equiv \mathcal{SU}(1),$$

one obtains from (6.7) the following isomorphism of (abelian) groups:

(6.11) $$\Phi^1_{\mathcal{A}}(X) = H^1(X, \mathcal{SU}(1)),$$

the first member of (6.11) being the Picard group of X (see Section 2). So, by analogy with (6.9), one concludes that

(6.12)   every line sheaf on X admits an $\mathcal{SU}(1)$-structure,

the last conclusion being an equivalent formulation of (6.11) (see also the ensuing comments): Thus, specializing (7.18) of Chapter II to the present case, and for $n = 1$ as in (6.11), one gets at the following bijection

(6.13) $$\mathcal{L} \equiv [\mathcal{L}] \longleftrightarrow (g_{\alpha\beta})$$

such that

(6.14) $$(g_{\alpha\beta}) \in Z^1(\mathcal{U}, \mathcal{SU}(1));$$

namely, one has by definition

(6.15) $$(g_{\alpha\beta}) \in Z^1(\mathcal{U}, \mathcal{A}^{\cdot}),$$

such that

(6.16) $$|g_{\alpha\beta}| = 1 \in \mathcal{A}^{\cdot}(U_{\alpha\beta}) = \mathcal{A}(U_{\alpha\beta})^{\cdot},$$

where we set

(6.17) $$U_{\alpha\beta} = U_{\alpha} \cap U_{\beta} (\neq \emptyset),$$

while

(6.18) $$\mathcal{U} = (U_{\alpha})_{\lambda \in I}$$

stands for a given local frame of $\mathcal{L}$ (see also [VS: Chapt. V; p. 401f, (9.31) and (9.36)]). Our claim in (6.13)–(6.16) is a consequence of our hypothesis for X, as in (6.1), along with our previous conclusions in [VS: Chapt. V; p. 402, Theorem

9.2]; see also loc. cit., Chapt. VIII, p. 213, Proposition 5.1, as well as Note 6.1. This establishes completely our assertion in (6.11). ■

Note 6.1 Concerning (6.11), we further remark that in its proof we made use of Theorem 9.2 in [VS: Chapt. V; p. 402], as well as Proposition 5.1, ibid., Chapt. VIII, p. 213. One is based here on our hypothesis for X, as in (6.1), together with (6.5). So the assumption that X is a sheaf exponential space (cf. (6.1), along with [VS: Chapt. VII, p. 144; (7.1)]), enables one to prove that a line sheaf $\mathcal{L}$ for which Proposition 5.1 of [VS: Chapt. VIII, p. 213] holds, a fact that actually is here the case, is a flat $\mathbb{C}$-line sheaf (in fact, a consequence of (6.5)). In this connection, cf. also [VS: Chapt. VIII; p. 208, proof of Theorem 5.2, in particular p. 211, (5.48), as well as, Chapt. V; p. 370, Definition 5.1].

It is still of interest to look at (6.11) in comparison with the corresponding one, referring to the *Picard group* of X, as expressed through the coordinate 1-cocycles of the line sheaves involved; see Chapter II, (7.18). On the other hand, the new interpretation as given by (6.11) is due to our supplementary assumptions for X, the same point of view also being of importance in physics (cf., for instance, (6.2) below). Of course, these hypotheses for X (viz. cond. (6.6)) always hold in the classical case, while they still remain in force for various important particular examples in the abstract setting, as already mentioned (cf. (3.26) and comments following (3.29)).

Let us now denote by

$$(6.19) \qquad \Phi^1_{\mathcal{A}}(X)_{her}$$

the set of equivalence classes of Hermitian line sheaves on X, namely, those determined by (6.15) and (6.16). (We note that by virtue of (6.12), viz. as a consequence of our assumptions for X, cf. (6.1) and (6.5), the set (6.19) is not empty.) Now, by applying (6.15), (6.16), along with a similar argument, as in Note 6.1, based on the cited results in [VS], one can see that the same set (6.19) is an abelian group; see also [VS: Chapt. V; p. 358, Theorem 2.1, for $n = 1$]. Taking the present terminology into account, we give (6.11) the following more precise version; that is,

$$(6.20) \qquad \Phi^1_{\mathcal{A}}(X)_{her} = H^1(X, \mathcal{SU}(1)),$$

within an isomorphism of the (abelian) groups concerned. We call the first member of (6.20), viz. the set (6.19), the *Hermitian Picard group* of X, which always has a meaning for a space X, satisfying, for instance, (6.1) and (6.5).

6.1 Action of $H^1(X, S^1)$ on $\Phi^1_{\mathcal{A}}(X)^\nabla$

The action that is referred to in the heading of this subsection is the relative action of

$$(6.21) \qquad H^1(X, S^1) < H^1(X, \mathbb{C}^{\bullet})$$

on the Maxwell group of X, $\Phi^1_{\mathcal{A}}(X)^\nabla$, induced on it by (5.10.1), as defined by (5.10.2). Thus, one gets the group action

(6.22)
$$H^1(X, S^1) \times \Phi_{\mathcal{A}}^1(X)^\nabla \to \Phi_{\mathcal{A}}^1(X)^\nabla,$$

such that one has

(6.23)
$$\lambda \cdot [(\mathcal{L}, D)] = [(\lambda \cdot \mathcal{L}, D)]$$

for any $\lambda \equiv [(\lambda_{\alpha\beta})] \in H^1(X, S^1)$, viz. $(\lambda_{\alpha\beta}) \in Z^1(\mathcal{U}, \mathbb{C}^\cdot)$, with

(6.24)
$$|\lambda_{\alpha\beta}| = 1 \in S^1 \subseteq \mathbb{C}^\cdot,$$

so that by definition one obtains

(6.25)
$$(\lambda_{\alpha\beta}) \in Z^1(\mathcal{U}, S^1) \underset{\longrightarrow}{\subset} Z^1(\mathcal{U}, \mathbb{C}^\cdot)$$

and $[(\mathcal{L}, D)] \in \Phi_{\mathcal{A}}^1(X)^\nabla$ (see also (5.10.2), (5.10.3), and (5.12)).

On the other hand, by virtue of (5.104) and (6.5), one has

(6.26)
$$\ker \tilde{\partial} = \ker \partial \cap \mathcal{A}^\cdot = \mathbb{C} \cap \mathcal{A}^\cdot = \mathbb{C}^\cdot,$$

so that one concludes that (6.5) entails (5.14). Accordingly, based now on Lemma 5.1, one infers that

(6.27)   the action of $H^1(X, S^1)$ on the Maxwell group of X, $\Phi_{\mathcal{A}}^1(X)^\nabla$, as before, is free.

Furthermore, one can consider the action of $H^1(X, S^1)$ restricted to a given light bundle

(6.28)
$$\Phi_{\mathcal{A}}^1(X)_R^\nabla,$$

as before (cf., for instance, (5.65)), and look further for the transitivity of the action of $H^1(X, S^1)$ on (6.28), under supplementary conditions on X, as was the case in the previous Section 5 (see Theorem 5.1).

However, based on our hypothesis for X, as in (6.1), one can actually look at Maxwell fields on X possessing Hermitian $\mathcal{A}$-connections (thus, one considers Hermitian Maxwell fields), as we explain presently below in the next subsection.

6.2 Hermitian Maxwell Fields

We first remark that as a consequence of our hypothesis in (6.1) and (6.4), one concludes that

(6.29)   every vector sheaf, hence in particular any line sheaf on X, admits a Hermitian $\mathcal{A}$-connection.

Indeed, the previous claim follows from our relevant assumptions, as indicated above, according to [VS: Chapt. VII; p. 174, Theorem 10.1]; see also loc. cit. Chapt. IV; p. 336, Definition 10.1. In this connection, we remark that to prove (6.29) we do not actually need to consider any conditions pertaining to sheaf exponential items (as they are involved in (6.1)).

In other words (see also Chapter I; (9.27), along with the ensuing comments, therein), any line sheaf $\mathcal{L}$ on X admits an $\mathcal{A}$-connection D; hence (cf. (1.4)), one actually has a Maxwell field

$$(6.30) \qquad\qquad (\mathcal{L}, D)$$

whose $\mathcal{A}$-connection D satisfies the relation

$$(6.31) \qquad\qquad \partial(\rho(s, t)) = \rho(Ds, t) + \rho(s, Dt),$$

for any s, t in $\mathcal{L}(U)$, with U open in X. Thus, (6.31) characterizes, by definition, D in (6.30) as a Hermitian $\mathcal{A}$-connection, while the corresponding pair $(\mathcal{L}, D)$, as above, is called a *Hermitian Maxwell field*.

Note 6.2 By an obvious abuse of notation, we have used for simplicity in (6.31) the same symbol ρ for the corresponding Hermitian $\mathcal{A}$-metric on $\mathcal{L}$, viz. for an $\mathcal{A}$-valued Hermitian inner product on it (see [VS: Chapt. IV; p. 330, Section 9, and Chapt. VII; p. 171, Section 10]; cf. Chapter I).

On the other hand, the possibility of providing any line sheaf and, more generally, any vector sheaf on X with a Hermitian $\mathcal{A}$-metric is, according to our hypothesis on X, as in (6.1) (see also, however, the relevant comments after (6.29) above), an outcome of [VS: Chapt. IV; p. 333, Theorem 9.1]. This settles entirely the present situation concerning (6.31). The above was our main motivation, pertaining to the framework described by our previous conditions in (6.6).

Henceforth, we denote by

$$(6.32) \qquad\qquad \Phi^1_{\mathcal{A}}(X)^{\nabla_{her}}$$

the set of equivalence classes of Maxwell fields on X that are endowed with a Hermitian $\mathcal{A}$-connection (cf. (6.29)), thus, by definition, a subset of the Maxwell group of X, $\Phi^1_{\mathcal{A}}(X)^{\nabla}$.

In this regard, we further note that

(6.33) a gauge equivalence between Maxwell fields (see (1.16.1)) preserves the hermiticity (viz. (6.31)) of a given $\mathcal{A}$-connection.

One proves that the set (6.32) is actually a subgroup of the Maxwell group of X, that is,

$$(6.34) \qquad\qquad \Phi^1_{\mathcal{A}}(X)^{\nabla_{her}} < \Phi^1_{\mathcal{A}}(X)^{\nabla}.$$

Here we used in anticipation (6.38′), along with [VS: Chapt. VIII; p. 233, (9.8)]. The same relation below has already been applied in (6.33). ∎

Scholium 6.1 Before we proceed further, we comment upon the previous relation (6.31), supplying another equivalent local formulation: Assume that we are given a Hermitian line sheaf

$$(6.35) \qquad\qquad (\mathcal{L}, \rho)$$

on a topological space X satisfying (6.1); viz. a line sheaf $\mathcal{L}$ on X endowed with a Hermitian $\mathcal{A}$-metric ρ. (In this connection, see also [VS: Chapt. IV; p. 333, Theorem 9.1].) Suppose that $\mathcal{L}$ is equipped with an $\mathcal{A}$-connection D (as follows from our hypothesis in (6.1); see [VS: Chapt. VI; p. 85, Theorem 16.1, together with Chapt. II; p. 247, (8.56)]. Assume that we are given a 0-cochain of 1-forms, say

$$(6.36) \qquad\qquad \theta \equiv (\theta_\alpha) \in C^0(\mathcal{U}, \Omega^1),$$

characterizing D "locally" (see Chapt. I; (2.45), for $n = 1$). So one concludes that

(6.37) the given $\mathcal{A}$-connection D on $\mathcal{L}$ as above is Hermitian (i.e., (6.31) is in force with respect to (6.35)), if and only if the following Ricci's identity holds; cf. (6.36):

$$(6.37.1) \qquad\qquad \theta + \bar\theta = \tilde\partial(\rho).$$

See [VS: Chapt. VII; p. 173, (10.18.1)]. In fact, (6.37.1) is valid more generally for any given Hermitian $\mathcal{A}$-module $(\mathcal{E}, \rho)$ on X (loc. cit.). One can employ the Gram–Schmidt orthonormalization process (cf. Chapt. I; Scholium 9.1) with respect to a given orthonormal frame of $\mathcal{L}$,

$$(6.38) \qquad\qquad \mathcal{U} = (U_\alpha)_{\alpha \in I}$$

(cf. also [VS: Chapt. IV; p. 337, (10.18)], applied locally). Thus, one obtains that

(6.38′) (6.37.1) is equivalent to the following relation:

$$(6.38'.1) \qquad\qquad \bar\theta = -\theta.$$

Indeed, one infers within the previous framework that

$$(6.39) \qquad\qquad \tilde\partial(\rho) = 0,$$

which proves our assertion in (6.38′), in view of (6.37.1); see M. Postnikov [1: p. 174, in particular, Proposition 4]. ∎

In this connection, we finally remark that more generally, (6.38′) is valid for any Hermitian vector sheaf $(\mathcal{E}, \rho)$ on X, as is also the case with (6.37).

Looking at the group action (6.22), taking also (6.34) into account, one gets the following free group action:

$$(6.40) \qquad\qquad H^1(X, S^1) \times \Phi^1_{\mathcal{A}}(X)^{\nabla_{her}} \to \Phi^1_{\mathcal{A}}(X)^{\nabla_{her}},$$

relativized one from (6.22) on the set (group) (6.32). Indeed, our claim is a consequence of (6.23), (6.38'), and (6.27). ∎

We next consider the previous action, as in (6.40), restricted in particular to a light bundle, as we did in Section 5, however, now to a light bundle consisting of Hermitian Maxwell fields (Hermitian light bundle).

6.3 Hermitian Light Bundles

By looking at the map τ, as in (3.12), we further consider, by virtue of (6.34), its restriction to the set (group) (6.32). Thus, having now the map

$$(6.41) \qquad \tau : \Phi^1_{\mathcal{A}}(X)^{\nabla_{her}} \to \Omega^2(X),$$

as given by (cf. also (3.13))

$$(6.42) \qquad \tau([(\mathcal{L}, D)]) := R(D) \equiv R,$$

where D in (6.42) stands for a Hermitian $\mathcal{A}$-connection on the line sheaf $\mathcal{L}$ (cf. (6.29)), one further defines a *Hermitian light bundle* on X by the relation

$$(6.43) \qquad \Phi^1_{\mathcal{A}}(X)^{\nabla_{her}}_R := \tau^{-1}(R),$$

whenever this has meaning; viz., in view of (6.41), we assume that

$$(6.44) \qquad R \in \Omega^2(X), \text{ in such a manner that it is the curvature of Hermitian Maxwell fields on } X.$$

Thus, a Hermitian light bundle, as in (6.43), is just, by definition, a fiber of the map τ in (6.41), as defined by (6.42), at a point of the set

$$(6.45) \qquad \operatorname{im} \tau \subseteq \Omega^2(X),$$

viz. of the image of τ in $\Omega^2(X)$.

On the other hand, we now give conditions on X guaranteeing that (6.44) is in force: Thus, according to Weil's integrality theorem (cf. Theorem 3.1), one has

$$(6.46) \qquad \frac{1}{2\pi i} R(D) \equiv \frac{1}{2\pi i} R \in H^2(X, \mathbb{Z})$$

(see (3.60)), such that

$$(6.47) \qquad R \in \Omega^2(X)^{int}_{cl},$$

while we still assume that

$$(6.48) \qquad X \text{ is a Bianchi–Weil space}$$

(cf. also (3.66)). So, apart from (6.6), we further suppose, in view of (6.48), that

(6.49)
$$X \text{ is a Bianchi space (cf. (3.17)) such that one has}$$

(6.49.1) $\ker d = \operatorname{im} \partial.$

As already remarked (6.49.1) together with the commutativity of the exponential sheaf triangle (cf. (5.109)) entails that

(6.50) $\ker d = \operatorname{im} \tilde{\partial}.$

The previous condition is what one really needs for our next main result. Thus, one gets the following structural information pertaining to a *Hermitian light bundle* on X.

Theorem 6.1 Suppose we are given a topological space X, satisfying (6.6) and (6.50). Then any given Hermitian light bundle on X,

(6.51) $\Phi^1_{\mathcal{A}}(X)^{\nabla_{her}}_R$

(cf. (6.43)), is an affine space with structure group $H^1(X, S^1)$. That is, one concludes that

(6.52)   the group $H^1(X, S^1)$ provides a free and transitive action on the set (6.51) (cf. also (6.53) below).

Proof. First, based on (6.23), we see that one can restrict the action of the group $H^1(X, S^1)$, as in (6.40), to the set (6.51), so that one then gets the following (relative) group action:

(6.53) $H^1(X, S^1) \times \Phi^1_{\mathcal{A}}(X)^{\nabla_{her}}_R \to \Phi^1_{\mathcal{A}}(X)^{\nabla_{her}}_R.$

Therefore, since (6.40) refers to a free action, one gets that the above action (6.53) is free as well. (In this connection, we remark that (6.50) was not applied in the previous argument; cf. also Lemma 5.1). So it remains to prove that

(6.54)   the group action (6.53) is transitive; equivalently, the set (6.51), viz. a light bundle on X, with a given field strength (color), that is, curvature R, is a homogeneous space with respect to the group $H^1(X, S^1)$, or simply an $H^1(X, S^1)$-homogeneous space.

Thus, taking two elements of the set (6.51), i.e., two light rays of the same color (cf. also (1.19)), hence two Maxwell fields with a given field strength (curvature) R, see e.g. (6.93), and by applying also a local description of the fields concerned (cf., for instance, (2.26)), we know (see Theorem 5.1) that there exists a 1-cocycle

(6.55) $\lambda \equiv (\lambda_{\alpha\beta}) \in Z^1(\mathcal{U}, \mathbb{C}^{\bullet})$

such that

(6.56) $$\lambda_{\alpha\beta} := g''_{\alpha\beta} \cdot g^{-1}_{\alpha\beta}$$

(cf. (5.98), (5.99)). One thus obtains that

(6.57) $$[(\mathcal{L}', D')] = \lambda \cdot [(\mathcal{L}, D])$$

(see also (5.102)). On the other hand, by virtue of our hypothesis on X and (6.12), we may assume that we are dealing with Hermitian Maxwell fields, so that in view of (6.19), (6.16), in conjunction with (6.56), one obtains that

(6.58) $$|\lambda_{\alpha\beta}| = 1;$$

that is, finally, one has

(6.59) $$(\lambda_{\alpha\beta}) \in Z^1(\mathcal{U}, S^1),$$

so that (by an obvious and usual abuse of notation concerning (6.55)) one gets

(6.60) $$\lambda \equiv [(\lambda_{\alpha\beta})] \in H^1(X, S^1),$$

which by virtue of (6.57) settles what we wanted to prove in (6.54), and this also finishes the proof of the theorem. ■

In this connection, we also refer to Note 5.5 for the terminology we applied in the statement of Theorem 6.1. We refer to the same note for the terminology employed in our next result, pertaining to a reformulation of Theorem 6.1, according to (5.91). One gets the following equivalent form of (6.52). Namely,

(6.61)

suppose we have a topological space X satisfying (6.6) and (6.50). Then, any Hermitian light bundle

(6.61.1) $$\Phi^1_{\mathcal{A}}(Z)^{\nabla_{her}}_R$$

is a principal homogeneous $H^1(X, S^1)$-space.

On the other hand, keeping the same hypothesis for X as in (6.61), and following the terminology applied in (5.111), one obtains the subsequent equivalent formulation of Theorem 6.1, hence of (6.61) too. Thus, we have the following relation:

(6.62) $$\Phi^1_{\mathcal{A}}(X)^{\nabla_{her}}_R = H^1(X, S^1) \cdot [(\mathcal{L}, D)],$$

within a bijection of the corresponding sets, and for any Hermitian Maxwell field $[(\mathcal{L}, D)]$ having

(6.63) $$R(D) = R,$$

as in (6.62) (see also (6.46)).

Furthermore, analogous statements with those in (5.115) and (5.116) are still in force within the same framework as that in (6.61). Thus, (6.62) is equivalent to the bijection

$$(6.64) \qquad \Phi^1_{\mathcal{A}}(X)^{\nabla_{her}}_R = H^1(X, S^1),$$

defined by any given Hermitian Maxwell field, as in (6.63).

We come finally, in the next subsection, to a particular important case of the above, namely, to that in which the space X is path-connected.

6.4 Hermitian Light Bundles over Path-Connected Spaces

As already mentioned, our aim in this final subsection is to consider a special case of (6.61), where the space X involved is also path-connected. However, we have first to cite the following general result, which will be of use presently.

Lemma 6.1 Suppose we are given a path-connected topological space X and a group G. Then, one has the following bijection of the sets concerned:

$$(6.65) \qquad H^1(X, G) = Hom(\pi_1(X), G)/G.$$

Proof. See, for instance, R.C. Gunning [1: p. 186, Lemma 27]. Concerning the first member of (6.65), see also [VS: Chapt. III; p. 274, Definition 11.1] for the particular case considered here, namely of a constant group sheaf G. In this connection, cf. loc. cit., p. 275; (11.26.1), along with Chapt. II; p. 90, Example 1.1. ∎

For convenience, we further comment on the notation applied in (6.65): We have already referred to the 1st cohomology set appearing in (6.65) (see the previous proof). On the other hand, the second member of (6.65) concerns the set of group morphisms of $\pi_1(X)$ into G, i.e., of the fundamental (or Poincaré) group of X,

$$(6.66) \qquad \pi_1(X),$$

into an arbitrary group G, modulo the equivalence relation defined on the latter set by G; that is, one looks at the same set

$$(6.67) \qquad Hom(\pi_1(X), G)$$

as a G-set, that is, one defines an action of the group G on it,

$$(6.68) \qquad G \times Hom(\pi_1(X), G) \to Hom(\pi_1(X), G),$$

given by the relation

$$(6.69) \qquad a \cdot \phi := i_a \circ \phi \equiv Ad(a) \circ \phi,$$

for any $a \in G$ and $\phi \in Hom(\pi_1(X), G)$, where i_a stands for the *inner automorphism* of G defined by $a \in G$. Equivalently, one has

$$(6.70) \qquad \begin{aligned} (a \cdot \phi)(\gamma) &\equiv (i_a \circ \phi)(\gamma) = i_a(\phi(\gamma)) := a\phi(\gamma)a^{-1} \\ &\equiv Ad(a) \cdot \phi(\gamma) = (Ad(a) \circ \phi)(\gamma), \end{aligned}$$

for any $\gamma \in \pi_1(X)$, which also explains the notation in (6.69).

The second member of (6.65) stands for the *set of equivalence classes* in (6.67) defined by the above group action, or equivalently, the space of orbits (orbit space) of the action. (In this regard, cf. also Scholium 6.2.)

In the particular case of an abelian group G, since

$$(6.71) \qquad i_a = id_G, \quad \text{for any } a \in G,$$

one gets, in view of (6.69), that (6.65) is reduced to the relation

$$(6.72) \qquad H^1(X, G) = Hom(\pi_1(X), G),$$

within an (abelian) group isomorphism. (See also [VS: Chapt. III; p. 273, Remark 11.1], concerning now the (abelian) 1st cohomology group in the first member of (6.72).) For the important special case of the above, that $G = S^1$ (circle group), one gets from (6.72)

$$(6.73) \qquad H^1(X, S^1) = Hom(\pi_1(X), S^1) \equiv (\pi_1(X))^* \equiv \pi_1(X)^*,$$

within an isomorphism of (abelian) groups, where

$$(6.74) \qquad \pi_1(X)^*$$

stands by definition for the *character group* of $\pi_1(X)$.

We note (cf. Lemma 6.1) that (6.72) and (6.73) are in force for X a path-connected topological space.

Our previous conclusion in (6.61) reads thus:

(6.75)

> Suppose we are given a path-connected space X that also satisfies (6.6) and (6.50). Then every Hermitian light bundle on X,
>
> $$(6.75.1) \qquad \Phi^1_{\mathcal{A}}(X)_R^{\nabla_{her}},$$
>
> is a principal homogeneous $\pi_1(X)^*$-space.

On the other hand, by still employing the terminology of (6.62), one concludes that the same light bundle is given by the relation

$$(6.76) \qquad \Phi^1_{\mathcal{A}}(X)_R^{\nabla_{her}} = \pi_1(X)^* \cdot [(\mathcal{L}, D)],$$

within a bijection of the sets involved. Equivalently, one has

$$(6.77) \qquad \Phi^1_{\mathcal{A}}(X)_R^{\nabla_{her}} = \pi_1(X)^*,$$

up to a bijection of the corresponding sets, established by any given Hermitian Maxwell field $[(\mathcal{L}, D)]$ having $R(D) = R$ as in (6.77).

The preceding will find a particular application in Chapter V, pertaining to a parametrization (classification) of the so-called *geometric prequantizations* of the space X.

As a consequence of (6.77) (or (6.64)), we still remark that

(6.78)   all the (Hermitian) light bundles on X are of the same capacity (cardinality, number of light rays $[(\mathcal{L}, D)]$), independently of the color (viz. of the particular curvature R). (See also (5.115).)

We end up with the ensuing scholium, referring to further standard interpretations of the first member of (6.64), hence to its physical counterpart as well (cf. (6.64) or (6.77)).

Scholium 6.2 By looking further at (6.65), we recall that classically, its first member, viz. the 1st cohomology set,

(6.79) $H^1(X, G),$

classifies the so-called *principal G-bundles* over X when appropriately specialized concerning the space X and the group G; see, for instance, N. Steenrod [1: p. 66, Classification theorem 13.9]: Indeed, any group G in the discrete topology (thus, a totally disconnected space) can be considered, with X a suitable path-connected space (loc. cit.). See also R.C. Gunning [1: p. 189, Remark].

In view of the preceding relations, e.g., in (6.64), (6.77), or (5.115), one gets here too a corresponding physical interpretation of the above important notion of algebraic topology, as in (6.79). The significance, in either discipline, of the relevant notions here are indeed very indicative.

On the other hand, within the same vein of ideas, pertaining in particular to algebraic topology, cf. also R.M. Switzer [1: p. 191ff]. Concerning another point of view of the same notion viz. of the 1st cohomology set in terms of topological algebras and vector sheaves on their (global) spectra (thus, a potential relevance of all the preceding disciplines), see A. Mallios [5: p. 412, Theorem 2.1], together with [VS: Chapt. XI; p. 431, (8.12) and p. 344, Theorem 8.2].

Finally, the same set as above (cf. (6.79)), however with $G \equiv \mathcal{G}$ now a sheaf of groups (nonabelian in general) on X has been employed by Yu.I. Manin [1: p. 117ff] in studying extensions (of analytic spaces) and obstructions thereof, through cohomological methods. An analogous treatment in the case of $\mathcal{A}$-modules, with applications to the problem of existence of $\mathcal{A}$-connections, has been given in [VS: Chapt. III; p. 260, Section 10. and Chapt. V, VII]. Elements of the set

(6.80) $H^1(X, \mathcal{G})$

(or their corresponding cocycles) are also called $\mathcal{G}$-*torsors*; cf. Yu.I. Manin (loc. cit.).

We continue further in the next two sections to consider other aspects of the same action of the group $H^1(X, \mathbb{C}^{\bullet})$ as before (see Section 5), along with their Hermitian counterparts, which will also be of use in our subsequent discussion.

7 Equivariant Actions of H^1 $(X, \mathbb{C}^\cdot)$ (Continued)

As already pointed out (see (5.8) and (5.10)), the abelian group

(7.1) $$H^1(X, \mathbb{C}^\cdot)$$

acts on the abelian group

(7.2) $$H^1(X, \mathcal{A}^\cdot)$$

(or the so-called *Picard group* of X, (cf. (2.3)), as well as on the corresponding Maxwell group of X,

(7.3) $$\Phi^1_{\mathcal{A}}(X)^\nabla,$$

also an abelian group (cf. Theorem 2.1). Indeed, the action of (7.1) on the latter group is the relative action of the first one on (7.3) (see (5.9), (5.10)).

According to what we have already seen in previous sections of the present chapter,

(7.4)
> the above are valid, for any $\mathbb{C}$-algebraized space
>
> (7.4.1) $(X, \mathcal{A})$
>
> (cf. e.g. (5.3.1)) on which there is given a basic differential triad (cf. (0.1)),
>
> (7.4.2) $(\mathcal{A}, \partial, \Omega^1)$.

Thus, our aim in by the subsequent discussion is virtually to show that

(7.5)   the previous two actions of $H^1(X, \mathbb{C}^\cdot)$ are (canonically) interrelated.

Keeping in mind (7.4) assume further that one has the relation

(7.6) $$H^1(X, \mathcal{A}^\cdot) = H^2(X, \mathbb{Z}),$$

within an isomorphism of the (abelian) groups involved. It concerns here the so-called *Chern isomorphism*, which we have already considered. (See (3.45) in conjunction with (3.26), as well as Scholium 3.1 pertaining to a physical interpretation of (7.6)). We recall here that one gets (7.6) by taking, for instance,

(7.7)
> a sheaf exponential $\mathbb{C}$-algebraized space
>
> (7.7.1) $(X, \mathcal{A})$
>
> (cf. Chapt. I; (1.4)), while X is also a paracompact (Hausdorff) space and $\mathcal{A}$ a fine ($\mathbb{C}$-algebra) sheaf on X. (So we can look, e.g., at a Weil space see (3.26).)

Accordingly, by virtue of the above identification (7.6) and of what also has been said in the beginning of this section, one obtains

(7.8)   a simultaneous action of $H^1(X, \mathbb{C}^{\cdot})$ on the (abelian) groups $H^2(X, \mathbb{Z})$ (cf. (7.6)) and $\Phi^1_{\mathcal{A}}(X)^{\nabla}$.

The corresponding group spaces of the previous actions of $H^1(X, \mathbb{C}^{\cdot})$, viz. the spaces (groups for the case at hand) on which $H^1(X, \mathbb{C}^{\cdot})$ acts, are interconnected through the following diagram:

(7.9)

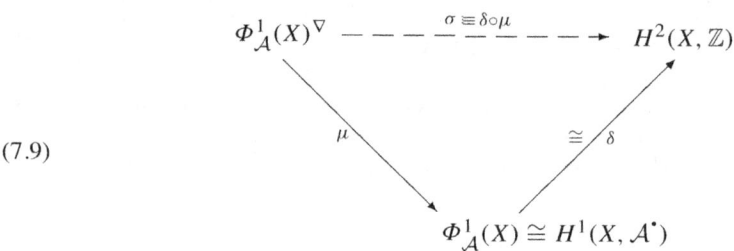

(concerning the (group) isomorphisms indicated in the previous diagram, see (2.3) and (7.6)). Hence by definition, (7.9) is a commutative diagram. Furthermore,

(7.10) $$\mu : \Phi^1_{\mathcal{A}}(X)^{\nabla} \to \Phi^1_{\mathcal{A}}(X)$$

in (7.9) stands for the (canonical) *forgetful map*; that is, one sets

(7.11) $$\mu([(\mathcal{L}, D)]) := [\mathcal{L}]$$

for any $[(\mathcal{L}, D)] \in \Phi^1_{\mathcal{A}}(X)^{\nabla}$, the same being a well-defined map according to (2.45) in conjunction with (2.48), (2.49); in this connection, see also [VS: Chapt. V; p. 353, Lemma 2.1].

In other words, and still employing physical language, one concludes that

to every Maxwell field (light ray)

(7.12.1) $$[(\mathcal{L}, D)] \in \Phi^1_{\mathcal{A}}(X)^{\nabla}$$

(7.12)   (see (1.19)) one can associate, unambiguously, its carrier, viz. the corresponding line sheaf (photon)

(7.12.2) $$\mathcal{L} \equiv [\mathcal{L}] \in \Phi^1_{\mathcal{A}}(X)$$

on X.

See also Definition 1.1, together with (3.55) in the foregoing.

On the other hand, the map δ in (7.9) denotes, as already mentioned, the Chern (abelian group) isomorphism (see (7.6)). Thus, our aim now is to prove that

the map

(7.13)   (7.13.1) $\sigma := \delta \circ \mu,$

as given by (7.9), is equivariant, alias a $H^1(X, \mathbb{C}^\cdot)$-map.

Equivalentl, this means that one has the relation

(7.14) $\sigma \circ \tau_\alpha = \tau'_\alpha \circ \sigma$

for any clement (cohomology class) $\alpha \in H^1(X, \mathbb{C}^\cdot)$. That is,

(7.15)   the map σ (cf. (7.13.1)) commutes with the two actions of $H^1(X, \mathbb{C}^\cdot)$, as indicated in (7.8).

The above can also be depicted by the following commutative (to be proved, cf. (7.2) below) diagram:

(7.16)

$$
\begin{array}{ccc}
\Phi^1_{\mathcal{A}}(X)^\nabla & \xrightarrow{\quad \sigma \quad} & H^2(X, \mathbb{Z}) \\[2em]
\Big\downarrow{\tau_\alpha} & & \Big\downarrow{\tau'_\alpha} \\[2em]
\Phi^1_{\mathcal{A}}(X)^\nabla & \xrightarrow{\quad \sigma \quad} & H^2(X, \mathbb{Z}).
\end{array}
$$

Concerning the terminology employed above, we also refer to P. Tondeur [1: p. 7, Section 1.2], pertaining to the general theory of transformation groups. Before we proceed to the proof of (7.13), we first explain the relevant terminology as used in the foregoing: Namely, for any cohomology class

(7.17) $\alpha \in H^1(X, \mathbb{C}^\cdot),$

we denote by

(7.18) $\tau_\alpha : \Phi^1_{\mathcal{A}}(X)^\nabla \to \Phi^1_{\mathcal{A}}(X)^\nabla$

the corresponding (group) automorphism of the Maxwell group, as in (7.18), defined by the action on the latter group of the group $H^1(X, \mathbb{C}^\cdot)$ (ibid.), according to (5.10), so that one has

(7.19) $\tau_\alpha([(\mathcal{L}, D)]) := \alpha \cdot [(\mathcal{L}, D)] := [(\alpha \cdot \mathcal{L}, D)] \in \Phi^1_{\mathcal{A}}(X)^\nabla,$

for any $[(\mathcal{L}, D)] \in \Phi^1_{\mathcal{A}}(X)^\nabla$ (cf. (5.10.2)).

On the other hand, in view of (5.8), one has an action of $H^1(X, \mathbb{C}^{\boldsymbol{\cdot}})$ on $H^1(X, \mathcal{A}^{\boldsymbol{\cdot}})$. However, by virtue of the Chern isomorphism (7.6), we can transfer the same action as before to the (abelian) group $H^2(X, \mathbb{Z})$: That is for any element $\alpha \in H^1(X, \mathbb{C}^{\boldsymbol{\cdot}})$, one defines a (group) automorphism of $H^2(X, \mathbb{Z})$,

$$(7.20) \qquad\qquad \tau'_\alpha : H^2(X, \mathbb{Z}) \longrightarrow H^2(X, \mathbb{Z}),$$

by the relation

$$(7.21) \qquad\qquad \tau'_\alpha(z) := z + \delta(\alpha) \in H^2(X, \mathbb{Z}),$$

for any $z \in H^2(X, \mathbb{Z})$, where

$$(7.22) \qquad\qquad \delta : H^1(X, \mathbb{C}^{\boldsymbol{\cdot}}) \longrightarrow H^2(X, \mathbb{Z}),$$

as in (7.21), stands for the corresponding Bockstein operator (cf. (7.24)). That is, one looks first at the (classical exponential) short exact sequence (of constant sheaves)

$$(7.23) \qquad 0 \longrightarrow \mathbb{Z} \xrightarrow{\ \varepsilon\ } \mathbb{C} \xrightarrow{\ \exp \equiv e\ } \mathbb{C}^{\boldsymbol{\cdot}} \longrightarrow 1.$$

Therefore, since, according to our hypothesis (cf. (7.7)) X is a (Hausdorff) paracompact space, one gets the corresponding long exact sequence (of abelian groups) in cohomology

$$(7.24) \qquad \cdots \longrightarrow H^1(X, \mathbb{C}) \longrightarrow H^1(X, \mathbb{C}^{\boldsymbol{\cdot}}) \xrightarrow{\ \delta\ } H^2(X, \mathbb{Z}) \longrightarrow \cdots,$$

thus in particular at the map δ as in (7.22) (the same map being by definition a morphism of the groups concerned; in this connection see also [VS: Chapt. III; p. 207, Theorem 5.3, along with p. 234, Theorem 8.1]). Thus, the preceding now explains completely our notation in the (7.21). [We note here that the analogous Bockstein operator that is associated with the general short exact exponential sheaf sequence, as in (3.50), (cf. (3.51)), becomes the Chern isomorphism, as in (7.6), under suitable conditions for $(X, \mathcal{A})$; see (3.26) in conjunction with Scholium 3.1.]

Based on (7.14), or on (7.16), we also express these two equivalent statements, through another equivalent one by saying that

> the above (group) automorphisms τ_α and τ'_α, as defined by (7.19) and (7.21), respectively, are σ-related for any element (cohomology class)

(7.25) $(7.25.1)$ $\alpha \in H^1(X, \mathbb{C}^{\boldsymbol{\cdot}}).$

> Thus the above is an equivalent way (by definition) to (7.13), of saying that σ, as defined by (7.9), is an equivariant map, or a $H^1(X, \mathbb{C}^{\boldsymbol{\cdot}})$-map.

Thus, based now on the definitions as given, we come to the proof of (7.13), in fact of (7.14). That is, we have

$$
\begin{aligned}
(\sigma \circ \tau_{\alpha})([(\mathcal{L}, D)]) &= \sigma(\tau_{\alpha}([(\mathcal{L}, D)]) = \sigma(a \cdot [(\mathcal{L}, D)]) \\
&= \sigma([(a \cdot \mathcal{L}, D)]) = (\delta \circ \mu)([(a \cdot \mathcal{L}, D)]) \\
&= \delta(\mu([(a \cdot \mathcal{L}, D)]) = \delta([a \cdot \mathcal{L}]) = \delta(a \cdot [\mathcal{L}]) \\
&= \delta(a) + \delta([\mathcal{L}]) = \tau'_{\alpha}(\delta([\mathcal{L}])) = \tau'_{\alpha}(\delta(\mu([(\mathcal{L}, D)]))) \\
&= (\tau'_{\alpha} \circ \sigma)([(\mathcal{L}, D)]),
\end{aligned}
$$

(7.26)

for any Maxwell field $[(\mathcal{L}, D)]$, which thus proves (7.14), or, equivalently, the commutativity of the diagram (7.16). ∎

To recapitulate our discussion thus far, we have proved that

(7.27) $H^1(X, \mathbb{C}^{\bullet})$ entails an equivariant action on both the Maxwell group of X, $\Phi^1_{\mathcal{A}}(X)^{\nabla}$, as well as on the group $H^2(X, \mathbb{Z})$, this being defined by the equivariant map σ, or $H^1(X, \mathbb{C}^{\bullet})$-map, as given by (7.9). Thus, equivalently, one gets the commutative diagram (7.16) (for any $\alpha \in H^1(X, \mathbb{C}^{\bullet})$). [We recall that X is assumed to satisfy (7.7).]

Our next objective is to relate the map σ, as above, with the map τ that we already know (cf. (3.12)). This will also be of use in the Section 8. Thus, we are going to prove that

(7.28)   the maps σ and τ are (essentially) the same (viz. modulo a phase factor). That is,

(7.28.1) $\tau = 2\pi i \cdot \sigma.$

To verify (7.28.1), we have first to resort to certain appropriate identifications (in effect, group isomorphisms) that are provided by the preceding discussion. To start with, by looking at the map σ, one has

(7.29) $\sigma([(\mathcal{L}, D)]) = \delta(\mu([(\mathcal{L}, D)])) = \delta([\mathcal{L}])$
         $= \delta([(g_{\alpha\beta})]) \equiv [(g_{\alpha\beta})] \in H^2(X, \mathbb{Z}),$

for any $[(\mathcal{L}, D)] \in \Phi^1_{\mathcal{A}}(X)^{\nabla}$ (cf. (7.9)), where apart from the Chern isomorphism δ (see (7.6) and (7.9)) in the last relation of (7.29), we have employed the following identification (Picard isomorphism, see (2.2) or (2.3) and (3.55.1)) for the map μ (cf. also (7.11)):

(7.30) $\mu([(\mathcal{L}, D)]) = [\mathcal{L}] \equiv [(g_{\alpha\beta})] \in H^1(X, \mathcal{A}^{\bullet}) \underset{\delta}{\cong} H^2(X, \mathbb{Z}).$

One can further look at (7.29) as a concrete evaluation of (7.9), which will also be applied presently in proving (7.28.1).

By taking further the map τ, one gets (cf. also (3.55.2) and (3.59), as well as, (3.13))

(7.31) $\tau([(\mathcal{L}, D)]) = R(D) \equiv R \leftrightarrow [R] = 2\pi i \cdot [(g_{\alpha\beta})] \in H^2(X, \mathbb{C}).$

Therefore, in view of (7.29), one finally has

$$(7.32) \qquad \tau([(\mathcal{L}, D)]) = [R] = 2\pi i \cdot [(g_{\alpha\beta})] = 2\pi i \cdot \sigma([(\mathcal{L}, D)]),$$

for any Maxwell field $[(\mathcal{L}, D)]$, which is, of course, the desired (7.28.1). ∎ In other words, one can remark here that

(7.33)   (7.28.1) is just another formulation of (3.57.1), in terms of the maps σ and τ. Moreover, both of these relations are essentially a consequence of the Chern isomorphism, which thus acquires an especially interesting physical significance as well. (Cf. the same remark (3.57), as well as the relevant comments before it, along also with Scholium 3.1.)

By looking at (7.28.1), and in conjunction with the map τ and its kernel $\ker\tau$, one concludes that

(7.34)   the map σ along with the map τ becomes one-to-one onto the set (in fact, abelian group)

$$(7.34.1) \qquad\qquad \Phi^1_{\mathcal{A}}(X)^\nabla / \ker \tau.$$

Accordingly, in that sense, and also by taking (7.29) into account, one can look at the map σ as another interpretation of the Chern isomorphism concerning the restriction of the latter to the Maxwell group of X, $\Phi^1_{\mathcal{A}}(X)^\nabla$, the same isomorphism being, in principle, defined on the Picard group of X,

$$(7.35) \qquad\qquad \Phi^1_{\mathcal{A}}(X) \cong H^1(X, \mathcal{A}^\cdot)$$

(cf. (2.2), (2.3), and (2.4)). Further comments on (7.34.1) will be made in Section 7.(a), where we identify the (abelian) group $\ker\tau$ (cf. (4.3)), and also continued in Section 8.

Before we come to the following subsection, we remark that our previous calculations in (7.31) were actually rooted in (3.84). Hence, by virtue of (7.7), one further concludes, concerning the proper setting of what has been said, that

(7.36)   an appropriate framework for (7.28.1) is an exact Bianchi–Weil space (see (3.83), along with (3.26)).

7.1 The Kernel of the Map τ

By looking at the map τ, as in (3.12), (3.20), and (7.31), one gets the following diagram:

(7.37)

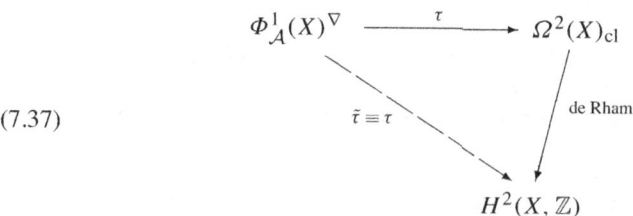

so that by virtue of (7.28.1) and (7.29), one finally obtains the map

$$(7.38) \qquad \frac{1}{2\pi i}\tilde{\tau} \equiv \frac{1}{2\pi i}\tau = \sigma,$$

having values in $H^2(X, \mathbb{Z})$, viz., by definition the cohomology class of the curvature (field strength) of the Maxwell field $[(\mathcal{L}, D)]$, or the corresponding Chern (characteristic) class of $\mathcal{L}$ (modulo a sign; see also Note 3.2, along with (3.55.2)). [The previous argument is based on Weil's integrality theorem (see Theorem 3.1 in the preceding). Thus, our framework is still as described by (7.36).]

On the other hand, by further considering the image of τ as given by (3.68.1), one has

$$(7.39) \qquad \operatorname{im} \tau \equiv \tau(\Phi^1_{\mathcal{A}}(X)^\nabla) = \Omega^2(X)^{\mathrm{int}}_{\mathrm{cl}} \lhd \Omega^2(X)_{\mathrm{cl}}$$

(see also (4.11), as well as the subsequent comments). So in view of the short exact sequence of (abelian) groups

$$(7.40) \qquad 1 \longrightarrow \ker \tau \longrightarrow \Phi^1_{\mathcal{A}}(X)^\nabla \xrightarrow{\tau} \operatorname{im} \tau \longrightarrow 0,$$

one obtains (cf. also (7.34))

$$(7.41) \qquad \operatorname{im} \tau = \Omega^2(X)^{\mathrm{int}}_{\mathrm{cl}} = \Phi^1_{\mathcal{A}}(X)^\nabla / \ker \tau,$$

within an isomorphism of the (abelian) groups concerned, regarding the second equality in (7.40). [That is, (7.39) is split exact, according to the first isomorphism theorem; cf., for instance, J.J. Rotman [1: p. 21, Theorem 2.12].]

Furthermore, we note for later use that an equivalent version of (7.41) (viz. of the splitting of (7.40)) is given by the relations

$$(7.42) \qquad \Phi^1_{\mathcal{A}}(X)^\nabla = \operatorname{im} \tau \cdot \ker \tau = \ker \tau \cdot \operatorname{im} \tau,$$

modulo isomorphisms of groups. One can consider above, in place of $\operatorname{im} \tau$, an isomorphic image of it by means of a section of τ in (7.40), as, for instance, that provided through Weil's theorem; viz., by means of the "if" part of it in the sense that (cf. Theorem 3.1, along with Note 3.2)

$$(7.43) \qquad \begin{array}{l} \text{the (only) integral closed 2-forms on } X \text{ are the Chern classes of carriers} \\ \text{(line sheaves) of Maxwell fields on } X. \end{array}$$

(In this connection, see thus the proof of the second (viz. the "if") part of the aforesaid theorem, starting with (3.44); see Section 3.(d), along with Remark 3.1(ii).)

We come now to our main objective in this subsection, that is, to show that

$$(7.44) \qquad \ker \tau = H^1(X, \mathbb{C}^{\cdot}),$$

modulo an isomorphism of the (abelian) groups involved. Thus, taking in anticipation the validity of (7.44), one obtains, in view of (7.41), that

$$(7.45) \qquad \operatorname{im} \tau = \Omega^2(X)_{\mathrm{cl}}^{\mathrm{int}} = \Phi_{\mathcal{A}}^1(X)^{\nabla}/H^1(X, \mathbb{C}^{\cdot}),$$

the last equality being valid up to an isomorphism of the (abelian) groups considered. Furthermore, we can supplement our last information in (7.44) by remarking that

$$(7.46) \qquad \operatorname{im} \tau = H^2(X, \mathbb{Z}).$$

Indeed, the last relation can be construed, as already explained (see Section 3.(d)), as another version of Weil's integrality theorem (see (7.42) or Theorem 3.1). Thus (7.41) now takes the form

$$(7.47) \qquad \Phi_{\mathcal{A}}^1(X)^{\nabla} = H^2(X, \mathbb{Z}) \cdot H^1(X, \mathbb{C}^{\cdot}),$$

or, by virtue of (7.44), one obtains

$$(7.48) \qquad H^2(X, \mathbb{Z}) = \operatorname{im} \tau = \Omega^2(X)_{\mathrm{cl}}^{\mathrm{int}} = \Phi_{\mathcal{A}}^1(X)^{\nabla}/H^1(X, \mathbb{C}^{\cdot}),$$

modulo the pertinent (abelian) group isomorphisms, whenever needed. So one gets, through (7.47), a complete cohomological description of the Maxwell group of X, $\Phi_{\mathcal{A}}^1(X)^{\nabla}$. This, in anticipation of a cohomological classification of the latter group that will be given in the sequel, supplying a more intrinsic description in terms of Maxwell's equations (in vacuo), formulated within our abstract setting (see the next Chapter IV; Sections 5, 6). Finally, we can further supplement (7.48) by resorting to the Chern (along with the Picard) isomorphism (cf. (7.6) and (2.3) respectively), so that one obtains

$$(7.49) \qquad \begin{aligned} \Phi_{\mathcal{A}}^1(X) &= H^1(X, \mathcal{A}^{\cdot}) = H^2(X, \mathbb{Z}) = \operatorname{im} \tau \\ &= \Omega^2(X)_{\mathrm{cl}}^{\mathrm{int}} = \Phi_{\mathcal{A}}^1(X)^{\nabla}/H^1(X, \mathbb{C}^{\cdot}), \end{aligned}$$

modulo the standard (abelian) group isomorphisms. In that sense, (7.47) can be written in the form of the following isomorphism (of the abelian groups concerned):

$$(7.50) \qquad \Phi_{\mathcal{A}}^1(X)^{\nabla} = H^1(X, \mathcal{A}^{\cdot}) \cdot H^1(X, \mathbb{C}^{\cdot}),$$

the second member of (7.50) denoting the direct product of (abelian) groups. More on the preceding relations will be considered in Section 8, pertaining to further useful interpretations of the same relations.

We come now to the proof of (7.44). We first prove the following (abelian) group isomorphism (into):

$$(7.51) \qquad H^1(X, \mathbb{C}^{\cdot}) \underset{\nu}{\subseteq} \Phi_{\mathcal{A}}^1(X)^{\nabla},$$

as given by the map

$$(7.52) \qquad \nu : \lambda \equiv [(\lambda_{\alpha\beta})] \longmapsto \nu(\lambda) := [(\lambda, 0)],$$

where we set (cf. also (2.26))

(7.53) $$(\lambda, 0) \equiv ((\lambda_{\alpha\beta}), (\theta_\alpha \equiv 0)).$$

Since we assume (cf. (7.51)) that

(7.54) $$(\lambda_{\alpha\beta}) \in Z^1(\mathcal{U}, \mathbb{C}^{\cdot}),$$

we obtain the relation

(7.55) $$\delta(\theta_\alpha) = 0 = \tilde{\partial}(\lambda_{\alpha\beta})$$

(see also Chapter I; (1.34)), so that the map ν, as above, is well defined (cf. Lemma 2.1). Of course, the same map is a group morphism according to the definition (see (7.51), along with [VS: Chapt. V; p. 367, (4.16)]); we finally prove that ν is one-to-one. Supposing that

(7.56) $$[(\lambda, 0)] = [(\lambda', 0)],$$

so that by definition one has (cf. (1.16))

(7.57) $$(\lambda, 0) \sim (\lambda', 0),$$

one obtains (Lemma 2.2, in particular, (2.45))

(7.58) $$\lambda'_{\alpha\beta} = \delta(s_\alpha^{-1}) \cdot \lambda_{\alpha\beta},$$

for some 0-cochain

(7.59) $$(s_\alpha) \in C^0(\mathcal{U}, \mathbb{C}^{\cdot}) \underset{\longrightarrow}{\subset} C^0(\mathcal{U}, \mathcal{A}^{\cdot}).$$

In this connection, by virtue of (7.52), (7.56), and (2.46), one concludes that

(7.60) $$\tilde{\partial}(s_\alpha^{-1}) = 0;$$

that is, one finally obtains that (see also (7.36))

(7.61) $$(s_\alpha) \in C^0(\mathcal{U}, \ker \tilde{\partial} = \mathbb{C}^{\cdot}),$$

which justifies (7.59). Therefore, one has, by (7.58),

(7.62) $$\lambda'_{\alpha\beta} \cdot \lambda_{\alpha\beta}^{-1} = \delta(s_\alpha^{-1}) \in \delta(C^0(\mathcal{U}, \mathbb{C}^{\cdot})) \equiv B^1(\mathcal{U}, \mathbb{C}^{\cdot}),$$

so that one gets

(7.63) $$\lambda \equiv [(\lambda_{\alpha\beta} = [(\lambda'_{\alpha\beta})] \equiv \lambda',$$

which also proves our claim about the injectivity of ν. As a result of the preceding, one thus obtains the following sequence of (abelian) groups:

(7.64) $$0 \longrightarrow H^1(X, \mathbb{C}^{\cdot}) \overset{\nu}{\longrightarrow} \Phi^1_{\mathcal{A}}(X)^\nabla \overset{\tau}{\longrightarrow} \operatorname{im} \tau \longrightarrow 0,$$

being also exact at $H^1(X, \mathbb{C}^{\cdot})$ (injectivity of the map v) as just proved, while we have

(7.65) $H^1(X, \mathbb{C}^{\cdot}) \cong \operatorname{im} v \subseteq \ker \tau,$

as follows straightforwardly from the definitions of the maps v and τ (cf. (7.52) and (3.13), respectively). Indeed, one actually proves the equality in the last relation of (7.65), so that one concludes that

(7.66)   (7.64) is a short exact sequence of (abelian) groups, in fact, split exact by definition.

Consider an element in $\ker \tau$, viz. a Maxwell field

(7.67) $[(\mathcal{L}, D)] \in \Phi^1_{\mathcal{A}}(X)^{\nabla},$

such that (cf. (3.13))

(7.68) $\tau([(\mathcal{L}, D)]) := R(D) \equiv R = 0,$

so that one obtains

(7.69) $[(\mathcal{L}, D)] \in \Phi^1_{\mathcal{A}}(X)^{\nabla}_{R=0} \approx \tau^{-1}(0) \equiv \ker \tau$

(see also (3.71)). Therefore (see (5.111)), one concludes that there exists a (uniquely defined) cohomology class, element of $H^1(X, \mathbb{C}^{\cdot})$, say

(7.70) $\lambda \equiv [(\lambda_{\alpha\beta})] \in H^1(X, \mathbb{C}^{\cdot}),$

such that

(7.71) $[(\mathcal{L}, D)] = \lambda \cdot [(\mu, 0)],$

where we have (cf. (7.52) and (7.65))

(7.72) $[(\mu, 0)] \in \operatorname{im} v \subseteq \ker \tau = \Phi^1_{\mathcal{A}}(X)^{\nabla}_{R=0}$

(see also (7.69)). Thus, one obtains from (7.71) the relation

(7.73) $[(\mathcal{L}, D)] = \lambda \cdot [(\mu, 0)] = [(\lambda\mu, 0)] \in \operatorname{im} v \cong H^1(X, \mathbb{C}^{\cdot}),$

which also proves the desired equality in (7.65), so that one finally obtains

(7.74) $H^1(X, \mathbb{C}^{\cdot}) \cong \operatorname{im} v = \ker \tau,$

that is, the exactness of (7.64), as well, thus the desired (7.44) too. ∎ Further use of the foregoing will be made in Section 8.

7.2 Hermitian Counterpart (Continued)

We first note that all the preceding material of this section can be formulated within the Hermitian framework that has been considered in Section 6. To this end, one has to be given the appropriate set-up. Thus, one gets, for instance, the following Hermitian analogue of (7.13):

(7.75)

> suppose that we are given the framework of (6.6). Then, the (abelian) group $H^1(X, S^1)$ has an equivariant action on the (abelian) group $H^2(X, \mathbb{Z})$, as well as on the set
>
> (7.75.1) $$\Phi^1_{\mathcal{A}}(X)^{\nabla_{her}}$$
>
> (cf. (6.32)), so that the corresponding map σ, as in (7.9), is an $H^1(X, S^1)$-map viz. an equivariant map.

In this connection, by further commenting on our previous terminology in (7.75), we note for convenience that we have already considered (cf. (6.40)) the action of the (abelian) group $H^1(X, S^1)$ on the set (7.75.1) (an (abelian) group as well, cf. (6.34)). On the other hand, by using the obvious induced map (group morphism) in cohomology,

(7.76) $$i^* : H^1(X, S^1) \longrightarrow H^1(X, \mathbb{C}^{\cdot}),$$

that derives from the natural injection

(7.77) $$S^1 \underset{\longrightarrow{i}}{\subseteq} \mathbb{C}^{\cdot},$$

one gets, by analogy to (7.20), an action of $H^1(X, S^1)$ on the (abelian) group $H^2(X, \mathbb{Z})$ as well. Accordingly, one finally comes to an analogous commutative, by definition, diagram, as in (7.9), defining in the present case the sought-for map σ:

(7.78)

$$
\begin{array}{ccc}
\Phi^1_{\mathcal{A}}(X)^{\nabla_{her}} \subseteq \Phi^1_{\mathcal{A}}(X)^{\nabla} & \xrightarrow{\ \sigma = \delta \circ \mu\ } & H^2(X, \mathbb{Z}) \\
\Big\downarrow{\mu} & & \Big\downarrow{\cong} \\
\Phi^1_{\mathcal{A}}(X)_{her} \cong H^1(X, \mathcal{SU}(1)) & \xrightarrow{\ i^*\ } & H^1(X, \mathcal{A}^{\cdot})
\end{array}
$$

which also fully clarifies our relevant terminology in (7.75).

Concerning the Hermitian counterparts of relations, like (7.50), for instance, one can look at and also obtain such results in light of (6.11) and (7.76). Furthermore, one can have a Hermitian analogue, in the appropriate sense, of Weil's integrality theorem (cf. Theorem 3.1) by taking also (6.12) into account. In this connection, we further remark that one has at one's disposal the Chern-Weil theorem (see [VS: Chapt. IX; p. 258, Theorem 3.1]), pertaining, for the case at issue, to the definition of the *Chern class-curvature* of a given Maxwell field $(\mathcal{L}, D)$ independently of the (gauge) potential ($\mathcal{A}$-connection). See also Remark 3.1(ii), in particular (3.127).

8 The Maxwell Group $\Phi^1_{\mathcal{A}}(X)^{\nabla}$ as a Central Extension (Continued)

We consider in the ensuing discussion one further application of Theorem 5.1 in conjunction with certain of our main conclusions in Section 7, as, for instance, those in (7.45) and (7.49). So to fix the terminology,

(8.1)   assume that we are given the set-up of (7.36).

Now, by employing the language of group extensions (see, for instance, J.J. Rotman [1: p. 127ff]), we see that the short exact sequence

$$(8.2) \qquad 1 \longrightarrow H^1(X, \mathbb{C}^\bullet) \xrightarrow{\ \upsilon\ } \Phi^1_{\mathcal{A}}(X)^{\nabla} \xrightarrow{\ \tau\ } H^2(X, \mathbb{Z}) \longrightarrow 0$$

(cf. (7.44) and (7.46)) can be viewed as a central (we are dealing here with abelian groups) extension of $H^2(X, \mathbb{Z})$ by $H^1(X, \mathbb{C}^\bullet)$. Hence, by virtue of the Chern isomorphism (cf. (7.6)), one further concludes that

the Maxwell group of X,

(8.3.1) $\Phi^1_{\mathcal{A}}(X)^{\nabla}$,

can be construed as a central extension of the Picard group of X,

(8.3)   (8.3.2) $H^1(X, \mathcal{A}^\bullet)(\cong H^2(X, \mathbb{Z}))$,

by

(8.3.3) $H^1(X, \mathbb{C}^\bullet)$,

viz. by the group of flat $\mathbb{C}$-line sheaves on X (cf. also (7.51)).

Accordingly, the whole situation is displayed by (7.49) and (7.50). The same can be put in the form of the following short exact sequence:

$$(8.4) \qquad 1 \longrightarrow H^1(X, \mathbb{C}^\bullet) \xrightarrow{\ \upsilon\ } \Phi^1_{\mathcal{A}}(X)^{\nabla} \xrightarrow{\ \tau\ } \Omega^2(X)^{\mathrm{int}}_{\mathrm{cl}} \longrightarrow 0.$$

Thus, one can restate (8.3), by saying that

the Maxwell group of X, $\Phi^1_{\mathcal{A}}(X)^\nabla$ can be viewed as a central extension
of the image of τ, viz. of the (abelian) group

(8.5)

(8.5.1) $\Omega^2(X)^{\text{int}}_{\text{cl}} \lhd \Omega^2(X)$

(Weil group of X) by $H^1(X, \mathbb{C}^\bullet)$.

Indeed, the above is actually another version of (7.64), or even of (7.40), by taking
also into account (7.41) along with (7.44). ∎

On the other hand, let us now restrict ourselves to the set

(8.6) $\Phi^1_{\mathcal{A}}(X)^\nabla_R$,

that is, to all those Maxwell fields on X having a given (fixed) field strength (curvature) R. Thus, as already known, based on Weil's integrality theorem (cf. Theorem
3.1), one further concludes, pertaining to the given curvature R as in (8.6), that

(8.7) $\frac{1}{2\pi i}[R] \in H^2(X, \mathbb{Z})$,

viz. one thus obtains a 2-dimensional integral cohomology class of X (cf., for instance, (3.59) and (3.60)). Here, by an obvious abuse of language, one usually speaks
just of R, in place of (8.7), when referring to the respective cohomology class of X.

Furthermore, by virtue of the same theorem, the set (8.6) is nonempty, that is,
one always gets the relation

(8.8) $\Phi^1_{\mathcal{A}}(X)^\nabla_R := \tau^{-1}(R) \neq \emptyset$,

within, of course, the appropriate context, as, for example, that of (8.1) (see also
(3.13) and (3.16)). So in this case, our conclusion in (8.3) is simply reduced to the
following relation, which actually we already know (cf. (5.111)); that is,

(8.9) $\Phi^1_{\mathcal{A}}(X)^\nabla_R = H^1(X, \mathbb{C}^\bullet) \cdot [(\mathcal{L}, D)]$,

for any $[(\mathcal{L}, D)]$ an element of (8.6). (In this connection, see also (5.115) or (5.116)).
Based on (8.8) and the fact that

(8.10) $H^1(X, \mathbb{C}^\bullet) \underset{\nu}{\subseteq} \Phi^1_{\mathcal{A}}(X)^\nabla$

(cf., for example, (8.4)), one has

(8.11) $H^1(X, \mathbb{C}^\bullet) = \Phi^1_{\mathcal{A}}(X)^\nabla_0 = \tau^{-1}(0)$,

so that the previous relation (8.9) can still be expressed by

(8.12) $\Phi^1_{\mathcal{A}}(X)^\nabla_R = \Phi^1_{\mathcal{A}}(X)^\nabla_0 \cdot [(\mathcal{L}, D)]$,

or even just by the relation

(8.13) $\Phi^1_{\mathcal{A}}(X)^\nabla_R = \Phi^1_{\mathcal{A}}(X)^\nabla_0$,

a bijection established through any Maxwell field $[(\mathcal{L}, D)]$ such that $R(D) = R$,
the given (fixed) curvature R. Equation (8.13) gives us additional nice (physical)
information pertaining to the inherent structure of a given "light bundle" (cf. also
(8.23)).

8.1 The Hermitian Counterpart (Continued)

Based on our discussion in Section 6, we can formulate here too all the preceding material of the present section within a Hermitian set-up by one considering, for instance, the framework of (6.6), (6.50), along with (7.36): Thus, by virtue of (6.12), one can then look at a

(8.14)   Hermitian analogue of Weil's integrality theorem,

in the sense that for any

(8.15) $$z \in H^2(X, \mathbb{Z}),$$

one can find a Hermitian Maxwell field (see (6.29) and (6.31)), whose curvature (field strength) corresponds to z. (In this connection, see also [VS: Chapt. IX; p. 258, Theorem 3.1] or Remark 3.1(ii), in particular (3.127).) In other words, there exists an element (cf. (6.32) and (6.34))

(8.16) $$[(\mathcal{L}, D)] \in \Phi^1_{\mathcal{A}}(X)^{\nabla_{her}} \lhd \Phi^1_{\mathcal{A}}(X)^{\nabla}$$

(viz. a Hermitian Maxwell field), such that

(8.17) $$\frac{1}{2\pi i}[R] = z \in H^2(X, \mathbb{Z}),$$

where $R \equiv R(D)$ (cf. also (6.32)).

Similarly, one obtains a Hermitian short exact sequence like (8.4); that is,

(8.18) $$1 \longrightarrow H^1(X, S^1) \overset{\nu}{\longrightarrow} \Phi^1_{\mathcal{A}}(X)^{\nabla_{her}} \overset{\tau}{\longrightarrow} \Omega^2(X)^{int}_{cl} \longrightarrow 0.$$

Therefore, by analogy to (8.5), one concludes that

the Hermitian Maxwell group

(8.19.1) $$\Phi^1_{\mathcal{A}}(X)^{\nabla_{her}}$$

can be viewed as a central extension of the image of τ,

(8.19)   (8.19.2) $$\mathrm{im}\,\tau \cong \Omega^2(X)^{int}_{cl}$$

(cf. (3.68.1)), by

(8.19.3) $$H^1(X, S^1),$$

the (abelian) group of (flat) principal S^1-sheaves on X.

Scholium 8.1 Based on our terminology, as applied in [VS: Chapt. V; p. 370, Definition 5.1, along with p. 371, scholium, in particular (5.12), for $n = 1$], one considers the source of the map (in effect, abelian group morphism)

(8.20) $$H^1(X, \mathbb{C}^\cdot) \longrightarrow H^1(X, \mathcal{A}^\cdot)$$

as the (abelian) group of (isomorphism classes of) flat $\mathbb{C}$-line sheaves on X. Within this same vein of ideas, one can further consider, by (8.19.3),

(8.21) $$H^1(X, S^1)$$

as the (abelian) group of (isomorphism classes of) flat principal S^1-sheaves on X. (In this connection, we also refer to the relevant recent work of E. Vassiliou [1] on the theory of principal sheaves.) The previous term "flatness" is actually referred to the corresponding coordinate 1-cocycles of the principal/vector sheaves concerned (characterized by the former), these coordinates being (S^1- or) $\mathbb{C}^\cdot$-valued locally constant functions (sections) on X; namely, one has,

(8.22) $$(g_{\alpha\beta}) \in Z^1(\mathcal{U}, \mathbb{C}^\cdot) \underset{\longrightarrow}{\subseteq} Z^1(\mathcal{U}, \mathcal{A}^\cdot),$$

along with the analogous relation for S^1. See also (5.33.1) as well as [VS: Chapt. V; p. 371, Theorem 5.1]. In this regard, cf. also Scholium 6.2.

Finally, by looking at the relations (7.47) or (8.12) and/or (8.13) (cf. also (8.11)), we realize that

two elements of
$$\Phi^1_{\mathcal{A}}(X)^\nabla_R \subseteq \Phi^1_{\mathcal{A}}(X)^\nabla$$

(8.23)   differ by an element of $H^1(X, \mathbb{C}^\cdot)$ (or of $H^1(X, S^1)$), viz. by a "polarized" light beam, to apply a convenient physical language to the elements of the latter groups.

Applications of the preceding material will also be considered in Chapters IV and V, pertaining mainly to a (cohomological) classification of geometric prequantizations of (Hermitian) Maxwell fields.

Cohomological Classification of Maxwell and Hermitian Maxwell Fields

> " ... the structure underlying an intrinsic approach to physics is "essentially"
> de Rham cohomology."
>
> C. von Westenholz in *Differential Forms in Mathematical Physics* (North-Holland, 1981). p. 321.

The classification alluded to in the title of this chapter will be supplied by means of the so-called Čech hypercohomology. Thus, for convenience, we first give the highlights of that mechanism, which will be of importance in the sequel, while we further refer to [VS] for relevant details and proofs. (So see loc. cit., Chapt. III; p. 218, Section 7, in particular, p. 227, Example 7.1, as well as, Chapt. VI; p. 92, Section 18, especially, p. 96, Theorem 18.2, and Chapt. VII; p. 174, Section 11.) Therefore, we start with the following introductory material.

1 Hypercohomology with Respect to a (Differential) $\mathcal{A}$-Complex

The cohomology theory we are dealing with differs from the usual sheaf cohomology (cf., for instance, [VS: Chapt. III], concerning the latter theory) in that instead of employing a sheaf of $\mathcal{A}$-modules (see below for the terminology) or simply an $\mathcal{A}$-module, as a sheaf of coefficients for the cohomology groups (in effect, $\mathcal{A}(X)$-modules) involved, one actually considers a whole complex of (sheaves of) $\mathcal{A}$-modules, or just an $\mathcal{A}$-complex (see (1.2)). In this connection, we further note that every $\mathcal{A}$-module can be construed in a trivial way (namely, by adding zeros in the corresponding sequences, as in (1.2)) as an $\mathcal{A}$-complex, hence the terminology employed here, referring to hypercohomology.

1.1 Sheaf Cohomology

As usual, we start by considering a $\mathbb{C}$-algebraized space

(1.1) $$(X, \mathcal{A})$$

(see, for instance, Chapter II; beginning of Section 6), with respect to a given (arbitrary) topological space X. Consider next an $\mathcal{A}$-complex on X, of positive degree that is,

(1.2)
$$\mathcal{E}^{\bullet} : 0 \longrightarrow \mathcal{E}^0 \xrightarrow{d^0} \mathcal{E}^1 \xrightarrow{d^1} \mathcal{E}^2 \xrightarrow{d^2} \cdots ,$$

viz. we assume for the latter sequence that

(1.2')
$$\mathcal{E}^n = 0, \qquad \text{for any integer } n < 0.$$

In other words, we are given a sequence of $\mathcal{A}$-modules and $\mathcal{A}$-morphisms between them as indicated in (1.2), still denoted by

(1.3)
$$\mathcal{E}^{\bullet} \equiv \{(\mathcal{E}^n, d^n)\}_{n \in \mathbb{Z}_+}.$$

Concerning (1.2), we also refer to it as a differential $\mathcal{A}$-sequence. The respective maps

(1.4)
$$d^n : \mathcal{E}^n \longrightarrow \mathcal{E}^{n+1}, \qquad n \in \mathbb{Z}_+ \equiv \mathbb{N} \cup \{0\},$$

are $\mathcal{A}$-morphisms of the $\mathcal{A}$-modules (see also [VS: Chapt. III; (1.3)]) whose corresponding sequence is denoted simply by

(1.5)
$$d \equiv (d^n)_{n \in \mathbb{Z}_+},$$

while we also assume, by referring to the above, that

(1.6)
$$d^{n+1} \circ d^n = 0, \qquad n \in \mathbb{Z}_+$$

(which constitutes the differential condition, concerning (1.2)). In this connection, we also write (1.6), succinctly, in the form

(1.6')
$$d^2 \equiv d \circ d = 0.$$

Warning! The map d^2 in the last relation is different from the corresponding $\mathcal{A}$-morphism, as in (1.5)). Thus, (1.6) is equivalent with the following relation;

(1.7)
$$\operatorname{im} d^n \subseteq \ker d^{n+1}, \qquad n \in \mathbb{Z}_+.$$

The obstruction of (1.7) being an equality constitutes the study of the so-called cohomology of the $\mathcal{A}$-complex, or of the sequence (1.2), that is, the study of the sequence of $\mathcal{A}$-modules

(1.7')
$$\ker d^{n+1} / \operatorname{im} d^n, \qquad n \in \mathbb{Z}_+$$

(see also (1.10)).

Furthermore, we refer to the above $\mathcal{A}$-complex $\mathcal{E}^{\bullet}$, as in (1.3) simply as the $\mathcal{A}$-complex

(1.8)
$$(\mathcal{E}^{\bullet}, d)$$

on X, in fact an abbreviated expression of the more concrete terminology for (1.8), as a differential $\mathcal{A}$-complex on X.

For convenience in the subsequent discussion, we first highlight the fundamentals pertaining to the cohomology of a differential $\mathcal{A}$-complex $(\mathcal{E}^{\cdot}, d)$ on X. This lies, in turn, at the basis of the concept of sheaf cohomology of X, with respect to (or with coefficients in) a given $\mathcal{A}$-module $\mathcal{E}$ on X, while in the case of (sheaf) hypercohomology of X, which we are also going to employ in the sequel, one considers instead more generally (hence the term hyper here), as domain of coefficients an arbitrary (differential) $\mathcal{A}$-complex $\mathcal{E}^{\cdot}$ on X, as in (1.2), thus not just an $\mathcal{A}$-module $\mathcal{E}$ on X as before. (In this connection, cf. also the pertinent comments at the beginning of this section, along with, as particular examples, the next two sections.

Thus, as already hinted at, by considering the cohomology of an $\mathcal{A}$-complex $\mathcal{E}^{\cdot}$ on X, as in (1.2), one actually looks by definition at the following sequence of $\mathcal{A}$-modules

(1.9) $$h^{*}(\mathcal{E}^{\cdot}) := \{h^{n}(\mathcal{E}^{\cdot})\}_{n \in \mathbb{Z}_{+}},$$

where we set

(1.10) $$h^{n}(\mathcal{E}^{\cdot}) := \ker d^{n} / \operatorname{im} d^{n-1} \equiv Z^{n}(\mathcal{E}^{\cdot}) / B^{n}(\mathcal{E}^{\cdot}),$$

for any $n \in \mathbb{Z}_{+}$, while we have

(1.11) $$h^{n}(\mathcal{E}^{\cdot}) = 0, \qquad \text{for any } n < 0,$$

since in view of (1.2), one has

(1.12) $$d^{n} = 0, \qquad \text{for any } n < 0.$$

(In this connection, see also [VS: Chapt. III; p. 146ff.]) On the other hand, the same sequence (1.10) can still be construed, as a sequence of $A(X)$-modules (cf. also loc. cit., p. 148 (1.13)); indeed, the assertion is an immediate consequence of the definition of a sheaf morphism (according to our hypothesis, (1.5) is a sequence of $\mathcal{A}$-morphisms, as in (1.4)), and the coincidence in turn of this notion with that one of a morphism between the corresponding complete presheaves of sections (loc. cit., Chapt. I; p. 75, (13.19)).

The *sheaf cohomology* of X, with coefficients in a given $\mathcal{A}$-module $\mathcal{E}$, is, by definition, the cohomology of a certain $\mathcal{A}$- complex on X, provided by the application of the (global) section functor Γ_X on any injective $\mathcal{A}$-resolution of the given $\mathcal{A}$-module $\mathcal{E}$; to be thus concrete, let us first recall that, for any given $\mathcal{A}$-module $\mathcal{E}$ on X, one defines the aforesaid functor, according to the relation

(1.13) $$\Gamma_X(\mathcal{E}) := \Gamma(X, \mathcal{E}) \equiv \mathcal{E}(X),$$

the result being, in view of the hypothesis for $\mathcal{E}$, an $A(X)$-module, viz. that of the (continuous) global sections of $\mathcal{E}$ (see also loc. cit., Chapt. II; p. 141, (6.41), along with Chapt. I; p. 72, (13.1)). On the other hand, for any given $\mathcal{A}$-module $\mathcal{E}$ on X, there always exists an injective $\mathcal{A}$-resolution of $\mathcal{E}$, say

(1.14) $$\mathcal{E}^{\cdot} : 0 \longrightarrow \mathcal{E}^{0} \xrightarrow{d^{0}} \mathcal{E}^{1} \xrightarrow{d^{1}} \mathcal{E}^{2} \xrightarrow{d^{2}} \cdots,$$

viz. an injective (differential) $\mathcal{A}$-complex of positive degree (cf. (1.2′)), such that every $\mathcal{A}$-module in (1.14) is by definition *injective*, hence the terminology applied. See also [VS: Chapt. III; p. 155, Definition 1.2, and p. 157, Theorem 1.2]. Thus, one can finally consider the application of the functor Γ_X, as in (1.13) on the above $\mathcal{A}$-complex (1.14), so that one obtains the following $\mathcal{A}(X)$-complex (differential sequence of $\mathcal{A}(X)$-modules):

$$(1.15) \quad \Gamma_X(\mathcal{E}^{\bullet}) : 0 \longrightarrow \Gamma_X(\mathcal{E}^0) \xrightarrow{\Gamma_X(d^0)} \Gamma_X(\mathcal{E}^1) \xrightarrow{\Gamma_X(d^1)} \Gamma_X(\mathcal{E}^2) \xrightarrow{\Gamma_X(d^2)} \cdots .$$

Thus, as already said, one now defines

> the *sheaf cohomology of* X, *with coefficients in a given* $\mathcal{A}$-*module* $\mathcal{E}$, as the cohomology of the (differential) $\mathcal{A}(X)$-complex

$$(1.16) \qquad (1.16.1) \qquad\qquad\qquad \Gamma_X(\mathcal{E}^{\bullet}),$$

> as above, viz. that derived from the application of Γ_X to an injective resolution of $\mathcal{E}$, as in (1.14).

In other words, by virtue of (1.9), one defines

$$(1.17) \qquad h^*(X, \mathcal{E}) \equiv \{H^n(X, \mathcal{E})\}_{n \in \mathbb{Z}_+} := \{h^n(\Gamma_X(\mathcal{E}^{\bullet})\}_{n \in \mathbb{Z}_+},$$

that is, in particular, one sets (cf. (1.10), (1.15))

$$(1.18) \qquad \begin{aligned} H^n(X, \mathcal{E}) &\equiv h^n(\Gamma_X(\mathcal{E}^{\bullet})) \equiv h^n(\Gamma(X, \mathcal{E}^{\bullet})) \\ &:= \ker \Gamma_X(d^n)/\operatorname{im} \Gamma_X(d^{n-1}), \end{aligned}$$

for any $n \in \mathbb{Z}_+$.

Note 1.1 (Terminological) In accord with the previously employed terminology, referring to the notion of sheaf cohomology of X as that of an appropriate $\mathcal{A}$-complex on X, we further remark here that (1.16.1) can still be viewed as a (differential) complex consisting of sheaves of modules with respect to the constant $\mathbb{C}$-algebra sheaf $\mathcal{A}(X)$ on X. In this regard, cf. also [VS: Chapt. II; p. 97, (1.40)].

On the other hand,

$$(1.19)$$
> we also refer to (1.18), as the n dimensional (sheaf) cohomology $\mathcal{A}(X)$-module (or even, by abusing classical terminology, cohomology group) of X, with coefficients in the given $\mathcal{A}$-module $\mathcal{E}$.

In this connection, we further note that

$$(1.20)$$
> (1.18) is independent of the injective resolution of $\mathcal{E}$ considered, for instance, in (1.14).

The proof of our previous claim in (1.20) is essentially an immediate consequence of the so-called fundamental theorem of sheaf cohomology (see [VS: Chapt. III; p. 166, Theorem 3.1]). However, "is" in (1.20) actually means modulo an isomorphism of the (cohomology) functors involved (loc. cit.).

On the other hand, the same relations (1.18) as above give by definition the se-quence of the so-called *right derived functors of Γ_X*; i.e., one sets, for any given $\mathcal{A}$-module $\mathcal{E}$ on X,

$$(1.21) \qquad (R^n \Gamma_X)(\mathcal{E}) := h^n(\Gamma_X(\mathcal{E}^{\cdot})), \quad n \in \mathbb{Z}_+,$$

where $\mathcal{E}^{\cdot}$ stands for an injective resolution of $\mathcal{E}$ (cf. (1.14)). Accordingly, one thus obtains

$$(1.22) \qquad \begin{aligned} H^n(X, \mathcal{E}) &:= (R^n \Gamma_X)(\mathcal{E}) := h^n(\Gamma_X(\mathcal{E}^{\cdot})) \equiv h^n(\Gamma(X, \mathcal{E}^{\cdot})) \\ &:= \ker \Gamma_X(d^n)/\operatorname{im} \Gamma_X(d^{n-1}), \end{aligned}$$

for any $n \in \mathbb{Z}_+$ and any given $\mathcal{A}$-module $\mathcal{E}$ on X with $\mathcal{E}^{\cdot}$ as in (1.21). One thus concludes that

(1.23)   the sheaf cohomology of a topological space X with coefficients in a given $\mathcal{A}$-module $\mathcal{E}$ (or sheaf cohomology of X with respect to an $\mathcal{A}$-module $\mathcal{E}$) is the derived functor cohomology of the global section func-tor Γ_X.

Throughout the preceding we considered injective resolutions of $\mathcal{E}$; these are all Γ_X-acyclic, in the sense that every member of the resolution at issue has trivial derived functor cohomology. Thus, by considering, for instance, the injective resolution of $\mathcal{E}$, as in (1.14), one obtains (see also [VS: Chapt. III; p. 171, Lemma 3.1])

$$(1.24) \qquad (R^n \Gamma_X)(\mathcal{E}^m) = 0, \qquad \text{for any, } n, m \text{ in } \mathbb{Z}_+.$$

The above property of injective resolutions, as in (1.24), is a crucial one concerning their presence in (1.22), since according to the so-called abstract de Rham theorem (loc. cit., Chapt. III; p. 169, (3.24)), one actually gets the following result;

(1.25)   the sheaf cohomology of a topological space X with coefficients from a given $\mathcal{A}$-module $\mathcal{E}$ is the cohomology of any Γ_X-acyclic resolution of $\mathcal{E}$.

In fact, something more is essentially true, namely,

(1.26)   a more general form of *(1.25)* is valid by considering in place of Γ_X any covariant left exact $\mathcal{A}(X)$-linear functor T (in particular, the functor Γ_X), along with any acyclic resolution $\mathcal{E}^{\cdot}$ of $\mathcal{E}$ with respect to that functor.

See loc. cit., Chapt. III; p. 172, (3.38). An application of this even more general form of the abstract de Rham theorem is found, for instance, in the definition of $\mathcal{E}xt$ functors: ibid., p. 249, Section 9; in particular, see p. 253, (9.16) and (9.17).

All told, the above cohomology functors

(1.27) $\{H^n(X, \cdot)\}_{n \in \mathbb{Z}_+}$,

as given by (1.22), provide by definition the *sheaf cohomology of X with coefficients in any given $\mathcal{A}$-module $\mathcal{E}$ on X*. On the other hand, the same functors as above are characterized as well by the aforementioned (fundamental) theorem (of sheaf cohomology); loc. cit., as well as, p. 231, (8.4), along with the ensuing discussion, as, for instance, p. 232, (8.7), and p. 234, (8.11). Finally, our previous conclusions in (1.25) and (1.26) are also formulated, succinctly, by the following relations:

(1.28) $H^*(X, \mathcal{E}) \cong h^*(\Gamma_X(\mathcal{E}^{\cdot}))$

for any $\mathcal{A}$-module $\mathcal{E}$ on X, with $\mathcal{E}^{\cdot}$ a Γ_X-acyclic resolution of $\mathcal{E}$. More generally, one obtains

(1.29) $H^*(X, \mathcal{E}) \cong h^*(T(\mathcal{E}^{\cdot}))$,

with T and $\mathcal{E}^{\cdot}$ as in (1.26). (The isomorphisms appearing in the two last relations refer to isomorphisms of the respective cohomology theories, loc. cit.).

1.2 Hypercohomology

The notion in the heading of this subsection refers to the (sheaf) cohomology of certain $\mathcal{A}$-bicomplexes, or double $\mathcal{A}$-complexes, on X. However, as we shall see presently, the latter is reduced to the (sheaf) cohomology of an appropriate $\mathcal{A}$-complex, as before, that is canonically associated with the $\mathcal{A}$-bicomplex under discussion. So let us first fix the terminology.

By the term $\mathcal{A}$-bicomplex, or double $\mathcal{A}$-complex, on X, one means first a double sequence of $\mathcal{A}$-modules,

(1.30) $\mathcal{E}^{\cdot\cdot} \equiv \{\mathcal{E}^{n,m}\}_{(n,m) \in \mathbb{Z}_+^2}$,

where we set $\mathbb{Z}_+^2 \equiv \mathbb{Z}_+ \times \mathbb{Z}_+$; that is we also assume that $\mathcal{E}^{\cdot\cdot}$ is of positive bidegree, in the obvious sense of the word (see, for instance, (1.2′)). Second, we further assume that we are given on the above double sequence the following two types of differentials:

(i) the horizontal differentials ($\mathcal{A}$ morphisms)

(1.31) $\delta^{n,m} : \mathcal{E}^{n,m} \longrightarrow \mathcal{E}^{n+1,m}$,

(ii) the vertical differentials (viz. $\mathcal{A}$-morphisms)

(1.32) $d^{n,m} : \mathcal{E}^{n,m} \longrightarrow \mathcal{E}^{n,m+1}$,

with $(n, m) \in \mathbb{Z}_+^2$. In referring to the previous types of differentials, we usually write, in abbreviated form,

(1.33) $\delta \equiv (\delta^{n,m})$ and $d \equiv (d^{n,m})$,

with (n, m) as before. Third, we further suppose that the same differentials satisfy the relations

(1.34) $\delta^2 (\equiv \delta \circ \delta) = 0,$ and $d^2 (\equiv d \circ d) = 0,$

as well as

(1.35) $\delta d = d\delta,$ that is $\delta \circ d = d \circ \delta.$

Thus, the preceding data constitute the definition of a *differential $\mathcal{A}$-bicomplex*, or for simplicity's sake and by obviously abusing terminology, of an *$\mathcal{A}$-bicomples* (or, as already said of a *double $\mathcal{A}$-complex*). The same is denoted by

(1.36) $(\mathcal{E}^{\cdot\cdot}, \delta, d),$

while it is also depicted by the following cartesian diagram, or (n, m)-plane, in the form of the next diagram of the $\mathcal{A}$-modules appearing in (1.30), along with the connecting horizontal and vertical $\mathcal{A}$-morphisms (differentials), as in (1.31) and (1.32):

(1.37)

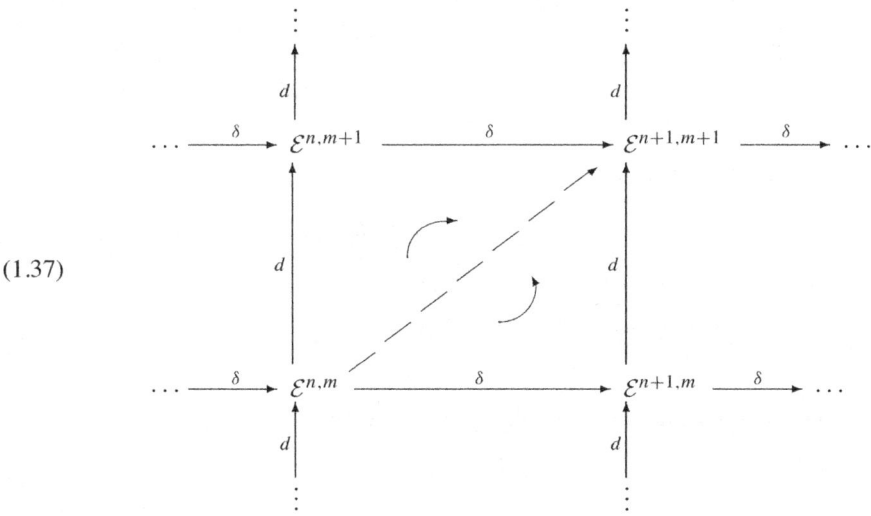

Now, by virtue of (1.35), each one of the rectangles in the previous diagram (1.37) is commutative; thus, expressing it "coordinate-wise" one has the relation (cf. also (1.33))

(1.38) $\delta^{n,m+1} \circ d^{n,m} = d^{n+1,m} \circ \delta^{n,m},$

for any $(n, m) \in \mathbb{Z}_+^2.$

Accordingly, for any given (differential) $\mathcal{A}$- bicomplex $\mathcal{E}^{\cdot\cdot}$, as in (1.36), one defines two sequences of ordinary (single) $\mathcal{A}$-complexes (see (1.2), (1.3), or (1.8)), as follows:

(i) the horizontal (alias row) $\mathcal{A}$-complex, associated with the horizontal differentials (cf. (1.31), (1.33)), viz.

$$(1.39) \qquad (\mathcal{E}^{\cdot,m}, \delta) \equiv \{(\mathcal{E}^{n,m}, \delta^{n,m} \equiv \delta)\}_{n\in\mathbb{Z}_+},$$

for any $m \in \mathbb{Z}_+$.

as well as

(ii) the vertical (or column) $\mathcal{A}$-complex connected with the vertical differentials (see (1.32) and (1.33)), that is,

$$(1.40) \qquad (\mathcal{E}^{n,\cdot}, d) \equiv \{(\mathcal{E}^{n,m}, d^{n,m} \equiv d)\}_{m\in\mathbb{Z}_+},$$

for every $n \in \mathbb{Z}_+$,

in such a manner that (1.34) and (1.35) (equivalently, (1.38)) are in force.

Now, starting from a given $\mathcal{A}$-bicomplex $\mathcal{E}^{\cdot\cdot}$, as above, one gets an ordinary $\mathcal{A}$-complex $\mathcal{E}^{\cdot}$ on X, the so-called total $\mathcal{A}$-complex (or else total complex) of $\mathcal{E}^{\cdot\cdot}$, denoted by

$$(1.41) \qquad tot(\mathcal{E}^{\cdot\cdot}) \equiv \mathcal{E}^{\cdot}.$$

That is, strictly speaking, one defines a (differential) $\mathcal{A}$-complex (cf. (1.8))

$$(1.42) \qquad \mathcal{E}^{\cdot} \equiv (\mathcal{E}^{\cdot}, D) \equiv \{(\mathcal{E}^p, D^p)\}_{p\in\mathbb{Z}_+},$$

such that in particular, one sets (see also (1.30))

$$(1.43) \qquad \mathcal{E}^p := \bigoplus_{n+m=p} \mathcal{E}^{n,m}, \qquad p \in \mathbb{Z}_+,$$

while concerning the corresponding total differential

$$(1.44) \qquad D \equiv (D^p)_{p\in\mathbb{Z}_+},$$

one defines (cf. (1.31) and (1.32))

$$(1.45) \qquad D^p := \sum_{n+m=p} \delta^{n,m} + (-1)^n d^{n,m}, \quad p \in \mathbb{Z}_+.$$

We come now to the following basic notion (cf. also [VS: Chapt. III; p. 222, Definition 7.1]).

Definition 1.1 Given an $\mathcal{A}$-bicomplex $\mathcal{E}^{\cdot\cdot}$ on X (see (1.36)), one defines, as the *cohomology of* $\mathcal{E}^{\cdot\cdot}$,

$$(1.46) \qquad\qquad h^*(\mathcal{E}^{\cdot\cdot}),$$

the cohomology of the corresponding total $\mathcal{A}$-complex of $\mathcal{E}^{\cdot\cdot}$ (cf. (1.41), (1.42)). Thus, one sets (see also (1.9)):

$$(1.47) \qquad\qquad h^*(\mathcal{E}^{\cdot\cdot}) := h^*(tot(\mathcal{E}^{\cdot\cdot})) \equiv h^*(\mathcal{E}^{\cdot}).$$

For any given $\mathcal{A}$-complex $\mathcal{E}^{\cdot}$ on X, there always exists an $\mathcal{A}$-bicomplex

$$(1.48) \qquad\qquad (\mathcal{F}^{\cdot\cdot}, \delta, d),$$

the so-called *(fully) injective $\mathcal{A}$-resolution of* $\mathcal{E}^{\cdot}$, supplied by considering suitable injective resolutions of the individual $\mathcal{A}$-modules of $\mathcal{E}^{\cdot}$; loc. cit., p. 222, (7.23)). On the other hand, we can further employ in (1.48) the (left exact global) section functor Γ_X, so that one gets the following $\mathcal{A}(X)$-bicomplex:

$$(1.49) \qquad \Gamma_X(\mathcal{F}^{\cdot\cdot}) \equiv \{\Gamma(X, \mathcal{F}^{n,m}), \Gamma_X(\delta), \Gamma_X(d)\}_{(n,m)\in\mathbb{Z}^2_+}$$

(see also (1.30), along with (1.31) and (1.32)). For convenience, we considered above total positive degree, while this will be the case as well in the particular instances, considered below.

Definition 1.2 Suppose we are given a $\mathbb{C}$-algebraized space $(X, \mathcal{A})$, and let $\mathcal{E}^{\cdot}$ be a (differential) $\mathcal{A}$-complex on X. One defines the *(sheaf) hypercohomology of X*, with respect to (or with coefficients from) $\mathcal{E}^{\cdot}$,

$$(1.50) \qquad\qquad h^*(X, \mathcal{E}^{\cdot}),$$

according to the following relations (cf. also (1.47)):

$$(1.51) \qquad h^*(X, \mathcal{E}^{\cdot}) := h^*(\Gamma_X(\mathcal{F}^{\cdot\cdot})) := h^*(tot(\Gamma_X(\mathcal{F}^{\cdot\cdot}))).$$

Consequently, one gets, through the last member of (1.51), the hypercohomology of X, as defined above, just as the sheaf cohomology of X with respect to the (differential) $\mathcal{A}$-complex $tot(\Gamma_X(\mathcal{F}^{\cdot\cdot}))$ (see also (1.41), (1.42)), that is, an instance that we already have dealt with in the first part of this section (cf. (1.28)). In this connection, we remark here that any given $\mathcal{A}$-module $\mathcal{E}$ on X, as already noted at the beginning of this section, can be viewed as a (trivial) particular case of a (differential) $\mathcal{A}$-complex (viz., by considering all differentials zero in (1.2) with $\mathcal{E}^0 \equiv \mathcal{E}$); hence, one looks at the hypercohomology of X as a generalization of the (usual) sheaf cohomology of X.

Note 1.2 The terminology applied in the foregoing connected with (differential) double $\mathcal{A}$-bicomplexes (see (1.36)) is reminiscent of the situation, one encounters in the heory of spectral sequences. Indeed, this actually happens in several of the notions that were employed in the preceding, while the proof of some of our previous conclusions is based on that theory as well; in this connection, we refer to [VS: Chapt. III; Section 7, p. 218ff], along with the references quoted therein, for relevant details. Nonetheless, for simplicity's sake, we systematically avoided any use of that more sophisticated technique. However, to quote, for instance, S. Lang [1: p. 163, §9], "... *the basic description of this gadget ... can be done in just a few pages.*" On the other hand, cf. J.J. Rotman [2: p. 366], "... *every purely homological result may be proved with spectral sequences ...,*" or still (loc. cit.), "... *spectral sequences offers a systematic approach in place of sporadic success,*" although "elementary" proofs may also exist, according to the same author (ibid.).

Another issue that we wish to point out here is that all the graded $\mathcal{A}$-modules considered in the foregoing were taken to be, for convenience, of positive degree (see, for instance, (1.2), or (1.49)). Indeed, a (partly) more general aspect has been adopted, for example, in [VS: Chapt. III; p. 222, (7.23.1)].

What we are going to employ in the sequel is Čech hypercohomology of X, with respect to a given (differential) $\mathcal{A}$-complex $\mathcal{E}^\bullet$ on X (cf. (1.2)), in fact, with respect to an appropriate 2-term such complex on X (see Sections 3, 4). Thus, as is usually the case with ordinary Čech cohomology, the corresponding hypercohomology is at least more transparent than the more general (abstract) sheaf hypercohomology of X, as described in the preceding. On the other hand, all these coincide in the particular (however, important) case that X is a paracompact (Hausdorff) space. See [VS: Chapt. III; p. 234, Theorem 8.1], or J.-L. Brylinski [1: p. 32, Theorem 1.3.13]. We come next to this item.

2 Čech Hypercohomology

As already mentioned, we specialize in the ensuing discussion to the case of Čech (sheaf) hypercohomology of X, with respect to (or with coefficients from) a given (differential) $\mathcal{A}$-complex $\mathcal{E}^\bullet$ on X (cf. (1.2)), instead of just an $\mathcal{A}$-module $\mathcal{E}$ on X, as is the usual Čech cohomology of X with coefficients in $\mathcal{E}$ (cf., for instance, [VS: Chapt. III; p. 182, (4.53′)]).

As usual, we start with a $\mathbb{C}$-algebraized space

$$(2.1) \qquad\qquad\qquad (X, \mathcal{A}),$$

and let

$$(2.2) \qquad\qquad\qquad \mathcal{U} = (U_\alpha)_{\alpha \in I}$$

be an open covering of X. Furthermore, assume that we are given a (differential) $\mathcal{A}$-complex on X,

$$(2.3) \qquad\qquad\qquad \mathcal{E}^\bullet \equiv (\mathcal{E}^\bullet, d)$$

(see (1.2), along with (1.8)). Thus, one now defines the following *Čech cochain* $A(X)$-bicomplex on X with respect to $\mathcal{E}^{\bullet}$ (associated with the given open covering $\mathcal{U}$ of X):

(2.4) $\check{C}^{\bullet}(\mathcal{U}, \mathcal{E}^{\bullet}) \equiv (\check{C}^{\bullet}(\mathcal{U}, \mathcal{E}^{\bullet}), \delta, d) := (\{\check{C}^{n}(\mathcal{U}, \mathcal{E}^{m})\}_{(n,m)\in\mathbb{Z}_{+}^{2}}, \delta, d).$

Here the differentials δ and d are defined, respectively, via the corresponding Bockstein (or coboundary) operators that are associated with the usual Čech cohomology of X (see loc. cit., Chapt. III; p. 176, (4.15)), and the differentials of the given A-complex $\mathcal{E}^{\bullet}$ on X (see (1.5)). By analogy with (1.30) and (1.37), here too we depict (2.4) by the following (n, m)-plane:

(2.5)

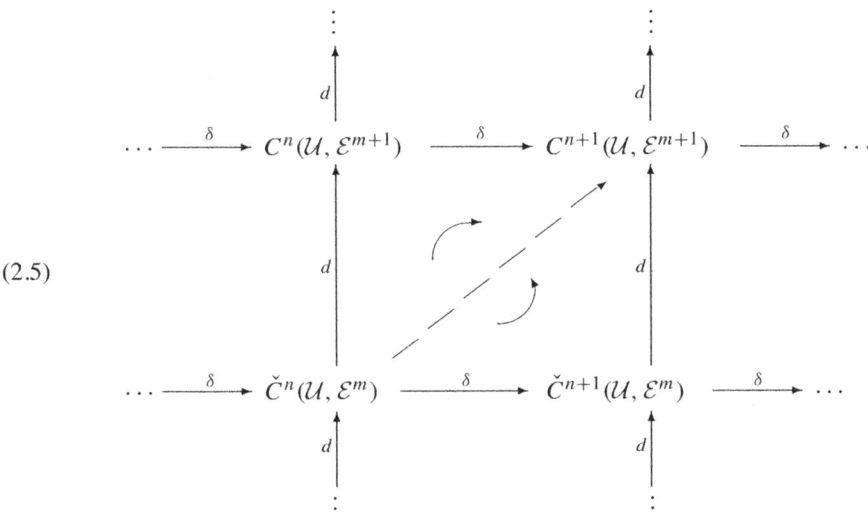

Furthermore, we remark that

(2.6)   the analogous relations to (1.34) and (1.35) are in force as well, hence the commutativity of the respective rectangle, as in (2.5), with (n, m) in $\mathbb{Z}_{+}^{2}$.

Indeed, concerning (1.34), see [VS: Chapt. III; p. 176, (4.18)], as well as (1.6). On the other hand, the analogue of (1.35) in the present case is easily verified, according to the definitions of the respective operators. See also loc. cit., p. 176, (4.15) and (4.16), still taking into account the linearity of the d's by virtue of their definition. ∎

Therefore, by analogy to (1.41) and (1.42), we can further consider the corresponding to (2.4) total $A(X)$-complex, while the respective total differential can be also defined, by virtue of the operators δ and d, as in (2.4) (cf. also (2.5)), and of (1.45). Consequently, one thus gets the (differential) $A(X)$-complex

(2.7) $(tot(\check{C}^{\bullet}(\mathcal{U}, \mathcal{E}^{\bullet})), D),$

viz. the total $\mathcal{A}(X)$-complex of (2.4), defined, for any open covering $\mathcal{U}$ of X and any given $\mathcal{A}$-complex $\mathcal{E}^{\bullet}$ on X. We come next to the following basic notion, being a prerequisite for the main one, throughout the sequence.

Definition 2.1 Suppose we are given a $\mathbb{C}$-algebraized space $(X, \mathcal{A})$, an open covering $\mathcal{U}$ of X, and a (differential) $\mathcal{A}$-complex $\mathcal{E}^{\bullet}$ on X. One defines the *Čech hypercohomology of $\mathcal{U}$ with respect to the $\mathcal{A}$-complex $\mathcal{E}^{\bullet}$*, denoted by

$$(2.8) \qquad\qquad h^*(\check{C}^{\bullet}(\mathcal{U}, \mathcal{E}^{\bullet})),$$

as the cohomology of the (differential) $\mathcal{A}(X)$-complex (2.7); that is, one has

$$(2.9) \qquad\qquad h^*(\check{C}^{\bullet}(\mathcal{U}, \mathcal{E}^{\bullet})) := h^*(tot(\check{C}^{\bullet}(\mathcal{U}, \mathcal{E}^{\bullet}))).$$

The graded cohomology $\mathcal{A}(X)$-module, that is, the sequence of cohomology $\mathcal{A}(X)$-modules alluded to by (2.8), is still denoted by:

$$(2.10) \qquad\qquad \check{\mathbb{H}}^*(\mathcal{U}, \mathcal{E}^{\bullet}) \equiv \{\check{\mathbb{H}}^n(\mathcal{U}, \mathcal{E}^{\bullet})\}_{n \in \mathbb{Z}_+},$$

so that one has, for any $n \in \mathbb{Z}_+$,

$$(2.11) \qquad \begin{aligned} \check{\mathbb{H}}^n(\mathcal{U}, \mathcal{E}^{\bullet}) &\equiv h^n(\check{C}^{\bullet}(\mathcal{U}, \mathcal{E}^{\bullet})) \\ &:= h^n(tot(\check{C}^{\bullet}(\mathcal{U}, \mathcal{E}^{\bullet}))) := \ker D^n / \operatorname{im} D^{n-1} \end{aligned}$$

(cf. also (1.9) and (1.10), along with (2.8)). On the other hand, concerning the previous (cohomology) $\mathcal{A}(X)$-modules, one proves that

(2.12)
> for every $n \in \mathbb{Z}_+$, the family
>
> $$(2.12.1) \qquad\qquad \check{\mathbb{H}}^n(\mathcal{U}, \mathcal{E}^{\bullet}),$$
>
> with $\mathcal{U}$ ranging over the (proper) open coverings of X, provides an inductive system of $\mathcal{A}(X)$-modules (the previous set of coverings being directed (upward), by "refinement").

In this connection, cf. also J.-L. Brylinski [1: p. 30, Lemma 1.3.8] concerning the independence of the inductive system considered by (2.12.1) from the "refinement maps" involved. (See also [VS: Chapt. III; p. 175, (4.9)].) We are now in a position to cope with the main issue of this section.

Definition 2.2 Assume that we have a $\mathbb{C}$-algebraized space $(X, \mathcal{A})$. Then, the *Čech hypercohomology of X with respect to a (differential) $\mathcal{A}$-complex $\mathcal{E}^{\bullet}$ on X* (of positive degree, cf. (1.2)), denoted by

$$(2.13) \qquad\qquad \check{\mathbb{H}}^*(X, \mathcal{E}^{\bullet}) \equiv \{\check{\mathbb{H}}^n(X, \mathcal{E}^{\bullet})\}_{n \in \mathbb{Z}_+},$$

is defined by the relation (see (2.12))

(2.14) $$\check{\mathbb{H}}^n(X, \mathcal{E}^{\cdot}) := \varinjlim_{\mathcal{U}} \check{\mathbb{H}}^n(\mathcal{U}, \mathcal{E}^{\cdot}),$$

for any $n \in \mathbb{Z}_+$.

In this connection, we further remark that the technique of spectral sequences is here too the proper device to look at (2.14); see also [VS: Chapt. III; p. 225, (7.36), along with the subsequent comments therein]. On the other hand, by still looking at (2.14), one gets, by virtue of the definitions that;

(2.15) $$\check{\mathbb{H}}^n(X, \mathcal{E}^{\cdot}) \equiv \varinjlim_{\mathcal{U}} \check{\mathbb{H}}^n(\mathcal{U}, \mathcal{E}^{\cdot}) := \left(\sum_{\mathcal{U}} \check{\mathbb{H}}^n(\mathcal{U}, \mathcal{E}^{\cdot}) \right) \Big/ \sim .$$

The equivalence relation that appears in the last member of the previous relation refers, by virtue of what was hinted at in (2.12), to the relation defined on the disjoint union, as indicated in (2.15), by refinement of the (proper) open coverings of X considered, indexing the corresponding inductive system, as in (2.12.1). Therefore, in view of (2.15), we further note that

(2.16)

with every element

(2.16.1) $$[z] \in \check{\mathbb{H}}^n(X, \mathcal{E}^{\cdot})$$

(viz. an n-dimensional hypercohomology class of X, cf. (2.11)), there is associated an open covering $\mathcal{U}$ of X such that one has

(2.16.2) $$[z] \in \check{\mathbb{H}}^n(\mathcal{U}, \mathcal{E}^{\cdot}),$$

where one also takes into account the aforementioned equivalence relation on the disjoint union; viz. that same definition of an inductive limit space.

In particular, for $n = 1$, (2.15) yields

(2.17) $$\check{\mathbb{H}}^1(X, \mathcal{E}^{\cdot}) := \varinjlim_{\mathcal{U}} \check{\mathbb{H}}^1(\mathcal{U}, \mathcal{E}^{\cdot}) = \sum_{\mathcal{U}} \check{\mathbb{H}}^1(\mathcal{U}, \mathcal{E}^{\cdot}).$$

Indeed, the last relation is an adaptation to the present setting (cf. also (2.11)) of [VS: Chapt. III; p. 185, Lemma 4.3], in conjunction with N. Bourbaki [2: p. 92, Remarque 1]. ∎

We are going to apply our previous conclusion in (2.17) straightforwardly in Section 3 by considering in particular a 2-term $\mathcal{A}$-complex on X. In turn, this particular instance is actually at the basis of our main result in Section 5, referring to the classification, via Čech hypercohomology, of the Maxwell fields on X, this being also our principal objective of this chapter. As an application, one can get, for instance, within the present abstract set-up (cf. Section 6 below), Maxwell's equations (in vacuo).

3 Čech Hypercohomology Relative to a Two-Term $\mathcal{A}$-Complex

As already mentioned, we specialize in the present section to a particular (differential) $\mathcal{A}$-complex on X consisting just of two (non zero) $\mathcal{A}$-modules, $\mathcal{E}^0$ and $\mathcal{E}^1$, along with the corresponding differential, say d, between them (all the rest being zero). That is, we suppose that we are given, as usual, a $\mathbb{C}$-algebraized space,

(3.1) $$(X, \mathcal{A}),$$

as well as the following 2-term (differential) $\mathcal{A}$-complex on X:

(3.2) $$\mathcal{E}^\bullet : 0 \longrightarrow \mathcal{E}^0 \xrightarrow{d^0 \equiv d} \mathcal{E}^1 \xrightarrow{d^1 = 0} 0 \longrightarrow \cdots \quad .$$

By applying the notation of (1.3), we assume that we have the following $\mathcal{A}$-complex on X:

(3.3) $$\mathcal{E}^\bullet \equiv (\mathcal{E}^\bullet, d) = \{(\mathcal{E}^0, d), (\mathcal{E}^1, 0)\}.$$

By further employing the terminology of Section 2, let us first consider an open covering of X, say

(3.4) $$\mathcal{U} = (U_\alpha)_{\alpha \in I},$$

so that we can next look at the Čech cochain $\mathcal{A}(X)$-bicomplex with respect to $\mathcal{E}^\bullet$ as in (3.3). One obtains (see also (2.4))

(3.5) $$\check{C}^\bullet(\mathcal{U}, \mathcal{E}^\bullet) \equiv (\check{C}^\bullet(\mathcal{U}, \mathcal{E}^\bullet), \delta, d) := (\{\check{C}^n(\mathcal{U}, \mathcal{E}^m)\}, \delta, d),$$

such that one has

(3.6) $$n \in \mathbb{Z}_+ \ \text{and} \ m = 0, 1.$$

Furthermore, according to (2.4), one defines that

(3.7)

in (3.5),

(3.7.1) $$\delta \equiv (\delta^n)_{n \in \mathbb{Z}_+}$$

stands for the respective Bockstein operators between the cochain $\mathcal{A}(X)$-modules concerned, as in (3.5) (cf. also (3.8) below, along with [VS: Chapt. III; p. 176, (4.15)]), while by virtue of (3.3), one has

(3.7.2) $$d \equiv (d^0, 0),$$

such that one further sets

(3.7.3) $$d \equiv (d^i)_{i \in \mathbb{Z}_+}$$

with

(3.7.4) $$d^0 \equiv d, \ \text{and} \ d^i = 0, \ \text{for} \ i \in \mathbb{N}.$$

Now, by analogy with (2.5), we depict the above by the following diagram that has (cf. (2.6)) commutative rectangles:

(3.8)

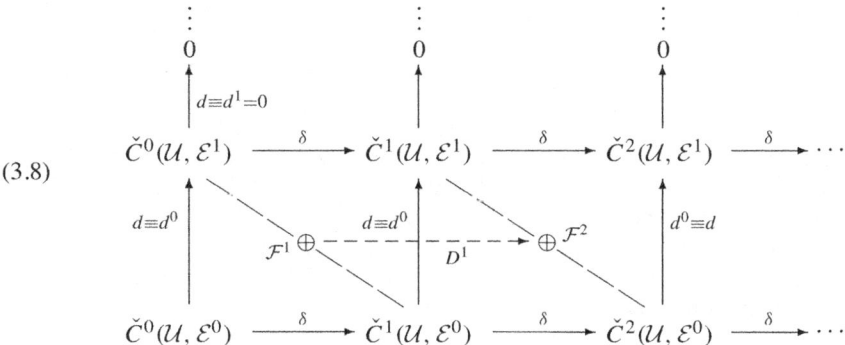

Consequently, based on (2.7), one can further consider the total $\mathcal{A}(X)$-complex of (3.5),

$$(3.9) \qquad (tot(\check{C}^{\bullet}(\mathcal{U}, \mathcal{E}^{\bullet})), D) \equiv (\mathcal{F}^{\bullet}, D) \equiv \{(\mathcal{F}^p, D^p)\}_{p \in \mathbb{Z}_+},$$

where,, we set

$$(3.10) \qquad tot(\check{C}^{\bullet}(\mathcal{U}, \mathcal{E}^{\bullet}))) \equiv \mathcal{F}^{\bullet}.$$

In the sequel we shall use details of the previous $\mathcal{A}$- complex $\mathcal{F}^{\bullet}$, in the particular $\mathcal{A}$-complexes we consider in Sections 4, 5. Therefore, for convenience, we exhibit in the sequel explicitly the individual issues of the same $\mathcal{A}$-complex as they are recorded in (3.9); thus, we look in particular at the three first $\mathcal{A}(X)$-modules of (3.9), which we shall also use in the sequel, while we give the general form of any one of them in (3.12), according to (1.43). So we have (cf. also (3.8))

$$(3.11) \qquad \begin{aligned} \mathcal{F}^0 &= \check{C}^0(\mathcal{U}, \mathcal{E}^0), \\ \mathcal{F}^1 &= \check{C}^1(\mathcal{U}, \mathcal{E}^0) \oplus \check{C}^0(\mathcal{U}, \mathcal{E}^1), \\ \mathcal{F}^2 &= \check{C}^2(\mathcal{U}, \mathcal{E}^0) \oplus \check{C}^1(\mathcal{U}, \mathcal{E}^1), \end{aligned}$$

where we have also taken into account that by our hypothesis (see (3.2) or (1.2)), all the cochain complexes involved are only of positive degree. On the other hand, by the same argument as before (cf. thus (1.43), along with (3.5), (3.6)), one obtains in full generality the following relation:

$$(3.12) \qquad \mathcal{F}^p = \check{C}^p(\mathcal{U}, \mathcal{E}^0) \oplus C^{p-1}(\mathcal{U}, \mathcal{E}^1), \quad p \in \mathbb{Z}_+,$$

of which equations (3.11) are a particular instance.

Furthermore, the respective differentials between the $\mathcal{A}(X)$-modules in (3.11) are given by the relations (see also (1.45), as well as (3.18))

$$D^0 = \delta^{0,0} + d^{0,0} \equiv \delta + d^0 \equiv \delta + d$$

(3.13)

$$: \mathcal{F}^0 = \check{C}^0(\mathcal{U}, \mathcal{E}^0) \longrightarrow \mathcal{F}^1 = \check{C}^1(\mathcal{U}, \mathcal{E}^0) \oplus \check{C}^0(\mathcal{U}, \mathcal{E}^1),$$

and also

(3.14)

$$D^1 = \delta^{1,0} + (-1)^1 d^{1,0} + \delta^{0,1} + (-1)^0 d^{0,1} = \delta + (\delta - d^0)$$

$$\equiv \delta \oplus (\delta - d) : \mathcal{F}^1 \longrightarrow \mathcal{F}^2.$$

Caution! By an obvious abuse of notation, we have employed for convenience the same symbol δ for various, different Bockstein operators ($\mathcal{A}(X)$-morphisms); similar identifications have been made for the various d's, but see also (3.2) or (3.7.4). In this connection, cf. also diagram (3.8) along with (1.31) and (1.32).

Based further on (1.45) and also taking into account (3.7.4), one gets the following general form of the D^p's.

(3.15)

$$D^p = \delta + (-1)^0 d^0 + \sum_{i=1}^{p} \delta^{p-1,i}, \qquad p \in \mathbb{Z}_+.$$

For convenience and applying the same abuse of notation as above, we also write

(3.16)

$$D^p \equiv \delta + d + \sum_{i=1}^{p} \delta^{p-1,i}, \qquad p \in \mathbb{Z}_+.$$

In practice we set, for convenience

(3.17)

$$\delta^{p,i} \equiv \delta, \quad \text{for any } p \in \mathbb{Z}_+, \text{ and } i = 0, 1, \dots, p.$$

On the other hand, according to the definitions (see (1.4) as applied to the (total) differential $\mathcal{A}(X)$-complex (3.9)), one has for the respective differentials ($\mathcal{A}(X)$-morphisms);

(3.18)

$$D^p : \mathcal{F}^p \longrightarrow \mathcal{F}^{p+1}, \qquad p \in \mathbb{Z}_+,$$

where the $\mathcal{A}(X)$-modules $\mathcal{F}^p$, $p \in \mathbb{Z}_+$, involved are given by (3.12), (which also explains the notation in (3.16) concerning the (Bockstein operators) $\delta^{p,i}$'s.

Applications of the preceding, as already mentioned, will be given throughout the remaining sections of this chapter. The above constitute the general theoretic point of view of the rest of our discussion in the present chapter. On the other hand, having in mind just this sort of applications, we are interested in particular in describing the corresponding 1-dimensional hypercohomology $\mathcal{A}(X)$-module of X with respect to the given $\mathcal{A}$-complex $\mathcal{E}^{\bullet}$ as in (3.2), something that we examine in the following subsection.

3.1 Identification of $\check{\mathbb{H}}^1(X, \mathcal{E}^0 \xrightarrow{d^0} \mathcal{E}^1)$

As the title indicates, our aim in this subsection is to point out the form of the elements of the (hyper)cohomology space ($\mathcal{A}(X)$-module). We are going to apply our conclusions to a similar description, hence, to a corresponding cohomological classification as well, of the Maxwell fields on X (cf. Section 5), the main objective of this chapter.

Suppose that we are given an arbitrary element of the space in the title of this subsection, that is, assume that we have (see also (3.2))

$$(3.19) \qquad [z] \in \check{\mathbb{H}}^1(X, \mathcal{E}^0 \xrightarrow{d^0} \mathcal{E}^1) \equiv \check{\mathbb{H}}^1(X, \mathcal{E}^{\cdot}).$$

Therefore, by virtue of (2.17), one concludes that

$$(3.20) \qquad [z] \in \check{\mathbb{H}}^1(\mathcal{U}, \mathcal{E}^{\cdot})$$

for some open covering $\mathcal{U}$ of X, so that (cf. (2.11)) one finally obtains that

$$(3.21) \qquad [z] \in \check{\mathbb{H}}^1(\mathcal{U}, \mathcal{E}^{\cdot}) = \ker D^1 / \operatorname{im} D^0,$$

where the D^i's, with $i = 0, 1$, are given by (3.13) and (3.14). In other words, one gets

$$(3.22) \qquad [z] = z + \operatorname{im} D^0,$$

such that one has (cf. also (3.14), along (3.8) and (3.11))

$$
(3.23) \qquad
\begin{aligned}
z \in \ker D^1 &= \ker(\delta \oplus (\delta - d^0)) = \ker \delta \oplus \ker(\delta - d^0) \\
&\equiv \ker \delta \oplus \ker(-d^0 + \delta) \equiv \ker \delta \oplus \ker(-d + \delta) \\
&\subseteq \check{C}^1(\mathcal{U}, \mathcal{E}^0) \oplus \check{C}^0(\mathcal{U}, \mathcal{E}^1).
\end{aligned}
$$

In this connection, cf. also N. Bourbaki [3: Chapt. II, p. 14, Corollaire 1; (25)].

Consequently, in view also of the last relation, one now concludes that (cf. (3.8) and (3.11), as well)

$$(3.24) \qquad z \equiv (g, \theta) \in \check{C}^1(\mathcal{U}, \mathcal{E}^0) \oplus \check{C}^0(\mathcal{U}, \mathcal{E}^1) = \mathcal{F}^1,$$

so that if in particular

$$(3.25) \qquad \mathcal{U} = (U_\alpha)_{\alpha \in I}$$

is the open covering of X under consideration, one obtains

$$(3.26) \qquad g \equiv (g_{\alpha\beta}) \in \check{C}^1(\mathcal{U}, \mathcal{E}^0)$$

as well as

$$(3.27) \qquad \theta \equiv (\theta_\alpha) \in \check{C}^0(\mathcal{U}, \mathcal{E}^1).$$

By still looking at (3.23) (cf. also (3.14), in that by hypothesis, one has $d^{0,1} = 0$), we further obtain that

(3.28) $g \in \ker \delta \subseteq \check{C}^1(\mathcal{U}, \mathcal{E}^0), \quad that\ is, \quad \delta(g) = 0;$

hence one finally gets

(3.29) $g \equiv (g_{\alpha\beta}) \in \check{Z}^1(\mathcal{U}, \mathcal{E}^0) \subseteq \check{C}^1(\mathcal{U}, \mathcal{E}^0).$

On the other hand, by (3.23), one further concludes that

(3.30) $(g, \theta) \in \ker(-d + \delta);$

namely, one has

(3.31) $(\delta - d)(\theta, g) = \delta\theta - dg = 0,$

or

(3.32) $\delta\theta = dg,$

so that by virtue of (3.26) and (3.27), one finally obtains

(3.33) $\delta(\theta_\alpha) = d(g_{\alpha\beta}).$

The preceding constitutes the proof of the following basic result.

Theorem 3.1 Suppose we are given a $\mathbb{C}$-algebraized space

(3.34) $(X, \mathcal{A})$

as well as the following (differential) 2-term $\mathcal{A}$-complex on X,

(3.35) $\mathcal{E}^{\bullet} : 0 \longrightarrow \mathcal{E}^0 \xrightarrow{d^0 \equiv d} \mathcal{E}^1 \longrightarrow 0 \longrightarrow \cdots \quad .$

Furthermore, let

(3.36) $\mathcal{U} = (U_\alpha)_{\alpha \in I}$

be an open covering of X.
    Then, a given pair,

(3.37) $(g, \theta) \in \check{C}^1(\mathcal{U}, \mathcal{E}^0) \times \check{C}^0(\mathcal{U}, \mathcal{E}^1)$

yields an element

(3.38) $[(g, \theta)] \equiv [z] \in \check{\mathbb{H}}^1(\mathcal{U}, \mathcal{E}^{\bullet}) \underset{\longrightarrow}{\subset} \check{\mathbb{H}}^1(X, \mathcal{E}^{\bullet})$

(cf. also (2.17)) if and only if one has (see (3.26), (3.27), as above)

(3.39) $\delta(g) \equiv \delta(g_{\alpha\beta}) = 0,$

along with the relation

(3.40) $\delta(\theta_\alpha) = d(g_{\alpha\beta}). \quad \blacksquare$

As we shall see, the preceding lie at the basis of our main result in Section 5. For reasons that will become clear and in view also of Theorem 3.1, we call the above $\mathcal{A}(X)$-module (cf. (3.21), (3.23))

$$(3.41) \qquad \ker D^1 \subseteq \check{C}^1(\mathcal{U}, \mathcal{E}^0) \oplus \check{C}^0(\mathcal{U}, \mathcal{E}^1)$$

the *abstract* (or *generalized*) *Maxwell space* of X with respect to the given (2-*term*) $\mathcal{A}$-complex $\mathcal{E}^\bullet$ on X, as in (3.35).

Within the same vein of ideas, *a pair*

$$(3.42) \qquad (g, \theta) \in Dom\, D^1 = \mathcal{F}^1$$

(cf. (3.24), (3.8), and (3.14)), as in (3.37), that also satisfies (3.39) and (3.40), is still called an *abstract* (or else *generalized*) *Maxwell field* on X. Thus, in other words,

(3.43)

the abstract Maxwell fields on X, with respect to given 2-term $\mathcal{A}$-complex $\mathcal{E}^\bullet$ on X, as in (3.29), constitute (or even, are characterized by) the solution space of the differential operator D^1 (cf. (3.14)), viz. the $\mathcal{A}(X)$-module

$$(3.43.1) \qquad \ker D^1,$$

as in (3.41).

Indeed, to put it differently,

(3.44)

a given pair

$$(3.44.1) \qquad (g, \theta)$$

as in (3.42) provides an abstract Maxwell field if and only if it is a solution (viz. a "zero place") of the differential operator) D^1, cf. (3.14); that is, if and only if one has

$$(3.44.2) \qquad D^1(g, \theta) = 0,$$

or equivalently, whenever one has

$$(3.44.3) \qquad (g, \theta) \in \ker D^1.$$

In fact, the preceding constitutes the abstract setting of our discussion in Section 5 in the sequel, which refers to the study of (concrete) Maxwell fields, viz. pairs $(\mathcal{L}, D)$ on X, as they are defined in Chapter III; Section 1. Indeed, a preamble to that material is already supplied by the next section.

4 Čech Hypercohomology, with Respect to the Two-Term Z-Complex $\mathcal{A}^{\cdot} \xrightarrow{\tilde{\partial}} \Omega^1$

We specialize below our previous considerations in Section 3 to the particular case of a (differential) 2-term $\mathbb{Z}$-complex of the form

$$(4.1) \qquad \mathcal{E}^{\cdot} : 0 \longrightarrow \mathcal{A}^{\cdot} \xrightarrow{\tilde{\partial}} \Omega^1 \longrightarrow 0 \longrightarrow \cdots .$$

So we first explain the relevant terminology, which just has employed for (4.1): Thus, suppose that we are given, as always, a $\mathbb{C}$-algebraized space

$$(4.2) \qquad\qquad (X, \mathcal{A}),$$

together with a differential triad

$$(4.3) \qquad\qquad (\mathcal{A}, \partial, \Omega^1)$$

on X (see Chapter I or Chapt. III; Introduction). Hence, according to the general theory (cf. Chapter I), one then defines the corresponding *logarithmic derivation*

$$(4.4) \qquad\qquad \tilde{\partial} : \mathcal{A}^{\cdot} \longrightarrow \Omega^1,$$

which is associated with the given $\mathbb{C}$-derivation (flat $\mathcal{A}$-connection) ∂, as in (4.3), such that one has, by definition

$$(4.5) \qquad\qquad \tilde{\partial}(\alpha) := \alpha^{-1} \cdot \partial(\alpha)$$

for any continuous (local) section

$$(4.6) \qquad\qquad \alpha \in \mathcal{A}^{\cdot}(U) = \mathcal{A}(U)^{\cdot}$$

with U open in X, where $\mathcal{A}^{\cdot}$ stands for the (abelian) group sheaf of units (invertible elements) of (the unital $\mathbb{C}$-algebra sheaf) $\mathcal{A}$; see also Chapter I or Chapt. II; (7.24), and (6.6). Indeed (see also [VS: Chapt. VI; p. 6, (1.27)]),

(4.7)
> $\tilde{\partial}$ is a morphism of the (abelian) group sheaves involved in (4.4), viz. one has
>
> $$(4.7.1) \qquad\qquad \tilde{\partial}(s \cdot t) = \tilde{\partial}(s) + \tilde{\partial}(t)$$
>
> for any s, t in $\mathcal{A}^{\cdot}(U)$ as in (4.6).

Consequently, the 2-term complex of the (abelian) group sheaves in (4.1) is a differential, in the sense that $\tilde{\partial}$ is, by virtue of (4.7.1), a $\mathbb{Z}$-morphism (viz. morphism of sheaves of groups), where each one of the group sheaves in (4.1) can be viewed as a $\mathbb{Z}$-module, with $\mathbb{Z}$ itself being considered as a constant sheaf (of rings) on X (cf.

also [VS: Chapt. I; p. 17, Section 4.1]). The above now justify completely the terminology, employed in (4.1); therefore, we can further apply all the mechanism that was already worked out in in Section 3, that is, employ Čech hypercohomology in the case of the particular $\mathbb{Z}$-complex at issue. (As already mentioned, it was the same complex as above that actually motivated our discussion in Section 3, the complex (3.2) being thus a direct generalization of the above $\mathbb{Z}$-complex (4.1).)

In view of (4.1), we have the (2-term) $\mathbb{Z}$-complex

$$(4.8) \qquad (\mathcal{E}^{\cdot}, d) \equiv \{(\mathcal{A}^{\cdot}, d^0 \equiv \tilde{\partial}), (\Omega^1, 0)\},$$

which we also denote, occasionally, by

$$(4.9) \qquad \mathcal{E}^{\cdot} \equiv (\mathcal{E}^{\cdot}, \tilde{\partial}) \equiv \{\mathcal{A}^{\cdot} \xrightarrow{\tilde{\partial}} \Omega^1\}$$

(see, for instance, the heading of the present section). Furthermore, consider an open covering of X, say

$$(4.10) \qquad \mathcal{U} = (U_\alpha)_{\alpha \in I}.$$

Therefore, we can next look at the Čech cochain $\mathbb{Z}$-bicomplex, that is, a bicomplex of (abelian) groups, associated with the $\mathbb{Z}$-complex $(\mathcal{E}^{\cdot}, \tilde{\partial})$, as in (4.9). Thus, one obtains (cf. also (2.4))

$$(4.11) \qquad \{\check{C}^{\cdot}(\mathcal{U}, \mathcal{E}^{\cdot}), \delta, d\} \equiv (\{\check{C}^n(\mathcal{U}, \mathcal{E}^m)\}_{(n,m) \in \mathbb{Z}_+^2}; \delta, d),$$

so that by analogy with (3.5), one has

$$(4.12) \qquad d^0 \equiv \tilde{\partial}, \quad \text{while} \quad d^i = 0, \qquad i \in \mathbb{N}.$$

Moreover, concerning our notation in (4.11), we have

$$(4.13) \qquad n \in \mathbb{Z}_+, \quad \text{while} \quad m = 0, 1,$$

so that according to (4.8) (or even (4.9)), one further sets

$$(4.14) \qquad \mathcal{E}^0 \equiv \mathcal{A}^{\cdot} \quad \text{and} \quad \mathcal{E}^1 \equiv \Omega^1.$$

For convenience, as we have already done, we depict the above $\mathbb{Z}$-bicomplex as in (4.11), specialized according to (4.12)–(4.14) by the following diagram, to the preceding ones, as in (1.37), (2.5), and (3.8). By virtue of what has been said in (2.6), the same diagram has the corresponding commutative rectangles. So we have

$$(4.15)$$

Hence, by following our previous argument in Section 3, we can further consider the total (differential) $\mathbb{Z}$-complex, corresponding to (4.11), specialized as above to (4.15), associated with the given open covering $\mathcal{U}$ of X, as in (4.10). Therefore, we have (see also (3.9))

$$(4.16) \qquad (tot(\check{C}^{\cdot}(\mathcal{U}, \mathcal{E}^{\cdot})), D) \equiv \{(\mathcal{F}^p, D^p)\}_{p \in \mathbb{Z}_+},$$

so that in particular one obtains, according to (3.11) and (4.14)

$$(4.17) \qquad \begin{aligned} \mathcal{F}^0 &= \check{C}^0(\mathcal{U}, \mathcal{A}^{\cdot}), \\ \mathcal{F}^1 &= \check{C}^1(\mathcal{U}, \mathcal{A}^{\cdot}) \oplus \check{C}^0(\mathcal{U}, \Omega^1), \\ \mathcal{F}^2 &= \check{C}^2(\mathcal{U}, \mathcal{A}^{\cdot}) \oplus \check{C}^1(\mathcal{U}, \Omega^1). \end{aligned}$$

The preceding (abelian) groups are actually those that we shall apply only in the sequel from those in (4.16). By further looking at the corresponding differentials (viz., for the case in hand, group morphisms, see (4.7.1)), one obtains, by virtue of (3.13), (3.14), and (4.12),

$$(4.18) \qquad D^0 = \delta + \tilde{\partial} : \mathcal{F}^0 \longrightarrow \mathcal{F}^1,$$

as well as

$$(4.19) \qquad D^1 = \delta \oplus (\delta - \tilde{\partial}) : \mathcal{F}^1 \longrightarrow \mathcal{F}^2.$$

Of course, we can still apply (3.12) and (3.16) to obtain the form of $\mathcal{F}^p$ and D^p for $p > 1$. (However, see also the comments following (4.17)). We come next to the particular abelian group from the (Čech hyper)cohomology of (4.16) (see (2.9) (2.11)), which will be of our concern in the sequel (cf. also (3.19) or (3.21)).

4.1 Characterization of the (Abelian) Čech Hypercohomology Group $\check{\mathbb{H}}^1(X, \mathcal{A}^{\cdot} \overset{\tilde{\partial}}{\to} \Omega^1)$

Following the general device as exhibited above in Section 2, one defines (cf. (2.14), (2.17))

$$(4.20) \qquad \check{\mathbb{H}}^1(X, \mathcal{A}^{\cdot} \overset{\tilde{\partial}}{\to} \Omega^1) \equiv \check{\mathbb{H}}^1(X, \mathcal{E}^{\cdot}) := \varinjlim_{\mathcal{U}} \check{\mathbb{H}}^1(\mathcal{U}, \mathcal{E}^{\cdot}) = \sum_{\mathcal{U}} \check{\mathbb{H}}^1(\mathcal{U}, \mathcal{E}^{\cdot})$$

(cf. also (4.9) concerning the abbreviated notation applied above), while one has, by definition (see (2.11)),

$$(4.21) \qquad \check{\mathbb{H}}^1(\mathcal{U}, \mathcal{E}^{\cdot}) := \ker D^1 / \operatorname{im} D^0.$$

Consequently, one gets the following basic result, being a straightforward consequence within the present particular case, of our previous conclusion in Theorem 3.1.

Lemma 4.1 Suppose we are given a differential triad

$$(4.22) \qquad\qquad (\mathcal{A}, \partial, \Omega^1)$$

on a topological space X, as well as an open covering of X

$$(4.23) \qquad\qquad \mathcal{U} = (U_\alpha)_{\alpha \in I}.$$

Then, a given pair

$$(4.24) \qquad \begin{aligned} (g, \theta) &\equiv ((g_{\alpha\beta}), (\theta_\alpha)) \in \check{C}^1(\mathcal{U}, \mathcal{A}^{\cdot}) \times \check{C}^0(\mathcal{U}, \Omega^1) \\ &\cong \check{C}^1(\mathcal{U}, \mathcal{A}^{\cdot}) \oplus \check{C}^0(\mathcal{U}, \Omega^1) \equiv \mathcal{F}^1 = Dom D^1 \end{aligned}$$

(cf. also (4.19)) belongs to the solution space of the differential operator D^1 (loc. cit.); that is, one has

$$(4.25) \qquad\qquad ((g_{\alpha\beta}), (\theta_\alpha)) \in \ker D^1$$

if and only if the following relations hold:

$$(4.26) \qquad\qquad \delta(g_{\alpha\beta}) = 0$$

and

$$(4.27) \qquad\qquad \delta(\theta_\alpha) = \tilde{\partial}(g_{\alpha\beta}).$$

Proof. See the comments before the statement of the lemma, along with (3.39) and (3.40) (cf. also (3.44)). ∎

On the other hand, by applying the terminology of the same Theorem 3.1 in conjunction with Lemma 4.1, we can obtain the following description of the cohomology group in the title of this subsection (cf. also (4.20)). Namely, one concludes that

> a given pair (g, θ), as in (4.24), yields an element (cf. also (4.21), along with (4.9))

$$(4.28) \qquad (4.28.1) \qquad \begin{aligned} &\equiv [z] \in \check{\mathbb{H}}^1(\mathcal{U}, \mathcal{E}^{\cdot}) \subset \check{\mathbb{H}}^1(X, \mathcal{E}^{\cdot}) \\ &\equiv \check{\mathbb{H}}^1(X, \mathcal{A}^{\cdot} \xrightarrow{\tilde{\partial}} \Omega^1) \end{aligned}$$

> if, and only if (4.26) and (4.27) are in force.

Furthermore, based on (4.24), one obtains;

$$(4.29) \qquad\qquad g \equiv (g_{\alpha\beta}) \in \check{C}^1(\mathcal{U}, \mathcal{A}^{\cdot}),$$

so that if (4.26) holds, then, in particular, one concludes that

$$(4.30) \qquad g \equiv (g_{\alpha\beta}) \in \check{Z}^1(\mathcal{U}, \mathcal{A}^{\cdot}) \subseteq \check{C}^1(\mathcal{U}, \mathcal{A}^{\cdot}).$$

In other words, one can then look at (4.30) as the coordinate 1-cocycle with respect to the open covering $\mathcal{U}$ of X, as in (4.23), of a line sheaf $\mathcal{L}$ on X, viz. one has

$$(4.31) \qquad \mathcal{L} \longleftrightarrow (g_{\alpha\beta}) \equiv g$$

(see Chapt. III; (2.2) or Chapt. II; (7.18) and (7.19), for $n = 1$). Therefore, in the case that *(4.26)* is in force, the relation *(4.24)* takes the form

$$(4.32) \qquad (g, \theta) \equiv ((g_{\alpha\beta}), (\theta_\alpha)) \in \check{Z}^1(\mathcal{U}, \mathcal{A}^{\cdot}) \times \check{C}^0(\mathcal{U}, \Omega^1).$$

In particular, one still obtains from (4.24),

$$(4.33) \qquad \theta \equiv (\theta_\alpha) \in \check{C}^0(\mathcal{U}, \Omega^1),$$

so that if, moreover, *(4.27)* is valid, one then concludes that

(4.34)
> the 0-cochain (θ_α), as in (4.33), may be construed (transformation law of potentials, see Chapt. III; (2.33), along with (2.34.2) and/or (2.36)) as yielding an $\mathcal{A}$-connection D on the line sheaf $\mathcal{L}$, the latter being defined by (4.31). Thus, one gets;
>
> $$(4.34.1) \qquad D \longleftrightarrow (\theta_\alpha) \equiv \theta.$$

See also Chapt. III; Lemma 2.1. Thus, all told, one concludes the following:

(4.35)
> any pair (g, θ), as in (4.24), which further supplies a solution of the equation
>
> $$(4.35.1) \qquad D^1 = 0,$$
>
> or equivalently, whenever one has
>
> $$(4.35.2) \qquad (g, \theta) \in \ker D^1,$$
>
> then one gets a Maxwell field $(\mathcal{L}, D)$ on X, in view of (4.31) and (4.34.1).

On the other hand,

(4.36)
> the converse of (4.35) holds, as well; viz. any Maxwell field $(\mathcal{L}, D)$ on X can be characterized by local data with respect to an open covering of X (or local frame of $\mathcal{L}$), that is, it is a pair, as in (4.24), that provides a solution of (4.35.1). (See Chapt. III; Section 2, in particular, Lemma 2.1.)

In fact, as we shall presently see in Section 5 (cf. Theorem 5.1), the equation

$$(4.37.1) \qquad\qquad D^1 = 0$$

(4.37)

characterizes the Maxwell group of X,

$$(4.37.2) \qquad\qquad \Phi^1_{\mathcal{A}}(X)^{\nabla}.$$

In other words, we can say that

the equation

$$(4.38.1) \qquad\qquad D^1 = 0$$

(4.38)   determines, through the cohomology group

$$(4.38.2) \qquad\qquad \check{\mathbb{H}}^1(X, \mathcal{A}^{\cdot} \xrightarrow{\tilde{\partial}} \Omega^1),$$

the Maxwell group of X, $\Phi^1_{\mathcal{A}}(X)^{\nabla}$.

5 Cohomological Wording of the Maxwell Group

We come now to discuss our main application of the preceding material, namely, the

(5.1)   coincidence of the Maxwell group of X with the 1st (Čech) hypercohomology group of X with respect to the 2-term $\mathbb{Z}$-complex

$$(5.2) \qquad\qquad \mathcal{E}^{\cdot} : 0 \longrightarrow \mathcal{A}^{\cdot} \xrightarrow{\tilde{\partial}} \Omega^1 \longrightarrow 0 \longrightarrow \cdots$$

(see also (4.1) along with Theorem 5.1). Indeed, our claim in (5.1) has been our
principal objective, the material being just the preamble, or even the abstraction, of
the relevant framework that one needs for our main conclusion in Theorem 5.1. In
this connection, we also refer to Section 4 for unexplained terms in the subsequent
discussion.

To start with, we assume that we are given again the framework of Section 4,
concerning in particular the terminology connected with (5.2). Our first goal is to
establish a map,

$$(5.3) \qquad\qquad \chi : \Phi^1_{\mathcal{A}}(X)^{\nabla} \longrightarrow \check{\mathbb{H}}^1(X, \mathcal{E}^{\cdot})$$

between the (abelian) groups concerned (cf. also (5.2), along with the ensuing discussion), which will be proved to be a group isomorphism:

So assume that we are given an element, say

$$(5.4) \qquad\qquad [(\mathcal{L}, D)] \in \Phi^1_{\mathcal{A}}(X)^{\nabla},$$

that is (see Chapt. III; (1.17), or even (1.19) therein) an isomorphism class of (gauge equivalent) Maxwell fields on X. Accordingly (loc. cit. Lemma 2.1), there exists a pair

$$(5.5) \qquad (g, \theta) \equiv ((g_{\alpha\beta}), (\theta_\alpha))$$

with respect to a given open covering of X,

$$(5.6) \qquad \mathcal{U} = (U_\alpha)_{\alpha \in I}$$

that is associated with a pair (Maxwell field) $(\mathcal{L}, D)$ on X, as in (5.4), viz. a local frame of $\mathcal{L}$ (ibid. (2.13)), in such a manner that one has;

$$(5.7) \qquad g \equiv (g_{\alpha\beta}) \in Z^1(\mathcal{U}, \mathcal{A}^{\boldsymbol{\cdot}})$$

as well as

$$(5.8) \qquad \theta \equiv (\theta_\alpha) \in \check{C}^0(\mathcal{U}, \Omega^1),$$

while one obtains the relation

$$(5.9) \qquad \delta(\theta_\alpha) = \tilde{\partial}(g_{\alpha\beta}).$$

The pair $(\mathcal{L}, D)$ was just a representative of the equivalence class (light ray), considered by (5.4).

Now, by virtue of (5.7), (5.8), as well as (5.9), one concludes, in view also of Lemma 4.1, that

(5.10)   the pair (g, θ), as in (5.5), which is associated with a given Maxwell field $(\mathcal{L}, D)$ on X through a local frame $\mathcal{U}$ of $\mathcal{L}$ (open covering of X, cf. (5.6)), satisfies the relation

$$(5.10.1) \qquad (g, \theta) \in \ker D^1.$$

Consequently,

the same pair (g, θ) as in (5.10) defines an element (cohomology class)

$$(5.11.1) \qquad [(g, \theta)] \in \check{\mathbb{H}}^1(\mathcal{U}, \mathcal{E}^{\boldsymbol{\cdot}}) \underset{\longrightarrow}{\subset} \check{\mathbb{H}}^1(X, \mathcal{E}^{\boldsymbol{\cdot}})$$

(cf. also (2.17)). That is, one sets

(5.11)
$$[(g, \theta)] := (g, \theta) + \operatorname{im} D^0$$
$$(5.11.2) \qquad \equiv (g, \theta) + D^0(\mathcal{F}^0) \in \ker D^1/\operatorname{im} D^0$$
$$\equiv \check{\mathbb{H}}^1(\mathcal{U}, \mathcal{E}^{\boldsymbol{\cdot}})$$

(see also (4.21)).

In this connection, we further note for immediate application that

(5.12) gauge equivalence of Maxwell fields (see Chapt. III; (1.16.1), as well as, Lemma 2.2) respects the equivalence relation in ker D^1, defined by

$$(5.12.1) \qquad \operatorname{im} D^0 = D^0(\mathcal{F}^0) \subseteq \mathcal{F}^1.$$

In fact, something much better is actually in force, in the sense that the equivalence relations at issue determine each other; namely, one has

(5.13) the relation

$$(5.13.1) \qquad (\mathcal{L}', D') \underset{u}{\sim} (\mathcal{L}, D)$$

(gauge equivalence of Maxwell fields, cf. Chapt. III; Lemma 2.2) is equivalent to

$$(5.13.2) \qquad (g', \theta') - (g, \theta) \in \operatorname{im} D^0.$$

Thus, by considering the above correspondence between the pairs $(\mathcal{L}, D)$ and (g, θ) (see (5.10), as well as, Chapt. III; (2.26)), suppose that (cf. (5.13.1))

$$(5.14) \qquad (\mathcal{L}', D') \sim (\mathcal{L}, D).$$

Hence, by looking at the respective pairs of cochains (cf. Chapt. III; Lemma 2.2), one obtains (see also 5.7), (5.8))

$$(5.15) \qquad g'_{\alpha\beta} = \delta(s_\alpha^{-1}) \cdot g_{\alpha\beta}, \quad \text{or} \quad g' = \delta(s^{-1}) \cdot g,$$

as well as

$$(5.16) \qquad \theta'_\alpha = \theta_\alpha + \tilde{\partial}(s_\alpha^{-1}), \quad \text{or} \quad \theta' = \theta + \tilde{\partial}(s^{-1}),$$

where (ibid.)

$$(5.17) \qquad s \equiv (s_\alpha) \in \check{C}^0(\mathcal{U}, \mathcal{A}^\cdot).$$

Consequently, one obtains;

$$(5.18) \qquad \begin{aligned} (g', \theta') - (g, \theta) &= (g'g^{-1}, \theta' - \theta) = (\delta(s^{-1}), \tilde{\partial}(s^{-1})) \\ &= (\delta, \tilde{\partial})(s^{-1}) \equiv (\delta \oplus \tilde{\partial})(s^{-1}) \equiv D^0(s^{-1}) \in \operatorname{im} D^0 \end{aligned}$$

(see also (4.17), (4.18), and (5.17)).
    On the other hand, by assuming (5.13.2), one has

$$(5.19) \qquad \begin{aligned} (g', \theta) - (g, \theta) &= (g'g^{-1}, \theta' - \theta) = D^0(s^{-1}) = (\delta \oplus \tilde{\partial})(s^{-1}) \\ &= (\delta, \tilde{\partial})(s^{-1}) = (\delta(s^{-1}), \tilde{\partial}(s^{-1})), \end{aligned}$$

so that one gets

(5.20) $$g'g^{-1} = \delta(s^{-1}), \quad \text{and} \quad \theta' - \theta = \tilde{\partial}(s^{-1}),$$

that is,

(5.21) $$g' = \delta(s^{-1}) \cdot g \quad \text{and} \quad \theta' = \theta + \tilde{\partial}(s^{-1}),$$

or even

(5.22) $$g'_{\alpha\beta} = \delta(s_{\alpha}^{-1}) \cdot g_{\alpha\beta}, \quad \text{and} \quad \theta'_{\alpha} = \theta_{\alpha} + \tilde{\partial}(s_{\alpha}^{-1}),$$

which was to be proved, viz. (Chapt. III; Lemma 2.2) (5.13.1) which thus proves completely our claim in (5.13). ∎

The situation we have had so far can be depicted, by the following diagram

(5.23)

$$
\begin{array}{ccc}
(\mathcal{L}, D) & \xleftarrow{\ \text{Chapt. III; (2.26)}\ } & (g, \theta) \in \ker D^1 \\
& \mathcal{U} & \\
& \searrow \quad \swarrow\ {\scriptstyle (5.11)} & \\
& & \\
& [(g, \theta)] \in \ker D^1/\mathrm{im}\, D^0 & \\
& = \check{\mathbb{H}}^1(\mathcal{U}, \mathcal{E}^{\cdot}) \subseteq \check{\mathbb{H}}^1(X, \mathcal{E}^{\cdot}), &
\end{array}
$$

which shows the final correspondence by means of the dashed line. By virtue of our previous conclusion in (5.13), the same correspondence passes to the quotient, viz. to the respective equivalence class of $(\mathcal{L}, D)$, so that one finally gets the well-defined map

(5.24) $$\chi : [(\mathcal{L}, D)] \longmapsto [(g, \theta)] : \Phi^1_A(X)^{\nabla} \longrightarrow \check{\mathbb{H}}^1(X, \mathcal{E}^{\cdot})$$

that we were looking for, as in (5.3). (In this regard, see also N. Bourbaki [1: Chapt. II; p. 118]). So we first prove that

(5.25)   the map χ, as in (5.24), is a morphism of (abelian) groups.

Indeed, one has (cf. also Chapt. III; Theorem 2.1)

(5.26)
$$
\begin{aligned}
\chi([(\mathcal{L}, D)] \cdot [(\mathcal{L}', D')]) &= \chi([(\mathcal{L}, D)] \otimes_A [(\mathcal{L}', D')]) \\
&= \chi([(\mathcal{L} \otimes_A \mathcal{L}', D \otimes D')]) \equiv [(h, \omega)],
\end{aligned}
$$

in the sense that one sets (see Chapt. III, Lemma 2.1, or even (2.26) therein)

(5.27) $$(\mathcal{L} \otimes_A \mathcal{L}', D \otimes D') \underset{\mathcal{U}}{\longleftrightarrow} (h, \omega).$$

On the other hand, by applying the notation of (5.7), one obtains

(5.28) $\mathcal{L} \otimes_A \mathcal{L}' \longleftrightarrow h \equiv g \otimes g' := g \cdot g' \equiv ((g_{\alpha\beta}) \cdot (g'_{\alpha\beta})) = (g_{\alpha\beta} \cdot g'_{\alpha\beta})$

(see also [VS: Chapt. V; p. 367, proof of Theorem 4.1, in particular (4.17)]). Furthermore, one has

(5.29) $$D \otimes D' \longleftrightarrow \omega \equiv (\omega_\alpha) \in \check{C}^0(\mathcal{U}, \Omega^1),$$

such that (cf. (5.8))

(5.30) $$\omega \equiv (\theta_\alpha) \otimes (\theta'_\alpha) = (\theta_\alpha + \theta'_\alpha) = (\theta_\alpha) + (\theta'_\alpha) \equiv \theta + \theta'$$

(loc. cit., Chapt. VIII; p. 233, (9.8)). Therefore, one gets (see (5.27), (4.17), along with Chapt. III; (2.28))

(5.31)
$$\chi([(\mathcal{L}, D)] \cdot [(\mathcal{L}', D')]) = [(h, \omega)] = [(gg', \theta + \theta')]$$
$$= [(g, \theta)] + [(g', \theta')] = \chi([(\mathcal{L}, D)]) + \chi([(\mathcal{L}', D')]),$$

which proves our assertion in (5.25). ∎

We finally prove that

(5.32)   the group morphism χ, as in (5.24), is an isomorphism of (abelian) groups.

We first show that χ is one-to-one: Namely, suppose that (cf. also (5.24))

(5.33) $$\chi([(\mathcal{L}, D)]) := [(g, \theta)] = 0 \in \check{\mathbb{H}}^1(X, \mathcal{E}^{\cdot})$$

for some $[(\mathcal{L}, D)] \in \Phi^1_A(X)^\nabla$. In view of (2.17), one has

(5.34) $$\chi([(\mathcal{L}, D)]) = [(g, \theta)] = 0 \in \check{\mathbb{H}}^1(\mathcal{U}, \mathcal{E}^{\cdot}) = \ker D^1/\operatorname{im} D^0.$$

Therefore (see also (5.11) and (4.17)), one obtains

(5.35) $$(g, \theta) \in \operatorname{im} D^0 = D^0(\mathcal{F}^0) = D^0(\check{C}^0(\mathcal{U}, \mathcal{A}^{\cdot})),$$

that is, one gets (see (4.18))

(5.36) $$(g, \theta) = D^0(s^{-1}) = (\delta \oplus \tilde{\partial})(s^{-1}) = (\delta, \tilde{\partial})(s^{-1}) = (\delta(s^{-1}), \tilde{\partial}(s^{-1})),$$

where

(5.37) $$s \equiv (s_\alpha) \in \check{C}^0(\mathcal{U}, \mathcal{A}^{\cdot}).$$

Hence, in view of (5.36), one has

(5.38) $$g = \delta(s^{-1}) \text{ and } \theta = \tilde{\partial}(s^{-1}),$$

that is (see Chapt. III; Lemma 2.2), one concludes that

$$(5.39) \qquad\qquad (g, \theta) \sim (1, 0);$$

that is (see also Chapt. III; (2.8), (2.32)),

$$(5.40) \qquad\qquad [(\mathcal{L}, D)] = [(\mathcal{A}, \partial)] \equiv \mathbf{1} \in \Phi^1_{\mathcal{A}}(X)^\nabla,$$

which proves, by virtue of (5.33) that χ is one-to-one. ∎ Finally, we also prove that χ is onto: Thus, given an element

$$(5.41) \qquad\qquad z \equiv [(g, \theta)] \in \check{\mathbb{H}}^1(X, \mathcal{E}^\bullet),$$

one concludes that (cf. also (4.17), (4.21), and (2.17))

$$(5.42) \qquad\qquad [(g, \theta)] \in \check{\mathbb{H}}^1(\mathcal{U}, \mathcal{E}^\bullet) = \ker D^1 / \operatorname{im} D^0,$$

so that by the same definitions, one has

$$(5.43) \qquad\qquad (g, \theta) \in \ker D^1.$$

Therefore (cf. also (4.24), (4.28)), one obtains

$$(5.44) \qquad\qquad \delta(g) = 0 \quad \text{and} \quad \delta(\theta) = \tilde{\partial}(g).$$

This means (cf. Chapt. III; Lemma 2.2) that there exists a *Maxwell field* $(\mathcal{L}, D)$ such that (see also loc. cit. (2.26))

$$(5.45) \qquad\qquad (\mathcal{L}, D) \longleftrightarrow (g, \theta),$$

that is, one finally obtains (cf. also (5.23), (5.25))

$$(5.46) \qquad\qquad \chi([(\mathcal{L}, D)]) = [(g, \theta)],$$

which was to be proved, in view of (5.41), which thus completely establishes our assertion in (5.32). ∎

All told, we can now recapitulate the foregoing into the form of the following basic result. That is, we have.

Theorem 5.1 Suppose that we are given a differential triad

$$(5.47) \qquad\qquad (\mathcal{A}, \partial, \Omega^1)$$

on a topological space X, and let

$$(5.48) \qquad\qquad \tilde{\partial} : \mathcal{A}^\bullet \longrightarrow \Omega^1$$

be the corresponding logarithmic derivation induced on $\mathcal{A}^\bullet$ by ∂, and let us further consider the following 2-term $\mathbb{Z}$-complex on X:

(5.49) $$\mathcal{E}^{\cdot} : 0 \longrightarrow \mathcal{A}^{\cdot} \xrightarrow{\tilde{\partial}} \Omega^{1} \longrightarrow 0 \longrightarrow \cdots .$$

Then one obtains the following isomorphism of (abelian) group:

(5.50) $$\Phi^{1}_{\mathcal{A}}(X)^{\nabla} \underset{\chi}{\cong} \check{\mathbb{H}}^{1}(X, \mathcal{E}^{\cdot}) \equiv \check{\mathbb{H}}^{1}(X, \mathcal{A}^{\cdot} \xrightarrow{\tilde{\partial}} \Omega^{1}),$$

such that one sets

(5.51) $$\chi([(\mathcal{L}, D)]) := [(g, \theta)],$$

for any $[(\mathcal{L}, D)] \in \Phi^{1}_{\mathcal{A}}(X)^{\nabla}$, where one has

(5.52) $$(g, \theta) \in \check{C}^{1}(\mathcal{U}, \mathcal{A}^{\cdot}) \times \check{C}^{0}(\mathcal{U}, \Omega^{1}),$$

with $\mathcal{U}$ an open covering of X (local frame of $\mathcal{L}$), such that

(5.53) $$\delta(g) = 0 \quad and \quad \delta(\theta) = \tilde{\partial}(g).$$

6 Abstract Maxwell Equations

We discuss in this section similarities that exist between the classical Maxwell's equations (in vacuo) and the relevant situation that appears through the present study pertaining in particular to the equation

(6.1) $$D^{1} = 0$$

as in the preceding; cf., for instance, (4.35.1) or (4.37).

   According to the classical framework, Maxwell's equations (in vacuo), which by definition describe the electromagnetic field, constitute two pairs of equations, the first of which is simply reduced to the classical ("second") Bianchi's identity (see Chapt. III; (3.19), and Note 3.1)

(6.2) $$d^{2}(R) \equiv dR = 0,$$

which is always in force! (See below the analogous, more general, abstract counterpart.) On the other hand, the rest of Maxwell's theory of electromagnetism, that is, the remaining second pair of Maxwell's equations, is subsumed into the following relation:

(6.3) $$d * R = 0.$$

Concerning the previous two relations (6.2) and (6.3) (Maxwell's theory of electromagnetism), R stands for the respective field strength, while "$*$" denotes the corresponding Hodge operator. On this matter we shall return, however, in Chapter I of

Vol. II of this treatise, by considering more generally Yang–Mills fields, the electromagnetism being, of course, a particular case thereof (abelian gauge theory). Regarding the classical theory within the framework of the standard differential geometry of (smooth) manifolds, we refer, for instance, to M. Postnikov [1: p. 328ff] or B.A. Dubrovin et al. [1: p. 194ff], as well as R.W.R. Darling [1: p. 50, §2.11, in particular p. 52, (2.74), along with the comments following it, and p. 194ff]. Therefore, as already hinted at above, one concludes that

(6.4)   the Maxwell's equations (6.2) and (6.3) are those relations within the classical theory that describe and characterize the electromagnetic field.

On the other hand, we have already established in Chapter III the form that several fundamental issues of that important classical theory acquire when one is working within the present abstract setting of differential geometry. For convenience, we recall that within this vein of ideas we have viewed the electromagnetic field, as a pair

$$(6.5) \qquad\qquad (\mathcal{L}, D),$$

where $\mathcal{L}$ is a line sheaf on a topological space X, the base space of a $\mathbb{C}$-algebraized space

$$(6.6) \qquad\qquad (X, \mathcal{A}),$$

and D an $\mathcal{A}$-connection on $\mathcal{L}$ defined through a given differential triad

$$(6.7) \qquad\qquad (\mathcal{A}, \partial, \Omega^1)$$

on X (see Chapt. III; Definition 1.1). As already mentioned , the present important particular example of a pair, as in (6.5), motivated the terminology of a Maxwell category and relevant notions as applied in the foregoing. (In this connection, see also loc. cit. Note 1.1 and comments following it.)

Thus, by analogy with the classical theory, that is, (6.2) and (6.3), we may call *abstract Maxwell's equations* those relations that characterize (viz. can locate or even detect) a Maxwell field, that is, a pair as in (6.5) (cf. also (6.8)). Indeed, as we have seen the preceding discussion, such a characterization can be achieved by means of (Čech hyper)cohomology of X.

In particular, by looking at the fundamental constituents of a given field, as above [that is (i) the carrier of the field, hence in our case the line sheaf $\mathcal{L}$, as in (6.5), thus locally described by the corresponding coordinate 1-cocycle, viz., primarily by a 1-cochain; and then (ii) the field itself, that is, the gauge potential, or the respective $\mathcal{A}$-connection D on $\mathcal{L}$, again locally determined by a 0-cochain of 1-forms], one is thus led to consider the description of the field, at issue by means of a pair

$$(6.8) \qquad (g, \theta) \equiv ((g_{\alpha\beta}), (\theta_\alpha)) \in \check{C}^1(\mathcal{U}, \mathcal{A}^{\boldsymbol{\cdot}}) \times \check{C}^0(\mathcal{U}, \Omega^1).$$

Of course, $\mathcal{U}$ stands here for an open covering of X, which provides local gauges of $\mathcal{L}$ as needed (viz., as we also say, a local frame of $\mathcal{L}$; cf., Chapter I). Thus, in view of our previous discussion (see Sections 4, 5), we can further note that a given pair

(6.9)   (6.9.1) $(g, \theta),$

as in (6.8), yields a Maxwell field $(\mathcal{L}, D)$, viz. one obtains

(6.9.2) $(\mathcal{L}, D) \longleftrightarrow (g, \theta)$

(see also Chapt. III; (2.26)), if and only if one has (cf. (4.19))

(6.9.3) $D^1(g, \theta) = 0.$

So, what amounts to the same thing, one concludes that

the equation

(6.10)   (6.10.1) $D^1 = 0$

characterizes the Maxwell field on X.

In this connection, see also our previous comments in Section 4; (4.35)–(4.37). Consequently, one may construe the relation

(6.11)   (6.11.1) $D^1 = 0$

as the abstract Maxwell's equation (in vacuo) on X.

Accordingly,

(6.12)   (6.11.1) (abstract Maxwell's equation), along with our previous conclusion in Chapt. III; (3.30) (viz., equivalently, Weil's integrality theorem, ibid.; Theorem 3.1), when one can afford for the latter the appropriate set-up (Weil space, loc. cit.; (3.26)), provides two criteria, both of a *cohomological* nature, for the existence of an electromagnetic field (photon).

Therefore, one gets here too a justification of the point of view that

(6.13)   cohomology provides always an intrinsic approach to physics, being thus the arithmetics of nowadays elementary particle physics.

In this regard, see also C. v. Westenlolz [1: p. 318ff]. On the other hand, within the same vein of ideas, the above pair (6.8), which under the condition (6.9.3), characterizes a Maxwell field, is but a special case, for $n = 1$ (line sheaves) of the more general situation, that one has, by considering, instead, a Yang–Mills field

(6.14) $(\mathcal{E}, D).$

We recall that $\mathcal{E}$ stands here for a vector sheaf on X with $rk\mathcal{E} = n > 1$, and D for an $\mathcal{A}$-connection on $\mathcal{E}$, these two objects being locally expressed by similar local data, as in (6.8), suitably adjusted to the case at issue; see Vol. II of this study, Chapter I, where an analogous condition to (6.9.3) is given for any $n \in \mathbb{N}$ in general, extending thus the situation of the present section. By still extending further the analogy with the classical case, one can say that

(6.15)

> the same pair (g, θ) as in (6.8) might also be conceived, as the analogous, in our case, generalized velocity, namely the standard concept of classical mechanics. Thus, the range of these pairs, viz. the set
>
> (6.15.1) $$\check{C}(\mathcal{U}, \mathcal{A}^{\boldsymbol{\cdot}}) \times \check{C}^0(\mathcal{U}, \Omega^1),$$
>
> may still be viewed as the corresponding generalized (abstract) phase space of the system (physical field), under consideration.

In concluding the present section, we can say that the classification of Maxwell fields, as given by Theorem 5.1 is but the cohomological analogue in terms of Čech hypercohomology, hence, via the theory of spectral sequences of the respective transformation law of potentials (see Chapt. III; (2.34.2) or (2.36)). In other words, the latter appear to be just the differential geometric-physical formulation of the matter (viz. of the field at issue), whose cohomological counterpart is thus supplied by the preceding (that is, by (6.9.3); cf. also (5.53) or (4.26), (4.27)).

Furthermore, the same formalism as above, in characterizing (physical) fields as pairs like these in (6.8) (cf. also Part II of this treatise), is still in accord with similar considerations of C. von Westenholz [1: p. 322ff] by trying to circumvent basic deficiencies occurring in the conventional quantum field theory of elementary particles. An analogous, more natural, point of view, through the same abstract approach, against the traditional one (loc. cit.), is also supplied, when one looks at the so-called second quantization, as we shall see below in Chapter V. Thus, the nature of the aforesaid applications indicates that

(6.16)   the cohomological description of elementary particles, viz., essentially, fields, fits nicely with quantization issues.

See Chapter V; Section 5.

7 The Hermitian Analogue

In this section we examine the Hermitian counterpart of Theorem 5.1. In other words, we are interested in the set

(7.1) $$\Phi_{\mathcal{A}}^1(X)^{\nabla_{her}},$$

viz. in the set of all Hermitian Maxwell fields, strictly speaking of all equivalence classes of such fields, on a given (Hausdorff) topological space X (see Chapt. III;

(6.32)). Therefore, to have the appropriate Hermitian framework, we assume henceforth that we are given

(7.2)   a (Hausdorff) topological space X that further satisfies the rest of the conditions set forth in Chapt. III; (6.1).

As a result, one concludes under the above hypothesis that (cf. thus Chapt. III; (6.12) and (6.29))

(7.3)   every line sheaf on X admits an $\mathcal{SU}(1)$-structure, as well as a Hermitian $\mathcal{A}$-connection.

Therefore, by analogy to Theorem 5.1, and taking also into account the relevant terminology of Chapter III; Section 6, one can obtain the following Hermitian analogue of Theorem 5.1.

Theorem 7.1 Suppose we are given a differential triad

$$(7.4) \qquad\qquad (\mathcal{A}, \partial, \Omega^1)$$

with respect to a $\mathbb{C}$-algebraized space

$$(7.5) \qquad\qquad (X, \mathcal{A}),$$

with X a topological space satisfying (7.2), and let

$$(7.6) \qquad\qquad \tilde{\partial} : \mathcal{A}^{\bullet} \longrightarrow \Omega^1$$

be the respective logarithmic derivation induced on $\mathcal{A}^{\bullet}$ by ∂. Moreover, let us further consider the following 2-term $\mathbb{Z}$-complex on X

$$(7.7) \qquad\qquad \mathcal{E}^{\bullet} : 0 \longrightarrow \mathcal{SU}(1) \overset{\tilde{\partial}}{\to} \Omega^1 \longrightarrow 0 \longrightarrow \cdots,$$

where we set

$$(7.8) \qquad\qquad \mathcal{SU}(1) \equiv \mathcal{SU}(1, \mathcal{A}) \triangleleft \mathcal{A}^{\bullet},$$

viz. the special unitary group sheaf of $\mathcal{A}$ of order 1 (thus, by definition, a (normal) group subsheaf of $\mathcal{A}^{\bullet}$, cf. Chapt. III; (6.10)). Then one obtains the following (set-theoretic) bijection:

$$(7.9) \qquad \Phi^1_{\mathcal{A}}(X)^{\nabla_{her}} = \check{\mathbb{H}}^1(X, \mathcal{E}^{\bullet}) \equiv \check{\mathbb{H}}^1(X, \mathcal{SU}(1) \overset{\tilde{\partial}}{\to} \Omega^1). \blacksquare$$

In this connection, denoting by ψ the above bijection (7.9), one further sets

$$(7.10) \qquad\qquad \psi([(\mathcal{L}, D^{her})]) := [(g, \theta)],$$

such that

(7.11) $(g, \theta) \in \check{C}^1(\mathcal{U}, \mathcal{SU}(1) \times \check{C}^0(\mathcal{U}, \Omega^1)$,

where $\mathcal{U}$ stands for an open covering of X, being a local frame of $\mathcal{L}$. Furthermore, by analogy with Theorem 5.1 the same pair (g, θ) as in (7.11) satisfies the following conditions, a consequence of our hypothesis (cf. (7.9), (7.10)) that

(7.12) $(g, \theta) \in \ker D^1_{her}$

(see (7.16) for the notation). Thus, regarding (7.11), one has

(7.13) $\delta(g) = 0$ and $\delta(\theta) = \tilde{\partial}(g)$,

such that

(7.14) $| g_{\alpha\beta} |= 1 \in \mathcal{SU}(1)(U_{\alpha\beta}) \lhd \mathcal{A}^\bullet(U_{\alpha\beta}) = \mathcal{A}(U_{\alpha\beta})^\bullet$,

along with the relation

(7.15) $\theta + \bar{\theta} = 0$

(see also Chapt. III; (6.38), for the notation employed in the previous relation; cf. [VS: Chapt. VII; p. 172, (10.7), and subsequent comments therein]). In this regard, cf. also (5.52)–(5.54) as well as, Chapt. III; (6.15)–(6.18). ■

Scholium 7.1 As an application of the above, and in conjunction with our discussion in Section 6, we also remark that

(7.16)
> we can further consider Maxwell's equations (in vacuo) within the Hermitian framework, as above, so that we may say that (cf. also (6.11) the equation
>
> (7.16.1) $D^1_{her} = 0$
>
> characterizes the Hermitian Maxwell fields on X.

We can further say that

(7.17)
> the Maxwell's equation (7.16.1), formulated within the Hermitian set-up as before is a particular case of a gauge field equation, with gauge group
>
> (7.17.1) $\mathcal{SU}(1) = \mathcal{U}(1) \lhd \mathcal{A}^\bullet$,
>
> the latter being presented in sheaf-theoretic terms (cf. also (7.8)).

On the other hand, we are going to see a similar situation to (7.17) in Vol. II of the present study, Chapters I and IV, by considering the Yang–Mills equations and Einstein's equations (in vacuo), respectively.

5

Geometric Prequantization

> " ... geometric quantization ... a geometric, coordinate-free construction for the Hilbert space and observables of the underlying quantum theory: with no explicit dependence on a particular coordinate system, such a construction can be expected to give a very clear insight into the ambiguities involved in passing from the classical to the quantum domain."
>
> N. M. J. Woodhouse in *Geometric Quantization* (Oxford University Press. 2nd Ed., 1991). pp. vi, vii.

> "Prequantization ... seems to solve the problem of associating a quantum mechanical system to every classical system, at least if that system is quantizable ... "
>
> R. J. Blattner in *On geometric quantization*. (Proceedings of "Non-linear Partial Differential Operators and Quantization Procedures," Clausthal 1981. LNM 1037. Springer-Verlag, 1983. pp. 209–241). p. 228.

Our aim in the ensuing discussion is to present the classical theory of geometric (pre)quantization in terms of the abstract differential geometry, to the extent that the former is entangled with, and is based on, the standard differential geometry of differential manifolds. In particular, we are especially interested in the classification problem of prequantizations, by extending to the present abstract setting the classical counterpart, according to B. Kostant [1] and J.-M. Souriau [1]. In fact, this essentialy has been done already in the preceding (cf. Chapters III, IV), by considering the classification of Maxwell fields with respect to their field strength (see Chapt. III; Section 5), along with the corresponding cohomological classification of the same fields (cf. Chapt. IV; Section 5, in this connection, see Chapt. III; Scholium 8.1). As will be explained in the pertinent places throughout the sequel, we put here, among other things, the aforementioned material into perspective by looking at the relevant results through the corresponding terminology, within the present abstract set-up, of the theory of geometric quantization.

1 Symplectic Sheaves

The current classical setting for geometric quantization is a symplectic manifold, viz. a pair

$$(1.1) \qquad\qquad (X, \omega),$$

consisting of a smooth (i.e., C^∞-)manifold X, whose dimension, as is proved, is even (say, $2n$, with $n \in \mathbb{N}$), and a (nondegenerate) closed 2-form ω on X.

On the other hand, a similar pair is the standard differential-geometric framework for classical mechanics. (In this regard, see N.M.J. Woodhouse [1: pp. 1, 156], along with the standard sources B. Kostant [1] and J.-M. Souriau [1]. See also R. Abraham–J.E. Marsden [1], or M. Puta [1], as well as V. Guillemin–S. Sternberg [1] and A. Weinstein [1].) In this context, the pair (X, ω) stands for the classical phase space of the physical system at issue; therefore, the geometric quantization procedure (or quantization functor, cf. A. Weinstein loc. cit.) appears to be just a way of establishing a correspondence between the classical phase space and the quantum phase space (Hilbert space, suitably curved) of the quantum-mechanical system we are looking for.

In this connection, it is worth noticing the (conceptual) resemblance of the phase spaces concerned (i.e., pairs, as in (1.1)) to those already considered as pairs; see Chapt. IV; (5.53), (6.8), or (7.11) (cf. also (6.9.2), as well as (6.15)) along with their subsequent cohomological formulation (loc. cit. (5.11.2) or (7.10)). See also (2.19), as well as the following diagram depicting the correspondence hinted at in the preceding:

(1.2)

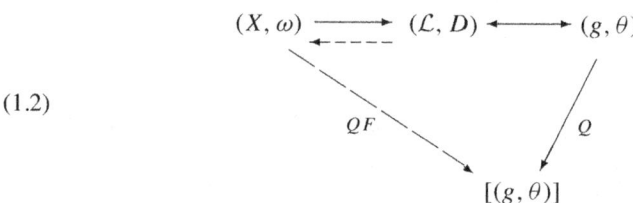

Concerning our notation in (1.2), Q stands there for quantization, in the sense that

(1.3)   "Quantization is provided by the Physical law itself."

See C. von Westenholz [1: p. 323; (ii′)], while according to the same author,

(1.4)   "Physical fields ... are defined as pairs"

(loc. cit., p. 322; (i′)), thus in our case, by a pair

(1.5) (g, θ),

as depicted in (1.2), representing a Maxwell field $(\mathcal{L}, D)$ (see also Chapt. III; (2.26); cf. (2.34) and/or (2.36), pertaining to a Yang–Mills field in general). Furthermore, QF in the same diagram stands for quantization functor (à la Weinstein, loc. cit.). The aforementioned part of axioms in (1.3), (1.4) belong to a set of postulates set forth by von Westenholz (ibid., p. 322f) and characterizing a geometrical approach to field theory, hence a geometrical approach (cf. also Chapt. IV; (6.8), as well as

(6.15)), which finally leads, through cohomology, to algebra, thus to quantization as well. The preceding support, by means of cohomology, something that, as we shall see later (cf. Vol. II, Chapter IV), contributes also to the question of quantizing gravity.

To put the classical set-up (cf. (1.1)) into perspective with the present abstract differential-geometric treatment, we first define, in terms of the language employed herewith, the relevant notion referred to in the title of this section. The applied terminology will be gradually justified in the subsequent discussion.

We start, as usual, by first considering a $\mathbb{C}$-algebraized space

$$(1.6) \qquad\qquad (X, \mathcal{A})$$

that we further assume to be a Bianchi space

$$(1.7) \qquad (\mathcal{A}, \partial, \Omega^1, d^1 \equiv d, \Omega^2, d^2 \equiv d) \equiv (\partial, d^1, d^2);$$

see Chapt. III; (3.17). In this connection, we also employ in the sequel, for convenience, as we did before occasionally, the notation d when referring to any one of the two differential operators d^1 and d^2. Another issue that will be of use presently is the notion of a

(1.8)   strictly exponential sheaf space, that is, a space $(X, \mathcal{A})$, as in (1.6), for which the following short exact exponential sheaf sequence, along with the associated commutative triangle, holds:

(1.8.1)

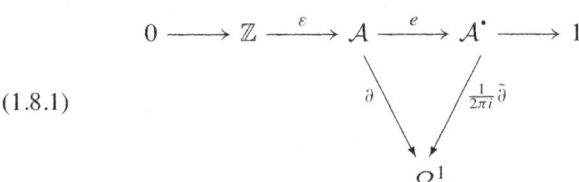

See also Chapt. III; (3.27). We shall also accept in the sequel that the topological space X, as in (1.6), is further a

generalized de Rham 2-space; viz. a (Hausdorff) paracompact space, for which the following sequence of $\mathbb{C}$-vector space sheaves is exact:

$$(1.9) \qquad (1.9.1) \qquad 0 \longrightarrow \ker \partial \longrightarrow \mathcal{A} \xrightarrow{\partial} \Omega^1 \xrightarrow{d^1} \Omega^2$$
$$\xrightarrow{d^2} d^2(\Omega^2) \equiv d\Omega^2 \longrightarrow 0.$$

(Thus, X is by definition a Bianchi space as well; cf. (1.7). In this regard, see also [VS: Chapt. IX; p. 254, Definition 3.1].)

In fact, the reason that we consider above that particular type of a ($\mathbb{C}$-algebraized) space is the following result, which in the classical case of smooth manifolds i in force, so that one then takes for granted its consequence; see (1.11) or (1.13).

Lemma 1.1 Suppose we are given a generalized de Rham 2-space X, along with a closed 2-form ω on X; that is, one has, by definition,

$$(1.10) \qquad \omega \in \Omega^2(X) \qquad \text{such that } d^2(\omega) \equiv d\omega = 0.$$

Then one obtains the following (2-dimensional Čech) cohomology class, which is supplied by ω:

$$(1.11) \qquad [\omega] \in \check{H}^2(X, \ker \partial).$$

Proof. See [VS: Chapt. IX; p. 256, Lemma 3.1, for $m = 2$]. ∎

Note 1.1 In the classical case (of smooth manifolds) one has the relation

$$(1.12) \qquad \ker \partial = \mathbb{C}$$

(Poincaré lemma; see also, for instance, Chapt. III; Scholium 2.1, or loc. cit; (3.27), (3.28).) Therefore, in that case (1.11) is then reduced to the 2-dimensional complex (Čech) cohomology class

$$(1.13) \qquad [\omega] \in \check{H}^2(X, \mathbb{C}),$$

hence the term generalized that may be attributed to (1.11).

The framework that is determined by the hypothesis of Lemma 1.1 provides the right setting for the formulation of the definition of our basic notion, which is also connected with the heading of this section.

Definition 1.1 We call a *symplectic sheaf space* X, or a *symplectic space*, a paracompact (Hausdorff) space that further fulfils the hypothesis of Lemma 1.1, denoted also by a pair

$$(1.14) \qquad (X, \omega).$$

We call the corresponding pair

$$(1.15) \qquad (\mathcal{A}, \omega)$$

a *symplectic sheaf* on X, while we also refer to the closed 2-form ω as the (*generalized*) *symplectic form* on X.

Note 1.2 (Terminological) The closed 2-form ω, as in (1.14) (or in (1.10)), which by definition is supposed to be defined on X (ibid.), is not necessarily nondegenerate, as happens in the classical case. On the other hand, the latter notion would require further (terminological) preparation if we needed to consider it within the present setting (see thus Vol. II; Chapt. I, Note 2.2, or Chapt. IV, (2.6)). However, this particular property of ω will not be necessary(!) for the sequel. Therefore, the epithet generalized for ω (cf. Definition 1.1) can be sustained in that sense.

The notion of a symplectic space, according to the previous definition, corresponds to what in the classical case is called a *"presymplectic space"* (*manifold*); see J.-M. Souriau [1: p. 82, Definition/Theorem]. The term *almost Hamiltonian* (*manifold*) is also used in the standard case concerning the analogous abstract setting; see W.A. Poor [1: p. 245, Definition 8.6].

We proceed now to consider the question of the prequantization of the spaces/ sheaves that appeared in the preceding, where the spaces introduced in (1.8) will be of particular importance (see the proof of Lemma 2.1).

2 Prequantizable Symplectic Sheaves

We come now to our main objective in this chapter, that is, to the classical notion of prequantization, within the present abstract setting, as concerns in particulat a Maxwell field (cf. Chapt. III; (1.4), (1.7)).

Assume that we are given a symplectic (sheaf) space X (cf. Definition 1.1, along with (1.9)), and let

$$(2.1) \qquad\qquad (\mathcal{A}, \omega)$$

be the corresponding symplectic sheaf on X (cf. (1.15)). We come now to the following basic notion.

Definition 2.1 A pair

$$(2.2) \qquad\qquad (\mathcal{A}, \omega)$$

is said to be a *prequantizable symplectic sheaf* whenever the corresponding closed 2-form ω is integral (cf. (2.3)).

According to Chapt. III; (3.33), the above requirement for ω means, by definition, that

$$(2.3) \qquad [\omega] \in \mathrm{im}(\zeta^*) \equiv \mathrm{im}(\check{H}^2(X, \mathbb{Z}) \xrightarrow{\zeta^*} \check{H}^2(X, \ker \partial)).$$

In this connection, we further remark that the previous map

$$(2.4) \qquad\qquad \zeta^* : H^2(X, \mathbb{Z}) \longrightarrow \check{H}^2(X, \ker \partial)$$

is provided, by virtue of the (canonical) embeddings, by the following diagram:

(2.5)

$$\ker \partial \subseteq \mathcal{A},$$

which in turn supplies a map in cohomology ζ^*, as in (2.4) corresponding to the map ζ. (In this regard, see also loc. cit., along with the subsequent comments.) Of course, $\mathbb{Z}$ and $\mathbb{C}$ stand here as usual for the respective constant sheaves on X, while the map ε ($\mathbb{C}$-vector space sheaf morphism) is given by Chapt. II; (6.6). See Chapt. III; (3.31), as well as [VS: Chapt. VII; pp. 150, 151: (7.29), (7.30), and (7.31)].

Note 2.1 The situation described by (2.3) is more general than that considered in Chapter III; (3.28), where we assumed that

(2.6) $\ker \partial = \mathbb{C}.$

However, this more stringent hypothesis (see also our previous Note 1.1) will not be necessary(!) for the subsequent discussion. As a consequence, one concludes that

> what one actually needs here (viz. by considering prequantizable symplectic sheaf spaces) is, first, to have the rel.

(2.7)   (2.7.1) $[\omega] \in H^2(X, \ker \partial),$

> which by virtue of Lemma 1.1, is always in force for any symplectic sheaf space X (cf. Definition 1.1).

On the other hand, (2.7.1) is more general, in principle, than its classical counterpart

(2.8) $[\omega] \in H^2(X, \mathbb{C}),$

concerning any closed 2-form ω on X (smooth manifold), in our case the given (generalized) symplectic form on X according to Definition 1.1. The guarantee of (2.7.1) was, in view of Lemma 1.1, the motivation of Definition 1.1.

Another item that will be of importance for the sequel is the relation

(2.9) $\check{H}^1(X, \mathcal{A}) = \check{H}^2(X, \mathcal{A}) = 0.$

The significance of the above relation for the ensuing discussion is that it actually supplies the so-called *Chern isomorphism*

(2.10) $\check{H}^1(X, \mathcal{A}^{\cdot}) = \check{H}^2(X, \mathbb{Z}).$

The last two relations have been considered already in the foregoing, within another context, pertaining in particular to a Weil space (see Chapt. III; (3.26), (3.27)), where the aforesaid relations are in force by virtue of the hypotheses for these spaces (loc. cit. (3.45) and (3.52)). Nevertheless, as we shall see, both contexts are akin to each other, the present one being more general. So we come now to the following basic conclusion of this section, which determines as well our subsequent terminology referring to the identification of the prequantizable symplectic sheaves. Indeed, the same can essentially be viewed as an extended half of Weil's integrality theorem (see Chapt. III; Theorem 3.1, along with Note 2.2 in this section).

Lemma 2.1 Suppose that X is a strictly exponential symplectic sheaf space satisfying (2.9) (cf. (1.8) and Definition 1.1), while we are also given a Maxwell field $(\mathcal{L}, D)$ on X. Then the corresponding field strength, that is, the curvature of the $\mathcal{A}$-connection D of $\mathcal{L}$,

$$(2.11) \qquad\qquad R(D) \equiv R,$$

yields an integral 2-dimensional cohomology class of X; namely, one has

$$(2.12) \qquad [R(D)] \equiv [R] \in \operatorname{im}(\check{H}^2(X, \mathbb{Z}) \longrightarrow \check{H}^2(X, \ker \partial)),$$

or, by an obvious abuse of notation,

$$(2.13) \qquad\qquad [R] \in \check{H}^2(X, \mathbb{Z}).$$

Proof. Based on Chapt. III; (3.9), (3.10), one has, as concerns the curvature of D,

$$(2.14) \qquad R(D) \equiv R = (d\theta_\alpha) \in Z^0(\mathcal{U}, d\Omega^1) \underset{\longrightarrow}{\subseteq} Z^0(\mathcal{U}, \Omega^2) = \Omega^2(X),$$

while one obtains (loc. cit. (2.17))

$$(2.15) \qquad\qquad \delta(\theta_\alpha) = \tilde{\partial}(g_{\alpha\beta})$$

(transformation law of potentials), so that

$$(2.16) \qquad\qquad (\theta_\alpha) \in \check{C}^0(\mathcal{U}, \Omega^1)$$

and

$$(2.17) \qquad\qquad (g_{\alpha\beta}) \in Z^1(\mathcal{U}, \mathcal{A}^\bullet)$$

are the local expressions of D and (a coordinate 1-cocycle) of $\mathcal{L}$, respectively, relative to a local frame of $\mathcal{L}$,

$$(2.18) \qquad\qquad \mathcal{U} = (U_\alpha)_{\alpha \in I}$$

(loc. cit., Section 2). On the other hand, in view of the short exact exponential sheaf sequence and the commutative triangle as in (1.8), one has (see also [VS: Chapt. III; p. 196, Lemma 5.2])

$$(2.19) \qquad\qquad \tilde{\partial}(g_{\alpha\beta}) = 2pi \cdot \partial(f_{\alpha\beta})$$

for some

$$(2.20) \qquad\qquad (f_{\alpha\beta}) \in \check{C}^1(\mathcal{U}, \mathcal{A}).$$

Furthermore (loc cit., p. 190, Lemma 5.1), one obtains

$$(2.21) \qquad\qquad \delta(f_{\alpha\beta}) \equiv (\lambda_{\alpha\beta\gamma}) \in \check{Z}^2(\mathcal{U}, \mathbb{Z})$$

in such a manner that one finally gets the following (defining) relation:

$$(2.22) \qquad [(g_{\alpha\beta})] \underset{\delta}{\cong} \frac{1}{2\pi i}[R] := [(\lambda_{\alpha\beta\gamma})] \in \check{H}^2(X, \mathbb{Z}),$$

δ being the Chern isomorphism, as in (2.10), an outcome of (2.9) and the short exact sequence in (1.8.1). See also Chapt. III; Scholium 3.1, in particular,(3.55.2) as well as Section 3.4, where, among other things, we explain the way that definition (2.22) becomes a theorem. Thus, (2.22), modulo the factor $1/2\pi i$, explains our abuse of notation in (2.13), and this also terminates the proof. ∎

To fix the terminology we are going to apply in the sequel, and still motivated by Lemma 2.1, we single out the following notion. Namely, we consider a

(2.23)   strictly exponential (cf. (1.8)) generalized de Rham 2-space X (see (1.9)) that also satisfies (2.9). For short (and for historical reasons as well), we call this type of space a *Kostant–Souriau space*.

In this connection, we thus now remark that according to Lemma 2.1 and Scholium 2.1 in the sequel, one obtains that

(2.24)   a Kostant–Souriau space provides a more general framework than an exact Bianchi–Weil space (cf. Chapt. III; (3.100)) within which Weil's integrality theorem holds. (We notice here that for a Kostant–Souriau space (2.22) is thus a theorem, viz. it can be proved; see also Scholium 2.1.)

Scholium 2.1 Commenting further on (2.24), we remark that in the case of a Kostant–Souriau space, as in (2.23), one realizes that (i) we do not necessarily assume the validity of (2.6) (but see Chapt. III; (3.83)), and (ii) $\mathcal{A}$ is not a fine sheaf on X, a fact that would imply (2.9): Namely, in that case $\mathcal{A}$ is a Γ_X-acyclic $\mathcal{A}$-module (cf. [VS: Chapt. III; p. 162, (2.20) and p. 238; (8.24)), so that one has, by definition,

$$(2.25) \qquad (R^n \Gamma_X)(\mathcal{A}) \equiv H^n(X, \mathcal{A}) = 0$$

for any $n \in \mathbb{N}$ (loc. cit., p. 162, (2.20) and pp. 164, 165; (3.3), (3.5)). Therefore, one concludes in particular that (2.9) is in force, a crucial issue in the proof of Lemma 2.1. (Furthermore, concerning our final comments in (2.24), cf. also Chapt. III; (3.125), as well as Remarks 3.1.)

On the other hand, by extending (and obviously abusing as well) classical terminology, a pair

$$(2.26) \qquad \{(\mathcal{L}, D); R(D) \equiv R\}$$

on a Kostant–Souriau space X with $(\mathcal{L}, D)$ a Maxwell field on it is called a *standard symplectic (sheaf) space*.

Indeed, name for the pair (2.26) is further justified by Lemma 2.1, in conjunction with the following more general version of it; viz.the closed 2-form ω that is involved in the hypothesis of that lemma is unnecessary (see also the above definition (2.23)). That is, one concludes that

> every standard symplectic sheaf space is prequantizable (see definition 2.1) in the sense that for the case at hand, the pair

(2.27)   (2.27.1) $(X, R(D) \equiv R)$

> (cf. also (2.26)) yields a prequantizable symplectic sheaf (same definition as above).

On the other hand, in view of Weil's integrality theorem (Chapt. III; Theorem 3.1; see also (2.24)), we already know that

(2.28)   the converse of (2.27) is still in force.

See also (2.32). All told, we can thus recapitulate the preceding by the following more general (cf. Scholium 2.1, ii)) version of Weil's integrality theorem:

> Within the framework of a Kostant–Souriau space X (see (2.23)),

(2.29)   (2.29.1)   the only prequantizable symplectic (sheaf) spaces are the standard ones,

> in the sense of (2.27).

Equivalently, and by slightly abusing our previous terminology, we can express the preceding by saying that

(2.30)   the only prequantizable Kostant–Souriau spaces X are the standard ones.

That is, if we are given a pair

(2.31) (X, ω)

with X a Kostant–Souriau space (cf. (2.23)) and ω a closed 2-form on X, or a Kostant–Souriau symplectic sheaf space, then

> the corresponding symplectic sheaf (cf. (2.1))

> (2.32.1) $(\mathcal{A}, \omega)$

> is prequantizable if and only if

(2.32)   (2.32.2) $\omega = R(D) \equiv R,$

> such that

> (2.32.3) $(\mathcal{L}, D)$

> is a Maxwell field on X.

Thus we can say that,

(2.33)   prequantization means that (2.3) is in force

with respect to a given closed 2-form ω on a space X, as in (1.14), or for the classical case, as in (1.1). Consequently, the moral here is that (cf. also (2.30))

(2.34)   the situation described by (2.3) can occur only within a framework like (2.32).

In conclusion, and paraphrasing a relevant comment in D.J. Simms–N.M.J. Woodhouse [1; p. 36], we can say that

(2.35)   Weil's integrality theorem is the starting point for geometric quantization while, as a result of the foregoing, a Kostant–Souriau sheaf space is proven to be (thus far!) the appropriate set-up for its abstract formulation concerning the present account.

3 The Hermitian Framework

Our aim in this section is to examine the analogous situation that one obtains when considering Hermitian $\mathcal{A}$-connections (cf. (3.7) below) of Maxwell fields. In other words, we have here the same scenario of geometric prequantization as before, however now with respect to Hermitian Maxwell fields as employed in the foregoing (cf. Chapt. III; Section 6). On the other hand, the present framework is virtually the direct analogue of the situation one has in the classical case (see Theorem 3.1 and subsequent comments), our previous discussion in Section 2 being thus a prelude as well as a guiding source to the present account.

Our task now is to establish that appropriate set-up, within which one could formulate the notion of a Hermitian symplectic (sheaf) space. Consequently, motivated, as mentioned above, by the relevant material in Chapter III; Section 6, we assume henceforth that we are given

(3.1)

an enriched ordered strictly involutive $\mathbb{C}$-algebraized space

(3.1.1) $(X, \mathcal{A})$

that is also a strictly exponential generalized de Rham 2-space such that

(3.1.2) $(\mathcal{A}, \rho)$

is a *Hermitian $\mathcal{A}$-module* on X with $\mathcal{A}$ a fine sheaf on X, while

(3.1.3) $(\mathcal{A}, \omega)$

is a symplectic sheaf on X (see Definition 1.1). Finally, we assume that the $\mathcal{A}$-module Ω^1 involved (see (1.9.1) in the foregoing) is also a vector sheaf on X.

Note 1.1 (Terminological) Before we proceed further, it is good first to point out certain consequences of the above framework as described by (3.1), items that will be applied presently. According to our hypothesis in (3.1), it follows that X is a paracompact (Hausdorff) space (cf. (1.9)). Therefore, since $\mathcal{A}$ is also by hypothesis a fine sheaf on X, it is then *a fortiori* Γ_X-acyclic (cf. [VS: Chapt. III; p. 238, (8.24), and p. 162, (2.20)]), so that

(3.2)   (2.9) holds true.

Accordingly, based on Definition 1.1, one thus concludes that

(3.3)   the space X, as in (3.1), is in particular a symplectic sheaf space. Hence, according to the preceding, Lemma 2.1 is also in force within the framework set forth by (3.1).

Some other important consequences of (3.1) will be pointed out subsequently that also will justify the following abbreviated name for a space X as in (3.1). Henceforth, we call such a space a

(3.4)   *Hermitian symplectic (sheaf) space.*

So our first important issue in this connection is that

(3.5)   every vector sheaf $\mathcal{E}$, hence in particular every line sheaf $\mathcal{L}$, on X, as in (3.1), admits a Hermitian $\mathcal{A}$-connection.

See [VS: Chapt. VII; p. 174, Theorem 10.1]. ∎

Accordingly, the above permits us to consider from the outset what we call in the sequel

(3.6)   *Hermitian Maxwell fields*, denoted by

(3.6.1) $(\mathcal{L}, D)_{her}$.

This is, in other words, pairs consisting of a line sheaf $\mathcal{L}$ on X, as in (3.4), endowed with a Hermitian $\mathcal{A}$-connection D, something that is of course always available, in view of (3.5). For convenience, we recall here that such a connection fulfils by definition the following (defining) relation:

(3.7) $\partial(\rho(s, t)) = \rho(Ds, t) + \rho(s, Dt)$

for any s, t in $\mathcal{L}(U)$, with U open in X. (For simplicity's sake, we retained in (3.7) the same symbol ρ for the $\mathcal{A}$-metric on $\mathcal{L}$, as was by hypothesis the case in (3.1.2); in this regard, see also [VS: Chapt. IV; p. 333, Theorem 9.1].)

We are now in a position to come to the main topic of the present section, that is, as hinted at above, to the precise transcription, within a Hermitian setting, of our previous account in Section 2: Indeed, by applying the terminology of Chapt. III;

Section 6 (ibid., (6.33), (6.34)), and taking also into account (3.2) in conjunction with Lemma 2.1, we first conclude that

(3.8)

every Hermitian Maxwell field

(3.8.1) $(\mathcal{L}, D)_{her}$,

as in (3.6.1), on a Hermitian symplectic (sheaf) space X (cf. (3.4)) supplies, by means of its field strength $R(D) \equiv R$ (curvature of the Hermitian $\mathcal{A}$-connection D, as in (3.8.1)), an integral 2-dimensional (Čech) cohomology class of X; that is,

(3.8.2) $[R] \in \check{H}^2(X, \mathbb{Z})$

(see also (2.12), (2.13)).

Accordingly, by adapting to the present Hermitian context the previous terminology of Section 2 we can say that

(3.9)

any Hermitian Maxwell field $(\mathcal{L}, D)_{her}$ provides a standard Hermitian symplectic (sheaf) space in the sense that one gets the pair

(3.9.1) $\{(\mathcal{L}, D)_{her}; R(D) \equiv R\}$

(cf. also (2.26)). Moreover, the same space in prequantizable (cf. (3.8.2) and (2.27)).

(3.9.2) In this connection, we further remark that any Hermitian symplectic sheaf space is in particular a Kostant–Souriau space,

according to the definitions, along with (3.2).

On the other hand, by analogy with (2.29), and also taking into account our conclusion in (3.9), we further obtain that

(3.10) the only prequantizable Hermitian symplectic (sheaf) spaces (see (3.4), as well as (3.8.2)) are the standard ones, that is, of the form (3.9.1) (cf. also (2.27)).

Equivalently, (3.10) can be stated in the following form (see also Theorem 3.1).

(3.11)

Suppose we are given a Hermitian symplectic sheaf space

(3.11.1) (X, ω).

Then X is prequantizable (i.e., ω satisfies (2.3)) if and only if

(3.11.2) $\omega = R(D)$

with respect to a pair (cf. (3.9.1))

(3.11.3) $\{(\mathcal{L}, D)_{her}, R(D) \equiv R\}.$

Thus, motivated now by the situation in Section 2 (see, for instance, (2.29)), one can say that

(3.12)   (3.10) (or equivalently, (3.11)) constitutes an equivalent version of what we may call an (abstract) Hermitian (form of) Weil's integrality theorem.

Indeed, the above abstract form of Weil's integrality theorem, formulated within the present Hermitian set-up as established in this section, is the standard situation corresponding to our abstract (axiomatic) setting, as found, for instance, in the classical work of B. Kostant [1] and J.-M. Souriau [1]. More precisely, one has, by analogy with Chapt. III; Theorem 3.1, the following fundamental result.

Theorem 3.1 (Hermitian Weil's integrality theorem). Let the space X be as in (3.4). Then a 2-dimensional complex (Čech) cohomology class, say

(3.13) $z \in \check{H}^2(X, \mathbb{C}),$

is integral, that is,

(3.14) $z \in \mathrm{im}(\check{H}^2(X, \mathbb{Z}) \longrightarrow \check{H}^2(X, \mathbb{C}))$

if and only if z (in point of fact, $2\pi i \cdot z$, cf. (2.22)) is the field strength (curvature) of a (Hermitian) $\mathcal{A}$-connection of a Hermitian Maxwell field on X; in other words,

(3.15) $z = 2\pi i \cdot R$

with respect to a pair (cf. also (3.9) or (3.11))

(3.16) $\{(\mathcal{L}, D)_{her}, R(D) \equiv R\}.$ ∎

Concerning the classical case alluded to above, see B. Kostant [1: p. 133, Proposition 2.1.1], J.-M. Souriau [1: p. 310; (18.15)], or D.J. Simms–N.M.J. Woodhouse [1: p. 36, Theorem] and N.M.J. Woodhouse [1: p. 160, Proposition 8.3.1].

Our next objective is to consider the cohomological classification of the previous abstract setting of (pre)-quantizations, following in this regard classical patterns (loc. cit.). In fact, this has been done in the foregoing, however, in another setting (see thus Chapt. III; Sections 5–8), referring to Maxwell fields only. It is now our task to transfer those results to the present geometric-prequantization jargon; for convenience, we shall restrict ourselves in the sequel to (prequantizable) Hermitian Maxwell fields. This is what we are going to discuss in the next section.

4 Cohomological Classification of (Abstract) Geometric Prequantizations of Hermitian Maxwell Fields with a Given Field Strength

As already said, our aim here is to look at the results of Chapter III, which refer to a cohomological classification of Maxwell and/or of Hermitian Maxwell fields by means of the language of geometric prequantization theory. On the other hand, we have also already considered in Chapt. IV; Section 5 the cohomological formulation of the Maxwell group in terms of (Čech) hypercohomology theory, along with the Hermitian analogue thereof (cf. Section 7), while in Chapter III; Section 5 we examined the cohomological expression of (the set of) Maxwell fields having a given field strength (ibid., (5.91) or (5.111)). It is these latter results of Chapter III that we are going to look at from the point of view of geometric prequantization.

To fix first the terminology,

(4.1)   we assume henceforth that we are given a Hermitian symplectic (sheaf) space X (cf. (3.4)).

The same space is going to be further, appropriately, specialized in the ensuing discussion to cope with the particular situation involved (see Theorem 4.1).

By analogy with the classical setting (cf. J.-M. Souriau, B. Kostant, op. cit.), working here within a Hermitian framework, we further consider the set

(4.2)
$$\Phi_{\mathcal{A}}^1(X)_R^{\nabla_{her}},$$

that is, all Hermitian Maxwell fields on X having a given field strength (curvature) R.

On the other hand, by applying physical parlance, we can also talk about the set (4.2) as

a Hermitian light bundle

(4.3.1)
$$\Phi_{\mathcal{A}}^1(X)_R^{\nabla_{her}}$$

(4.3)

on X having a certain color; the latter is by definition determined by the common field strength of the Hermitian Maxwell fields involved in (4.3.1), in fact of equivalence classes of such. We still call them (Hermitian) light rays.

In this regard, see also Chapt. III; (6.32), (6.43), where the same set (4.3.1) has been considered, however, within another perspective.

Referring to a Hermitian light bundle, as in (4.3.1), we further remark that this consists of light rays of the same color as that of the bundle itself, that is, of equivalence classes of Hermitian Maxwell fields (elements of the set (4.3.1))

(4.4)
$$[(\mathcal{L}, D)_{her}] \equiv [(\mathcal{L}, D^{her})]$$

(cf. also (3.6)), which, moreover, have a common field strength (curvature), which by definition characterizes the (common) color of the rays involved in (4.3.1), hence also of the light bundle itself; thus, one has

$$(4.5) \qquad\qquad R(D^{her}) = R,$$

for any light ray, as in (4.4) (cf. also Chapt. III; (6.43)). On the other hand, according to our hypothesis for X (cf. (4.1)), as well as Theorem 3.1 (see also (2.12), (2.13)), one obtains

$$(4.6) \qquad\qquad R(D^{her}) = R \in \check{H}^2(X, \mathbb{Z})$$

for any light ray (equivalence class that it determines), as in (4.4).

To proceed further, we need one more supplementary hypothesis for X, indeed a cohomological one, which, however, we adjust to the differential set-up (see (1.7) in the preceding), as has been our practice so far. In fact, the same refers in the classical case to the subtle question connected with the Poincaré lemma (again!). Precisely speaking, we shall need for the sequel the condition

$$(4.7) \qquad\qquad \ker \partial = \mathbb{C}.$$

In this regard, see also Note 2.1. In sum, we assume that

> the Hermitian symplectic (sheaf) space X, as in (4.1), fulfills the condition (4.7). Therefore (cf. also (1.9)), one gets the following exact sequence of $\mathbb{C}$-vector space sheaves:

$$(4.8) \qquad (4.8.1) \qquad\qquad 0 \longrightarrow \mathbb{C} \overset{\varepsilon}{\to} \mathcal{A} \overset{\partial}{\to} \Omega^1 \overset{d}{\to} \Omega^2 \overset{d}{\to} d\Omega^2 \longrightarrow 0.$$

> A topological space X, as in (4.1), that also satisfies (4.7) (hence equivalently (4.8.1)) will be called in the sequel a *strictly Hermitian symplectic (sheaf) space.*

Thus, the previous framework permits us now to employ for the present abstract setting our previous conclusions in Chapter III, adapted here to a cohomological classification of geometric prequantizations, which also was our initial objective. So we are now in a position to state the next fundamental result.

Theorem 4.1 Suppose that we are given a strictly Hermitian symplectic (sheaf) space

$$(4.9) \qquad\qquad (X, \omega)$$

(see (4.1), (4.8)). Then the set of equivalence classes of prequantizations of (X, ω) (or even, for simplicity, just of X), in other words, such classes of those Hermitian line sheaves on X (cf. also (4.4)),

$$(4.10) \qquad\qquad [(\mathcal{L}, D^{her})],$$

whose field strength (curvature) satisfies the relation

(4.11) $$R(D^{her}) = \omega,$$

that is, equivalently, the set

(4.12) $$\Phi^1_{\mathcal{A}}(X)^{\nabla_{her}}_{\omega},$$

with ω as in (4.9), given by the relation (cf. Chapt. III; (6.62))

(4.13) $$\Phi^1_{\mathcal{A}}(X)^{\nabla_{her}}_{\omega} = \check{H}^1(X, S^1) \cdot [(\mathcal{L}, D^{her})]$$

within a *bijection*. ■

Note 4.1 We express (4.13) by saying that

(4.14)   the possible (Hermitian geometric) prequantizations of (X, ω), as in (4.9), are parametrized by (the abelian group) $\check{H}^1(X, S^1)$.

In other words, if one is given a prequantizing (Hermitian) line sheaf (cf. also (3.6.1))

(4.15) $$(\mathcal{L}, D)_{her} \equiv (\mathcal{L}, D^{her}),$$

in the sense that (4.11) is in force, precisely speaking (cf. Chapt. III; (6.33)) a pre-quantizing (Hermitian) light ray

(4.16) $$[(\mathcal{L}, D^{her})],$$

then all the remaining possible ones are given by (4.13).

By further applying a more physical terminology for the set (4.12), one can give the following reformulation of the Theorem 4.1:

given a strictly Hermitian symplectic (sheaf) space

(4.17.1) $(X, \omega),$

(4.17)

any prequantizing (Hermitian) line bundle on X is a principal homogeneous $\check{H}^1(X, S^1)$-space (see also Chapt. III; (6.61.1)), or an affine space with structure group $(\check{H}^1(X, S^1)$.

On the other hand, employing our previous considerations in Chapt. III; Section 6.4, one gets the following important issues of (4.17), that is, of Theorem 4.1.

Suppose we are given a strictly Hermitian symplectic (sheaf) space

(4.18.1) (X, ω)

(4.18)

in such a manner that X is in particular a path-connected space. Then any prequantizing (Hermitian) light bundle on X is a principal homogeneous $\pi_1(X)^*$-space.

(See also loc. cit.; (6.75) and (6.76).) So one gets the following (set-theoretic) bijection:

$$(4.19) \qquad \Phi^1_{\mathcal{A}}(X)^{\nabla_{her}}_{\omega} = \pi_1(X)^* \cdot [(\mathcal{L}, D^{her})].$$

Here we recall that $\pi_1(X)$ stands for the Poincaré (or even fundamental) group of (the path-connected, by hypothesis, as in (4.18), space) X, while $\pi_1(X)^*$ denotes the corresponding character group of $\pi_1(X)$. (See also Chapt. III; Lemma 6.1, along with the subsequent discussion). More particularly, one gets now at the following result:

(4.20)

Suppose that we have a path-connected strictly Hermitian symplectic (sheaf) space

$$(4.20.1) \qquad\qquad (X, \omega),$$

being also a simply connected space (so that by definition, one has $\pi_1(X) = 1$. Then

the only prequantizing (Hermitian) light bundle on X is just

(4.20.2) a light ray,

$$[(\mathcal{L}, D^{her})], \qquad \text{with } R(D^{her}) = \omega.$$

That is,

$$(4.20.3) \qquad \Phi^1_{\mathcal{A}}(X)^{\nabla_{her}}_{\omega} = [(\mathcal{L}, D^{her})],$$

such that $R(D^{her}) = \omega$.

Equivalently, in a more standard way, we can say that

given a symplectic (sheaf) space (X, ω), as i (4.20.1), one concludes that

there exists just one (Hermitian) prequantizing line sheaf

$$(\mathcal{L}, D^{her})$$

(4.21)

(4.21.1)

on X such that (by definition)

$$R(D^{her}) = \omega.$$

In fact, one gets a unique (Hermitian) prequantizing light ray

$$(4.21.2) \qquad [(\mathcal{L}, D^{her})] \equiv [(\mathcal{L}, D)_{her}]$$

in such a manner that one has (prequantization condition)

$$(4.21.2) \qquad R(D^{her}) = \omega.$$

Concerning the classical counterpart of the preceding, pertaining to a standard symplectic (smooth) manifold X, phase space of a physical system, we refer, for instance, to A.A. Kirillov [1: p. 247, Theorem 2, along with the comments following it]. The latter are a very special case of the spaces considered by (4.18), or (4.20.1), as the case may be.

Consequently, by looking at an appropriate symplectic (sheaf) space of the pertinent homotopy type (see (4.20.1)), we are thus reminded of the famous apostrophe of R. Feynman, according to which one can say that

(4.22)   *"For all we know, there may well be* [just] *one electron in the Universe."*

See also J.-M. Souriau [1: p. 328, Theorem (18.130)].

5 Prequantization of Elementary Particles

> " ... to find a quantum model of ... an elementary relativistic particle it is unnecessary ... to quantize [first] the corresponding classical system."
>
> D.J. Simms–N.M.J. Woodhouse in *Lectures on Geometric Quantization* (Springer-Verlag, 1976) p. 86.

> " ... Quantization is provided by the Physical law itself."
>
> C. von Westenholz in *Differential Forms in Mathematical Physics.* (North-Holland, 1981). p. 323.

Our aim in the present section is to apply the preceding material to the particular case of elementary particles. In view of our discussion in Chapter II, elementary particles are classified, sheaf-theoretically, according to their spin-structure; thus, more precisely (loc. cit., (6.29): Selesnick's correspondence principle), we have seen that

(5.1)   the states of free elementary particles may be viewed as sections either of line sheaves or of vector sheaves, of rank greater than 1, in so far as the particles under consideration are either bosons or fermions, respectively.

On the other hand, the base space of the vector sheaves considered is (see, for instance, loc. cit.; (6.31)) a

(5.2)   topological space X, which in particular is a

(5.2.1)   compact connected complete flat 4-dimensional Lorentz manifold (hence, in effect, Minkowskian), representing thus an empty finite universe ("vacuum"; by hypothesis, all the elementary particles involved are "bare", viz. free, ones). More general topological spaces can still be considered, while our previous assumption is made only for technical reasons. In this regard, see also Chapt. II; Scholium 6.1, as well as Remark 6.2.

Accordingly, to achieve the desired flexibility of the language employed, as well as to keep track of the abstract setting that has been applied thus far in the previous sections of the present chapter, we assume (in general) that

(5.3) X is a topological space that further satisfies the above conditions in Section 1 (cf., for instance, (1.8), or Lemma 1.1 or Definition 1.1), along with (2.9) or (2.23) (Kostant–Souriau space), or those conditions in (3.1) when referring to the Hermitian case.

5.1 Bosonic Case

We first examine the case that the (free) elementary particle at issue is a boson, thus, by definition (cf. Chapt. II; (2.4)), an integral-spin (elementary) particle. This means that (ibid., (6.29))

(5.4)   the (free) elementary particle under consideration is represented by a line sheaf $\mathcal{L}$ on X. Here the space X is assumed to satisfy (5.3). On the other hand, the symplectic sheaf corresponding to X (see Definition 1.1)

(5.4.1) $$(\mathcal{A}, \omega)$$

is further assumed to satisfy the following data:

First,

(5.5)   our arithmetic, or structure sheaf, $\mathcal{A}$ on X is given conventionally by the relation

(5.5.1) $$\mathcal{A} \equiv^{\mathbb{C}} \mathcal{C}_X^{\infty},$$

viz., by the sheaf of germs of $\mathbb{C}$-valued "$\mathcal{C}^{\infty}$-functions" on X.

The quotation marks for the term $\mathcal{C}^{\infty}$-functions in (5.5) aim at pointing out the possibility, according to the present abstract setting, of considering in place of $\mathcal{A}$, as in (5.5.1), more general $\mathbb{C}$-algebra sheaves, provided, of course, the corresponding situation as described by (5.3) is in force. See Chapter IX; Section 5.

Second, by referring to the (generalized) symplectic form ω as in (5.4.1) (cf. also Definition 1.1), thus by definition a closed 2-form on X, we further make the following assumption:

whenever a line sheaf $\mathcal{L}$ on X (cf. (5.5), along with the comments following it) is present that (see Chapt. II; Section 6) represents by its (local) sections the states of a (bare, viz. free) boson on X, then

(5.6)

(5.6.1) the closed 2-form ω on X that we are looking for such that (5.4.1) is in force is the curvature form (field strength) $R(D)$ of the (boson, as before) standard Maxwell field naturally associated with $\mathcal{L}$; precisely speaking, by virtue of our previous terminology (cf. (2.26)), of a standard symplectic (sheaf) space, one has

(5.6.1.1) $\{(\mathcal{L}, D), R(D) \equiv \omega\}.$

In this connection, we first remark that in view of our hypothesis for X (cf. (5.3)),

(5.7)   every vector sheaf, hence in particular any line sheaf $\mathcal{L}$, on X admits an $\mathcal{A}$-connection, say D.

See our previous assumption in (5.3) about the space X in conjunction with [VS: Chapt. VI; p. 85, Theorem 16.1, and Chapt. III; p. 247, (8.56)]. ■ Consequently,

every (free) boson $\mathcal{L}$ (cf. (5.6)) can be construed as a standard symplectic (sheaf) space as in (5.6.1.1). By virtue of the same hypothesis about X as in (5.3), we may say (cf. (5.7)) that

(5.8)

(5.8.1) every (free) boson $\mathcal{L}$ on X provides by itself a standard symplectic (sheaf) space (cf. (5.6.1.1)),

or (same hypothesis as in (5.6) along with (3.5))

(5.8.2) a standard Hermitian symplectic (sheaf) space (Maxwell field)

$$\{(\mathcal{L}, D^{her}), R(D^{her}) \equiv \omega\}.$$

5.2 The Chern Isomorphism (Continued), and Consequences

We have already discussed (cf. Chapt. III; Scholium 3.1) the physical significance of the isomorphism in the title of this subsection. The same isomorphism is in force here as well by virtue of our hypothesis for the space X as exhibited in (5.3). Thus, one has here too the following (abelian) group isomorphism:

(5.9) $$\check{H}^1(X, \mathcal{A}^{\cdot}) = \check{H}^2(X, \mathbb{Z}).$$

Based on what has been said in Chapter III (loc. cit., in particular, (3.55)), one can say that the carrier that is the line shea $\mathcal{L}$, or the photon of the (Hermitian) electromagnetic field,

(5.10.1) $$(\mathcal{L}, D^{her}) \equiv (\mathcal{L}, D)_{her},$$

as appeared in (5.8.2), is actually identified with the effect itself of the field, namely, with the field strength (curvature) of the field under consideration, in view of the relation

(5.10)

(5.10.2) $$[(g_{\alpha\beta})] = \frac{1}{2\pi i}[R] \equiv [(\lambda_{\alpha\beta\gamma})] \in \check{H}^2(X, \mathbb{Z})$$

(see also Chapt. III; (3.43)); so the said "identification" is made in terms again(!) of cohomology theory. In this connection, we also recall here that one has

(5.10.3) $$\mathcal{L} \equiv [(g_{\alpha\beta})] \in \check{H}^1(X, \mathcal{A}^{\cdot}),$$

cf., for instance, Chapt. III; (2.2) and (2.14), or (2.33.3).

We discuss below certain consequences of the notion of the carrier of a line sheaf $\mathcal{L}$ on X, as expressed through a given coordinate 1-cocycle of $\mathcal{L}$; this is attained by an appropriate application of the concept of the Atiyah class of $\mathcal{L}$, the latter being formulated in terms of the aforementioned cocycle: Thus, given a line sheaf $\mathcal{L}$ on X, by looking at the corresponding Atiyah class of $\mathcal{L}$, $\mathfrak{a}(\mathcal{L})$, one sets

(5.11) $$\mathfrak{a}(\mathcal{L}) := [\tilde{\partial}(g_{\alpha\beta})] \in \check{H}^1(X, \Omega^1)$$

(see also [VS: Chapt. VI; p. 47, (9.14)]). Here

(5.12) $$(g_{\alpha\beta}) \in \check{Z}^1(\mathcal{U}, \mathcal{A}^{\cdot})$$

stands, as usual, for a coordinate 1-cocycle of $\mathcal{L}$ with respect to a given local frame

(5.13) $$\mathcal{U} = (U_\alpha)_{\alpha \in I}$$

of $\mathcal{L}$ (cf. Chapt. III; (2.13), (2.14), along with (2.18), (2.19)). In view of our hypothesis for X and $\mathcal{A}$ (see (4.1), together with (3.1), (3.4)), one concludes that

(5.14)   every vector sheaf on X is fine, therefore Γ_X-acyclic.

See [VS: Chapt. III; p. 238, (8,24), and p. 247, (8.56)]. ∎ Consequently, based further on (4.1) and (3.1), as well as on (5.11), one obtains

(5.15) $$\mathfrak{a}(\mathcal{L}) \equiv [\tilde{\partial}(g_{\alpha\beta})] \in \check{H}^1(X, \Omega^1) = 0,$$

so that (cf. (5.12), along with Chapt. III; (2.18) and Chapt. I; (4.9.3))

(5.16) $$\tilde{\partial}(g_{\alpha\beta}) \in \check{Z}^1(\mathcal{U}, \Omega^1)$$

is a coboundary; that is, one concludes that

$$(5.17) \qquad \tilde{\partial}(g_{\alpha\beta}) = \delta(\theta_\alpha)$$

for some 0-cochain of 1-forms

$$(5.18) \qquad (\theta_\alpha) \in \check{C}^0(\mathcal{U}, \Omega^1).$$

Thus, on the basis of our assumption in (5.11), we recast, via (5.17), the transformation law of potentials corresponding to (5.10.1) (see Chapt. III; Lemma 2.1, in particular (2.17); cf. also (2.33.2), as well as (2.36)). So one finally gets the conclusion that

(5.19)
> one can consider the (5.18) modulo an occasional translation in the affine space of $\mathcal{A}$-connections of $\mathcal{L}$,
>
> $$(5.19.1) \qquad Conn_{\mathcal{A}}(\mathcal{L}) = (\theta_\alpha) + \Omega^1(X)$$
>
> (see [VS: Chapt. VI; 33, Corollary 7.1]), as the (local form of the) gauge potential ($\mathcal{A}$-connection) of $\mathcal{L}$; moreover, in view of (5.17) and (5.18), the latter may be construed as (completely) determined by the carrier ((free) boson) itself,
>
> $$(5.19.2) \qquad \mathcal{L} \longleftrightarrow (g_{\alpha\beta}) \in \check{Z}^1(\mathcal{U}, \mathcal{A}^{\cdot})$$
>
> (Chapt. III; (2.26), (2.33)), or by the carrier of (the states of) the (bare) elementary particle at issue (see also Chapt. II; (6.29)).

Furthermore, as a consequence of the preceding, one concludes that:

(5.20)
> any free (bare) elementary particle here in particular a boson, provides by itself (precisely speaking by its carrier) the corresponding gauge potential ($\mathcal{A}$-connection of the carrier), as well as the field strength (curvature) associated with it. The latter are determined through our arithmetic (structure sheaf, or sheaf of coefficients) $\mathcal{A}$; the carrier itself, according to our assumption (loc. cit.), is similarly expressed by its corresponding "coordinate 1-cocycle"
>
> $$(5.20.1) \qquad (g_{\alpha\beta}) \in \check{Z}^1(\mathcal{U}, \mathcal{A}^{\cdot}).$$

The previous discussion, pertaining to (5.17) and its consequences, as exhibited by (5.19) and (5.20), also in agreement with the analogous situation one has within the classical set-up (cf., for instance, A. Weil [1: p. 88]): Thus, by considering a coordinate 1-cocycle of a line sheaf $\mathcal{L}$ on X,

$$(5.21) \qquad (g_{\alpha\beta}) \in \check{Z}^1(\mathcal{U}, \mathcal{A}^{\cdot})$$

(see also (5.13)), a given 0-cochain of 1-forms on X,

(5.22) $(\theta_\alpha) \in \check{C}^0(\mathcal{U}, \Omega^1)$

is said to be, according to the classical terminology (loc. cit.), a "*connection for the transition functions*" (the latter functions being, by definition, supplied, by (5.21)), whenever one has the relation

(5.23) $\delta(\theta_\alpha) = \dfrac{1}{2\pi i}\tilde{\partial}(g_{\alpha\beta}).$

Therefore, based on (5.23), one obtains

(5.24) $\tilde{\partial}(g_{\alpha\beta}) = 2\pi i \cdot \delta(\theta_\alpha) = \delta(2\pi i \cdot \theta_\alpha),$

so that the 1-cocycle, as defined by the first term of (5.24) (cf. also (5.16)), is actually a coboundary; hence, the Atiyah class of $\mathcal{L}$, as given by (5.11), is zero; that is, one has

(5.25) $\mathfrak{a}(\mathcal{L}) := [\tilde{\partial}(g_{\alpha\beta})] = 0 \in \check{H}^1(X, \Omega^1).$

Accordingly (see [VS: Chapt. VI; p. 54, Theorem 11.1, along with p. 52, (10.27), and (10.28)]), the given line sheaf $\mathcal{L}$, as represented by (5.21), admits an $\mathcal{A}$-connection D for which (5.22) provides the $\mathcal{A}$-connection 0-cochain of 1-forms on X (loc. cit. p. 112, (2.39), of course, for $n = 1$, concerning the case considered) corresponding to a local frame of $\mathcal{L}$ (cf., for instance, (5.13)). In this regard, see also our relevant comments in (5.19) pertaining to the affine space of $\mathcal{L}$-connections of $\mathcal{A}$, as in (5.19.1).

   Consequently, as an application of the notion of the Atiyah class of $\mathcal{L}$, as given by (5.11), one arrives at another proof of our previous claim, as in Lemma 2.1 of Chapter III (see in particular (2.17)). Indeed,

(5.26)   (2.17) of Chapt. III (viz., the transformation law of potentials) is equivalent, speaking in terms of (sheaf) cohomology, to the vanishing of the Atiyah class of $\mathcal{L}$ (as follows from the preceding).

In this connection, we know of course that

(5.27)   the vanishing of the Atiyah class of $\mathcal{A}$ is a criterion for the existence of an $\mathcal{L}$-connection of $\mathcal{L}$.

In other words, one concludes that

(5.28)   (2.17) in Chapt. III, that is, the transformation law of potentials (cf. also (5.17)) and the vanishing of the Atiyah class of $\mathcal{L}$, are all equivalent conditions for the realization of an $\mathcal{A}$-connection D of $\mathcal{L}$, which locally (in terms of a local frame of $\mathcal{L}$, cf. (5.13)), is given by (5.22). Of course, (5.23) is still another equivalent expression to the preceding ones (we recall here that by definition, $\Omega^1(X)$ is an $\mathcal{A}(X)$-module).

The preceding still justifies the classical terminology referring to (5.23) (see A. Weil, loc. cit.). See also [VS: Chapt.VI; p. 67, Corollary 12.1].

Finally, by further applying (5.23) (cf. also the comments in (5.28)), we still find the basic (standard) definition of the field strength of $(\mathcal{L}, D)$, in other words, the curvature of D. Thus, by considering the 0-cochain of 2-forms that corresponds to (5.22),

$$(5.29) \qquad (d\theta_\alpha) \in \check{C}^0(\mathcal{U}, d\Omega^1) \subsetarrow \check{C}^0(\mathcal{U}, \Omega^2),$$

one obtains (cf. also (5.23))

$$(5.30) \qquad \delta(d\theta_\alpha) = d(\delta(\theta_\alpha)) = d\left(\frac{1}{2\pi i}\tilde{\partial}(g_{\alpha\beta})\right)$$
$$= \frac{1}{2\pi i}d(\tilde{\partial}(g_{\alpha\beta})) = \frac{1}{2\pi i}(d \circ \tilde{\partial})(g_{\alpha\beta}) = 0$$

(one applies here the fact that $d \circ \tilde{\partial} = 0$, cf., for instance, Chapt. I; (7.5)). Therefore, (5.29) is a 0-cocycle of $d\Omega^1 \subseteq \Omega^2$; that is,

$$(5.31) \qquad (d\theta_\alpha) \in \check{Z}^0(\mathcal{U}, d\Omega^1) \cong (d\Omega^1)(X) \subsetarrow \Omega^2(X)$$

(cf. also [VS: Chapt. III; p. 234, Lemma 8.1, as well as, p. 183, (4.55)]). In other words, (5.31) entails a closed 2-form on X; that is, we have (we recall here that by definition, one has $d \circ d \equiv dd \equiv d^2 = 0$)

$$(5.32) \qquad (d\theta_\alpha) \in \Omega^2(X), \quad such\ that\ \ d(d\theta_\alpha) = 0.$$

The above 2-form, as given by (5.32), is further defined to be the curvature, denoted by

$$(5.33) \qquad R(D) \equiv R := (d\theta_\alpha) \in \Omega^2(X),$$

of the $\mathcal{A}$-connection

$$(5.34) \qquad D \longleftrightarrow (\theta_\alpha) \in \check{C}^0(\mathcal{U}, \Omega^1)$$

of $\mathcal{L}$; applying physical parlance, we still speak of the field strength (for R, as in (5.33)) of the gauge potential (connection D, as in (5.34)) of the (Maxwell) field $(\mathcal{L}, D)$ under consideration. See also the cited work of A. Weil.

5.3 Geometric Prequantization of Bosons (Continued)

Returning to our main objective of this part of the present section, namely, to the prequantization of (free) bosons, we further remark, as an outcome of the preceding discussion, that

> as far as, we dispose the appropriate "arithmetic" $\mathcal{A}$, in effect, $\mathbb{C}$-algebraized space

(5.33)

$$(5.35.1) \qquad\qquad (X, \mathcal{A})$$

(see for instance (5.3)),

then

(5.36) not only are the gauge potential and the corresponding field strength of the Maxwell field (viz. (free) boson) at issue (i.e., can be construed as) consequences of the carrier itself, through its realization via the associated with it coordinate 1-cocycle,

but in fact, one can say as a complementary remark to (5.8) that

(5.37) every (free) boson provides by itself its prequantization (see (5.8.1), in conjunction with (2.22)); therefore, any such elementary particle is prequantizable.

We now consider the case of (free) fermions, viz. of the other (spin-) subdivision of the world of (bare) elementary particles.

5.4 Fermionic Case

We have examined the (geometric) prequantization of (bare) particles of integer spin number, viz. of (free) bosons. We are going to look now at the case of (bare) elementary particles having half-integer spin number, that is, at (free) fermions in view of the (standard) classification of elementary particles according to their spin structure (cf. Chapt. II; (2.4)). We have also considered the sheaf-theoretic classification of elementary particles, in view of their spin structure, in terms of vector sheaves; thus, we have seen that (ibid., (6.29))

(5.38) fields of (states of bare) fermions correspond to (sections of) vector sheaves of rank greater than 1.

On the other hand, referring to our abstract set-up as expressed in general by a differential triad

$$(5.39) \qquad (\mathcal{A}, \partial, \Omega^1),$$

where the corresponding $\mathbb{C}$-algebraized space

$$(5.40) \qquad (X, \mathcal{A})$$

(cf., for instance, Chapt. III; (0.1)) can still be appropriately specialized according as a particular case might demand, we have seen that

(5.41) a (bare) boson can be identified (up to a (local) isomorphism of $\mathcal{A}$-modules) locally (viz. with respect to a given local gauge) with our "arithmetic" $\mathcal{A}$. This realization (identification) is achieved by means of a coordinate 1-cocycle that represents, in terms of $\mathcal{A}$, the carrier (line sheaf) $\mathcal{L}$ of the (free) boson at issue (see (5.19.2)).

Within the same vein of ideas, one can further realize (see also the comments following (5.44)) that

(5.42)   the $\mathcal{A}$-module Ω^1, as appears in (5.39), which here is supposed to be in particular a vector sheaf on X (hence, by definition, of rank greater than 1), may stand, through its sections, as the sheaf of states of a (bare) fermion, viz. of a spin-$\frac{1}{2}$ (free) elementary particle.

In the classical case, where the structure sheaf $\mathcal{A}$ is given by definition by (5.5.1), one further assumes that

$\Omega^1 :=$ sheaf of germs of sections of 1-forms on X, viz., strictly speaking, of the complexified cotangent bundle of X,

(5.43)   (5.43.1) $^{\mathbb{C}}\mathcal{T}^*(X),$

where the smooth (C^∞-)manifold X is specified by (5.2.1) (see also the relevant comments in (5.2)).

However, taking into account the relevant comments in (5.3) concerning the space X as well as those following (5.5), we assumed in (5.42) that

(5.44)   the $\mathcal{A}$-module Ω^1, as in (5.39), is a vector sheaf on X having the appropriate physical interpretation, as in (5.42).

Our previous assumption in (5.42) referring to the (physical) substance (function) of the $\mathcal{A}$-module (thus, by hypothesis, vector sheaf) Ω^1 on X is based on a suitable interpretation of the sections of Ω^1 as providing wave functions that transform as Lorentz spinors and therefore can be described as

(5.45)   single (bare) spin-$\frac{1}{2}$ fermion states (sections of Ω^1).

Indeed, the above is the outcome of a suitable dressing perturbation and switching it off again, on (and from) the states (-wave functions) represented by sections of Ω^1. In this regard, cf. also the comments following (4.3) in Chapter II pertaining to antisymmetric wave functions, as suitably described (states) sections of vector sheaves (of rank greater than 1); see the relevant discussion by S.A. Selesnick [1: p. 38f], the preceding being, in fact, an adaptation to the present abstract setting of the corresponding argument of his in the aforementioned work.

Thus, our objective in the sequel is to

(5.46)   provide for (the vector sheaf) Ω^1 as in (5.39), hence, according to (5.45), for the state space of a single (bare) fermion, a prequantizing line sheaf (see also the comments following (5.47) below).

Indeed, one actually proves that

(5.47)

Ω^1 can be endowed with an integral closed 2-form say, ω; yet by an (appropriate) application of Weil's integrality theorem (cf. Chapt. III; Theorem 3.1), we can look at ω as the field strength (curvature) of a Maxwell field $(\mathcal{L}, D)$ on Ω^1, that is, in such a manner that

(5.47.1) $$R(D) = \omega.$$

The preceding Maxwell field $(\mathcal{L}, D)$ we referred to in (5.47) becomes now, by definition, a prequantizing line sheaf for Ω^1, which we already looked for in (5.46), in fact, an appropriate, as we shall see (cf. (5.110) below), pull-back on Ω^1 of the aforementioned line sheaf.

However, before we come to the proof of our claim in (5.47), we first explain the technical part of the argument that we are going to employ in the sequel by making a slight aberration from the main stream of our exposition via the ensuing subsection, which is also of interest in itself.

5.5 Pull-Back of Maxwell Fields

Our aim in the present subsection is to point out, always within the abstract setting advocated thus far and in full generality as well, that

(5.48)   the pull-back functor preserves Maxwell fields.

For the terminology employed in (5.48), we refer to [VS: Chapt. I; p. 79, Section 14.2]. For convenience, we exhibit below the general set-up.

Suppose that we have a differential triad

(5.49) $$(\mathcal{A}, \partial, \Omega^1)$$

with respect to a given $\mathbb{C}$-algebraized space

(5.50) $$(Y, \mathcal{A}),$$

while we still assume that we are given a continuous map

(5.51) $$f : X \longrightarrow Y$$

from an arbitrary topological space X into Y. Furthermore, assume that we have a sheaf (of sets), say $\mathcal{S}$, on Y; that is, one has

(5.52) $$\mathcal{S} \equiv (S, \pi, Y).$$

Thus, we recall (loc. cit.) that the pull-back of $\mathcal{S}$ on the space X via f denoted by

(5.53) $$f^*(\mathcal{S}) \equiv f^{-1}(\mathcal{S}),$$

is the sheaf (of sets) on X given by the relation

(5.54) $$f^*(S) := \{(x, z) \in X \times S : f(x) = \pi(z)\} \subseteq X \times S.$$

That is, one has by definition

(5.55) $$f^*(S) \equiv (f^*(S),\ prx\big|_{f^*(S)},\ X),$$

where the corresponding (sheaf) projection onto X is by definition the restriction onto $f^*(S)$, in view of (5.54), of the first projection of $X \times S$ onto X.

On the other hand, $f^*(S)$ is endowed, by virtue of (5.54), with the relative topology from $X \times S$, so that one further proves that the map

(5.56) $$prx\big|_{f^*(S)} : f^*(S) \longrightarrow X : (x, z) \longmapsto x,$$

as in (5.55), is indeed a local homeomorphism (see [VS: Chapt. I; p. 79, (14.16), (14.17)]); hence (5.54), that is, (5.55), is a sheaf (of sets) on X, the so-called pull-back on X (alias the inverse image) through f (cf. (5.51)) of the given sheaf S on Y.

In this connection, it is useful to recall here the particular form that certain (defining) sections of $f^*(S)$ (see loc. cit., Chapt. I; p. 16, (3.17) have concerning the terminology employed here, along with (5.62)); thus, for any open set $V \subseteq Y$ and $t \in S(V)$ (viz. $V \equiv Dom(t)$), one has the relation

(5.57) $$f_V^*(t) \equiv f^*(t) = t \circ f,$$

where for convenience we applied have an obvious abuse of notation (see also (5.60)). That is, more precisely, one gets the relation

(5.58) $$f_V^*(t)(x) \equiv f^*(t)(x) := (x, t(f(x))) \in f^*(S)_x$$

for any $x \in f^{-1}(V)$, so that one has the corresponding adjunction map (or map of sections)

(5.59) $$f_V^* : S(V) \longrightarrow f^*(S(f^{-1}(V))$$

for any open $V \subseteq Y$, given (cf. (5.58)) by the relation

(5.60) $$f_V^*(t) \equiv f^*(t) := (id_U,\ t \circ f)$$

for any (local) section $t \in S(V)$ while we set in (5.60)

(5.61) $$U \equiv f^{-1}(V) = Dom(f_V^*(t)) \equiv Dom(f^*(t))$$

(see also loc. cit., Chapt. I; p. 80). On the other hand, as already hinted at, one further concludes, concerning the above type of sections of $f^*(S)$, that

(5.62)   sections of $f^*(S)$ over an open subset of X of the form $f^{-1}(V)$, with V open in Y, constitute a defining family of sections of $f^*(S)$

(ibid., Chapt. I; p. 81, (14.27), (14.28), and p. 16; (3.17)). ■ To put it more trans-
parently, we further remark that our previous conclusion in (5.62) means in practice
that

(5.63)   The sheaf $f^*(\mathcal{S})$ is determined through those sections of it that are taken
over open subsets of X of the form $f^{-1}(V)$, with V open in Y.

See also loc. cit., Chapt. I; Section 3, and p. 81; (14.28). ■

As an application of the above, pertaining to the fibers of the sheaves under consideration, one gets the relation

(5.64) $$f^*(\mathcal{S})_x = \mathcal{S}_{f(x)}$$

for any $x \in X$, within a (canonical, set-theoretic) bijection (loc. cit., p. 80; (14.26)).
In particular, as an outcome of (5.64), we know (ibid., Chapt. II; p. 119, (2.68)) that

(5.65)   the inverse image functor f^* associated with any given continuous map
f (cf., for example, (5.51)) is exact.

Referring in particular to (5.65), one considers sheaves of $\mathcal{A}$-modules (loc. cit., p.
118, (2.65)). Thus, according to the definitions, one further concludes that

(5.66)   the pull-back of any vector sheaf is a vector sheaf of the same (finite)
rank as the given one

(loc. cit., p. 128; (4.14); see also (5.64), concerning the last part of (5.66)). Thus, by
looking at a line sheaf $\mathcal{L}$ on Y and taking an open set $V \subseteq Y$, being a local gauge of
$\mathcal{L}$, one has the following commutative diagram:

(5.67)

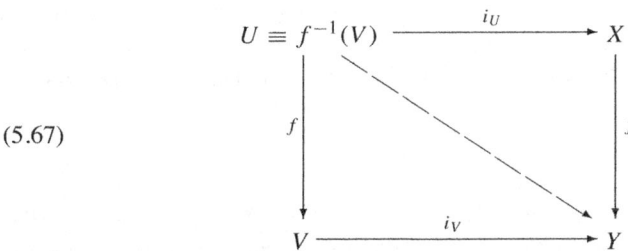

where i_U, i_V in (5.67) denote the *canonical inclusion (injection) maps*; therefore,
one has the relation

(5.68) $$i_V \circ f = f \circ i_{f^{-1}(V)} \equiv f \circ i_U.$$

Thus, referring to (5.66), and in particular to the line sheaf $\mathcal{L}$, one obtains

(5.69)
$$f^*(\mathcal{L})\big|_U = i_U^*(f^*(\mathcal{L})) = (f \circ i_U)^*(\mathcal{L}) = (i_V \circ f)^*(\mathcal{L})$$
$$= f^*(i_V^*(\mathcal{L})) = f^*(\mathcal{L}\big|_V) = f^*(\mathcal{A}\big|_V) = f^*(\mathcal{A})\big|_U,$$

which thus proves the assertion for the sheaf ($f^*(\mathcal{A})$-module) $f^*(\mathcal{L})$ on X (see also loc. cit., Chapt. I; p. 84, (14.46), as well as Chapt. II; p. 118, (2.66)). That is, we have actually proved, by (5.69), that

(5.70)   the pull-back $f^*(\mathcal{L})$ of a line sheaf $\mathcal{L}$ on Y by means of a (continuous) map f as in (5.51) is still a line sheaf on X (domain of definition of f).

A similar proof to (5.69) holds for any vector sheaf on Y by further taking into account that by virtue of (5.65) one always has the relation

$$(5.71) \qquad f^*(\mathcal{A}^n) = f^*(\mathcal{A})^n, \qquad n \in \mathbb{N},$$

within an isomorphism of $f^*(\mathcal{A})$-modules. This establishes completely our assertion in (5.66). ∎

Returning to the set-up of (5.49), we next remark that

(5.72)   the pull-back of a given differential triad as in (5.49) through a (continuous) map as in (5.51) entails a differential triad as well on the topological space, domain of definition, of the map at issue.

Namely, one sets

$$(5.73) \qquad f^*(\mathcal{A}, \partial, \Omega^1) := (f^*(\mathcal{A}), f^*(\partial), f^*(\Omega^1)),$$

where the second member of (5.72) denotes by definition the *differential triad on X, the pull-back*, according to (5.72), *of the given one on Y*, as in (5.49). In this regard, see also [VS: Chapt. VI; p. 25, (6.3)], along with subsequent comments therein concerning the justification of (5.73), hence the proof as well of (5.72). ∎

On the other hand, referring to the notation employed in the second member of (5.73), it is useful to recall here, for convenience, the notion of the pull-back, via f of an $\mathcal{A}$-connection D (loc. cit., p. 26, Definition 6.1): Thus, supposing again that we have the framework of (5.49)–(5.51), consider also a pair

$$(5.74) \qquad\qquad (\mathcal{E}, D)$$

consisting of an $\mathcal{A}$-module $\mathcal{E}$ on Y and an $\mathcal{A}$-connection D of $\mathcal{E}$. Then, by further looking at the differential triad on X, as given by (5.73), one defines the pull-back of D, through f, denoted by $f^*(D)$, as the sheaf morphism

$$(5.75) \qquad f^*(D) : f^*(\mathcal{E}) \longrightarrow f^*(\Omega^1(\mathcal{E})) \cong f^*(\mathcal{E}) \otimes_{f^*(\mathcal{A})} f^*(\Omega^1),$$

in such a manner that one sets

$$(5.76) \qquad\qquad f^*(D)(f_V^*(t)) := f_V^*(D(t)) = D(t) \circ f$$

for any (local) section $t \in \mathcal{E}(V)$ with V open in Y (cf. also (5.57), as well as, (5.60) in the preceding regarding the notation employed in (5.76); cf. [VS: Chapt. VI; p.

26, (6.11), along with Chapt. I; p. 83, (14.43), and p. 84, (14.44)]). Accordingly, one thus proves that

(5.77) $f^*(D)$, as defined by (5.76), yields an $f^*(\mathcal{A})$-connection of $f^*(\mathcal{E})$ (viz. the $f^*(\mathcal{A})$-module), pull-back, via f, of the given $\mathcal{A}$-module $\mathcal{E}$ on Y (see (5.74)).

See [VS: Chapt. VI; p. 27, (6.13)]. ■ In particular, by looking at the given standard (flat) $\mathcal{A}$-connection

(5.78) $$\partial : \mathcal{A} \longrightarrow \Omega^1$$

of $\mathcal{A}$ as in (5.49), one gets, by virtue of (5.76), the relation

(5.79) $$f^*(\partial)(f_V^*(t)) = f_V^*(\partial(t)) = \partial(t) \circ f$$

for any $t \in \mathcal{A}(V)$ and V open in Y.

Similarly, by considering the logarithmic derivation associated with ∂,

(5.80) $$\tilde{\partial} : \mathcal{A}^{\cdot} \longrightarrow \Omega^1$$

(cf. Chapt. I; (1.25)), and taking its pull-back via f, which also will be of use below, one obtains

(5.81) $$f^*(\tilde{\partial}) : f^*(\mathcal{A}^{\cdot}) = f^*(\mathcal{A}^{\cdot}) \longrightarrow f^*(\Omega^1),$$

such that one has (see 5.76), along with (5.57)),

(5.82) $$f^*(\tilde{\partial})(f^*(\alpha)) = f^*(\tilde{\partial}(\alpha)) = \tilde{\partial}(\alpha) \circ f,$$

for any $\alpha \in \mathcal{A}^{\cdot}(V) = \mathcal{A}(V)^{\cdot}$ and V open in Y. Concerning the domain of definition of $f^*(\tilde{\partial})$ as above, one has, as already noted, the relation

(5.83) $$f^*(\mathcal{A}^{\cdot}) = f^*(\mathcal{A})^{\cdot},$$

within an isomorphism of group sheaves (or of $\mathbb{Z}$-modules; regarding the latter terminology cf. also [VS: Chapt. II; p. 109, Remark 2.1]).

On the other hand, based further on (5.57) (cf. also (5.60)), one proves that

(5.84)   the pull-back of a 1-cocycle of $\mathcal{A}^{\cdot}$ yields a 1-cocycle of $f^*(\mathcal{A}^{\cdot})$. That is, formally speaking, one has the relation

(5.84.1)
$$f^*(Z^1(\mathcal{U}, \mathcal{A}^{\cdot})) \subseteq Z^1(f^{-1}(\mathcal{U}), f^*(\mathcal{A}^{\cdot}))$$
$$= Z^1(f^{-1}(\mathcal{U}), f^*(\mathcal{A})^{\cdot}).$$

Namely, for any $(g_{\alpha\beta}) \in Z^1(\mathcal{U}, \mathcal{A}^{\cdot})$, one has (cf. (5.57))

(5.85) $$\delta(f^*(g_{\alpha\beta})) = \delta(g_{\alpha\beta} \circ f) = \delta(g_{\alpha\beta}) \circ f = 0,$$

which proves our claim in (5.84.1). ■ (Of course, the previous assertion in (5.84) is actually valid for any $\mathcal{A}$-module $\mathcal{E}$, in general.)

We are now in a position to come to our main objective of the present subsection, that is, to the proof of (5.48):

Namely, suppose that we are given a Maxwell field on Y,

$$(5.86) \qquad (\mathcal{L}, D) \longleftrightarrow ((g_{\alpha\beta}), (\theta_\alpha))$$

(see Chapt. III; (2.26) concerning the notation applied in (5.86)), such that (ibid., Lemma 2.1)

$$(5.87) \qquad \delta(\theta_\alpha) = \tilde{\partial}(g_{\alpha\beta}).$$

Our task is to prove further that

the pull-back, via f, as in (5.51), of the Maxwell field $(\mathcal{L}, D)$ on Y, as in (5.86), that is (see also (5.73)), the object

(5.88)

$$(5.88.1) \qquad f^*((\mathcal{L}, D)) \equiv f^*(\mathcal{L}, D) := (f^*(\mathcal{L}), f^*(D)),$$

still yields a Maxwell field on X.

Of course, the pair

$$(5.89) \qquad (f^*(\mathcal{L}), f^*(D)) \equiv f^*(\mathcal{L}, D),$$

as appeared in (5.88.1), yields a Maxwell field on X, since by virtue of (5.70) and (5.77), it consists of a line sheaf $f^*(\mathcal{L})$ on X and an $f^*(\mathcal{A})$-connection $f^*(D)$ on it (cf. also Chapt. III; Definition 1.1 and (1.4), (1.5)), which thus proves our assertion in (5.88). ■ On the other hand, one can give an alternative to the previous proof, based on the preceding and the second member of (5.86) (local description of $(\mathcal{L}, D)$): Thus, one further concludes that

given a pair

$$(5.90.1) \qquad ((g_{\alpha\beta}), (\theta_\alpha))$$

determining a Maxwell field $(\mathcal{L}, D)$ on Y, as in (5.86) (see also Chapt. III; Lemma 2.1), its pull-back on Y, via f, viz. the pair

(5.90)

$$(5.90.2) \qquad (f^*(g_{\alpha\beta}), f^*(\theta_\alpha)),$$

provides a Maxwell field on X; that is (loc. cit.), (5.90.2) satisfies the relation (transformation law of potentials)

$$(5.90.3) \qquad \delta(f^*(\theta_\alpha)) = f^*(\tilde{\partial})(f^*(g_{\alpha\beta})).$$

We first remark that in view of (5.84.1), see also (5.59), the pair (5.90.2) consists of a 1-cocycle of $f^*(\mathcal{A}^\cdot) = f^*(\mathcal{A})^\cdot$ (cf. (5.83)) and a 0-cochain of $f^*(\Omega^1)$, viz. one has

(5.91) $(f^*(g_{\alpha\beta}), f^*(\theta_\alpha)) \in Z^1(f^{-1}(\mathcal{U}), f^*(\mathcal{A})^{\cdot}) \times C^0(f^{-1}(\mathcal{U}), f^*(\Omega^1)),$

while we assumed concerning (5.90.1), that

(5.92) $((g_{\alpha\beta}), (\theta_\alpha)) \in Z^1(\mathcal{U}, \mathcal{A}^{\cdot}) \times C^0(\mathcal{U}, \Omega^1)$

(cf. also Chapt. III; (2.16)). We next prove (5.90.3), that is, the transformation law of potentials for the pair (5.90.2); namely, one has

(5.93)
$$\delta(f^*(\theta_\alpha)) = \delta(\theta_\alpha \circ f) = \delta(\theta_\alpha) \circ f = \tilde{\partial}(g_{\alpha\beta}) \circ f$$
$$= f^*(\tilde{\partial}(g_{\alpha\beta})) = f^*(\tilde{\partial})(f^*(g_{\alpha\beta})),$$

which was to be proved, viz. (5.90.3). (In this regard, see also (5.57) and (5.82), along with Chapt. III; (2.17), in conjunction with our hypothesis for (5.90.1).) ∎
    On the other hand, as a byproduct of (5.93), one further obtains the relation

(5.94) $\delta(f^*(\theta_\alpha)) = \delta(\theta_\alpha \circ f) = \delta(\theta_\alpha) \circ f = f^*(\delta(\theta_\alpha));$

that is, one gets the relation

(5.95) $$\delta \circ f^* = f^* \circ \delta,$$

another instance of which has already been applied, in particular, to prove (5.84) (cf. also (5.85)). Thus, we can refer to (5.95) as the

(5.96)   commutativity of the pull-back (functor), with respect to the Bockstein (coboundary) operator.

Finally, in terminating the present subsection, we also want to point out another immediate byproduct of the preceding discussion. In considering an arbitrary topological space X, carrier of a given differential triad, as in (5.49) above, one easily sees that

the group of self-homeomorphisms of X,

(5.97.1) $Homeo(X) \equiv Aut(X),$

(5.97)   acts on the right on the Maxwell group of X,

(5.97.2) $\Phi_A^1(X)^\nabla$

(see Chapt. III; (2.5)), through the pull-back (functor).

Indeed, the assertion follows straightforwardly from (5.91) and (5.90.3), in conjunction with [VS: Chapt. I; p. 84, (14.46)]. Thus, one proves, for instance, that for any f, g in $Aut(X)$ (cf. (5.97.1)) and $(g_{\alpha\beta}) \in Z^1(\mathcal{U}, \mathcal{A}^{\cdot})$ (see (5.92)), one obtains

(5.96) $(g \circ f)^*(g_{\alpha\beta}) = (f^* \circ g^*)(g_{\alpha\beta}) = f^*(g^*(g_{\alpha\beta})),$

which vindicates our claim in (5.97). ∎

We can still express (5.97) by saying, in view of (5.98), that

(5.99)   the Maxwell group of X, $\Phi^1_{\mathcal{A}}(X)^\nabla$, is an $\mathcal{A}ut(X)^{op}$-set, the respective action being realized through the pull-back in fact, an

(5.99.1) $\mathcal{A}ut(X)^{op}$-group.

Here, we denote as usual by

(5.100) $\mathcal{A}ut(X)^{op}$

the opposite group of the group $\mathcal{A}ut(X)$, as in (5.97.1), viz. the same underlying set as $\mathcal{A}ut(X)$, but with multiplication reversed, viz. one sets

(5.101) $(g \circ f)^{op} := f \circ g,$ with f, g in $\mathcal{A}ut(X)$.

In this connection, see also, concerning the latter terminology P. Tondeur [1: p.2], or N. Bourbaki [3: Chapt. I; p. 49, Definition 1, or Chapt. II; p. 2, (M'_{III})].

Scholium 5.1 On the basis of the preceding discussion, we have concluded that (cf. (5.48) and (5.88))

(5.102)   the pull-back of a Maxwell field is still a Maxwell field.

In fact, instead of considering line sheaves, as in the case of a Maxwell field $(\mathcal{L}, D)$, one can actually prove, quite generally and within the general set-up as described by (5.49)–(5.51) that

(5.103)   the pull-back (functor) preserves Yang-Mills fields as well.

The justification of our previous claim can be based on a similar argument to that applied in the proof of (5.88). However, details of that proof will be given in Vol. II of this treatise, Chapter I; Section 9, in connection with relevant results pertaining to Yang–Mills fields.

We come now, in the following subsection, to deal with our previous considerations in (5.47), which are concerned with a (potential) prequantization of fermions, as was exactly the case for bosons (cf. (5.37)). This will complete our program of the present section.

5.6 Geometric Prequantization of Fermions (Continued)

The key notion for the response to (5.37) is to apply the so-called principle of

(5.104)   mediating forces, through the exchange of bosons.

In this connection, see also S.A. Selesnick [1: p. 43], whose relevant considerations were, as already mentioned, our motivation to the present abstract setting.
    In other words, one is led to the following situation:

(5.105)   by letting a boson act on Ω^1 (the state space of a (free) fermion, cf. (5.42)), one gets a new particle, which, however, locally coincides (modulo (natural) local $\mathcal{A}$-isomorphisms, cf. (5.106) below) with the initially given particle (fermion, or Ω^1).

Namely, assume that we have a line sheaf $\mathcal{L}$ on X, the carrier of a (free) boson (cf. Chapt. II; (6.29)), as in (5.105). Therefore, by looking at the action of $\mathcal{L}$ on Ω^1, as supposed in (5.105), one obtains the relations

(5.106)
$$\Omega^1(\mathcal{L})\big|_U \equiv (\mathcal{L} \otimes_\mathcal{A} \Omega^1)\big|_U = \mathcal{L}\big|_U \otimes_{\mathcal{A}\big|_U} \Omega^1\big|_U$$
$$= \mathcal{A}\big|_U \otimes_{\mathcal{A}\big|_U} \Omega^1\big|_U = (\mathcal{A} \otimes_\mathcal{A} \Omega^1)\big|_U = \Omega^1\big|_U,$$

modulo the obvious $\mathcal{A}\big|_U$-isomorphisms of the $\mathcal{A}\big|_U$-modules concerned, where U stands for an open subset of X, which one may take to be a common local gauge of $\mathcal{L}$ and Ω^1. (In this regard, see also [VS: Chapt. II; p. 125, (4.1), for $n = 1$, and p. 130, (5.15), as well as, p. 132, Lemma 5.1].) Of course, (5.106) proves already our assertion in (5.105); equivalently, one can express (5.105) by saying that

(5.107)   the action alluded to in (5.105) is locally undiscernible (viz., experimentally undetectable).

Accordingly, one realizes that

(5.108)   the aforementioned action, as in (5.105), serves, within the present context, to supply Ω^1 with a prequantizing line sheaf (see (5.46)), which we were looking for according to our claim in (5.47).

Consider now the sheaf on X defined by the given $\mathcal{A}$-module Ω^1 (see (5.49)), viz. one has by definition

(5.109) $$\Omega^1 \equiv (\Omega^1, \rho, X),$$

where ρ stands for the defining sheaf local homeomorphism (hence, continuous) projection of the sheaf (space) Ω^1 onto X. Therefore, by next looking at the line sheaf $\mathcal{L}$, one can further consider the pull-back $\rho^*(\mathcal{L}^1)$ of the sheaf ($\mathcal{A}$-module) $\mathcal{L}^1$

via ρ over (the topological space) Ω^1, so that one gets, by definition (cf. (5.55)), the following commutative diagram:

$$
\begin{array}{ccc}
\rho^*(\mathcal{L}) & \longrightarrow & \mathcal{L} \\
\downarrow & & \downarrow \pi \\
\Omega^1 & \underset{\rho}{\longrightarrow} & X
\end{array}
$$

(5.110)

On the other hand, by assuming for the initially given $\mathbb{C}$-algebraized space

(5.111) $(X, \mathcal{A})$

(see, for instance, (5.40)) the appropriate conditions, as for example that

(5.112) X is a paracompact (Hausdorff) space and $\mathcal{A}$ a fine ($\mathbb{C}$-algebra) sheaf on X,

then (cf. [VS: Chapt. VI; p. 85, Theorem 16.1], in conjunction with (5.112) and loc. cit., Chapt. III; p. 247, (8.56)), one concludes that

(5.113)   the line sheaf $\mathcal{L}$, as in (5.105), admits an $\mathcal{A}$-connection D,

so that one finally gets a Maxwell field

(5.114) $(\mathcal{L}, D)$

on X; in other words, the (free) boson we let act on Ω^1 in (5.105) (the state space of) a (free) fermion.

Accordingly, based now on (5.102), we conclude that the pull-back of $(\mathcal{L}, D)$ on Ω^1, via ρ (cf. (5.110)), that is, the pair

(5.115) $\rho^*((\mathcal{L}, D)) := (\rho^*(\mathcal{L}), \rho^*(D))$

(see also (5.88.1)), yields a Maxwell field on Ω^1. Thus, the field strength (curvature) of the latter is given by the relation

(5.116) $R(\rho^*(D)) = \rho^*(R(D))$.

See [VS: Chapt. VIII; p. 235, (9.24) and (9.25)].

On the other hand, by pulling back the curvature space structure of X to Ω^1, via ρ, one gets for Ω^1 a similar structure (cf. [VS: Chapt. VIII; p. 235, (9.24)]). Therefore,

(5.117) $R(\rho^*(D))$ is a closed 2-form on Ω^1.

See also Chapt. III; (3.19). Since

(5.118)   the pull-back functor is an exact functor

(cf. [VS: Chapt. II; p. 119, (2.68)], or (5.64)), one concludes, in particular, that

(5.119)   the pull-back of a Bianchi space, whichs is, by hypothesis, X (see Chapt. III; (3.17), for the terminology employed), is a Bianchi space as well. (Hence, this is the case for Ω^1 too, relative to the map ρ, as in (5.109).)

Thus, the above fully explains (5.117). ∎

Scholium 5.2 The question now arises whether the closed 2-form on Ω^1, as in (5.117), yields a 2-dimensional (integral) cohomology class of Ω^1. In fact, this can happen when (and this is the general moral of this treatise) one affords the appropriate set-up, as, for instance, a generalized de Rham 2-space (see [VS: Chapt. IX; p. 256, Lemma 3.1]); in this connection, concerning the exactness of the generalized de Rham sequence (loc. cit., p. 354, (3.1)), which we can assume for X, this can be transferred to Ω^1 by virtue of the exactness of the pull-back functor (see (5.118)). However, we still then need, by definition (ibid. p. 254, Definition 3.1), Ω^1 to be a paracompact (Hausdorff) space. Thus, in the classical case one takes as Ω^1 the (complexified) cotangent bundle of our base (space-time) manifold X, hence a type of space, as desired in view of our hypothesis, for X and Ω^1 (see also (5.43); cf. S.A. Selesnick [1: p. 43]). Accordingly,

(5.120)   on the (complexified) cotangent bundle of X (cf. (5.43.1)) one can define, by pull-back, a closed 2-form that yields a 2-dimensional (complex) cohomology class, in effect, by virtue of Weil's integrality theorem, integral (cf. Chapt. III; (3.33) or (3.34)).

Thus, the pull-back of the above 2-dimensional integral cohomology class on the respective sheaf of (germs of) sections of the bundle at issue, as in (5.43), will be, by definition, the desired cohomology class on the sheaf Ω^1. This is what we are doing below, within the present abstract setting.

Namely, by looking at the natural morphism in cohomology defined by the pull-back (functor) associated with ρ, as in (5.109), viz. the map

(5.121) $$\rho^* : H^2(X, \mathbb{Z}) \longrightarrow H^2(\Omega^1, \mathbb{Z}),$$

one defines (cf. also (5.116))

(5.122) $$[R(\rho^*(D))] := \rho^*([R(D)]).$$

In this connection, see also, for instance, R. Godement [1: 199;(1), and p. 200] or G.E. Bredon [1: p. 194] concerning (5.121) and the effect, in general, of a continuous map on the cohomology, that is, the behavior of the pull-back (functor) relative to the cohomology (functors).

Thus, based on (5.116), (5.121), and (5.122) (see also Chapt. III; (3.43)), one further obtains

(5.123)
$$\frac{1}{2\pi i}[R(\rho^*(D))] = \frac{1}{2\pi i}[\rho^*(R(D))] = \frac{1}{2\pi i}\rho^*([R(D)])$$

$$= \rho^*\left(\frac{1}{2\pi i}[R(D)]\right) \in \rho^*(H^2(X, \mathbb{Z})) \lhd H^2(\Omega^1, \mathbb{Z}),$$

which thus provides a 2-dimensional integral cohomology class of Ω^1. Consequently,

(5.124)

    the Maxwell field (cf. (5.115))

    (5.124.1) $\qquad\qquad\qquad (\rho^*(\mathcal{L}), \rho^*(D))$

    yields, by its field strength (curvature)

    (5.124.2) $\qquad\qquad\qquad R(\rho^*(D)),$

    a 2-dimensional integral cohomology class of Ω^1. Therefore, one can consider (5.124.1), in fact the line sheaf

    (5.124.3) $\qquad\qquad\qquad \rho^*(\mathcal{L}),$

    as a prequantizing line sheaf on Ω^1.

As a result of the preceding, one concludes that

(5.125)

    the pair

    (5.125.1) $\qquad\qquad\qquad (\Omega^1, \omega),$

    such that

    (5.125.2) $\qquad\qquad\qquad \omega := \frac{1}{2\pi i}R(\rho^*(D))$

    (see also (5.116), (5.117)), may now be considered as an integral symplectic (sheaf) space, hence prequantizable.

Indeed, the above terminology should essentially be taken in a generalized point of view concerning the previously applied terminology, as for instance in Definition 1.1. Thus, motivated by the situation we have in (5.125), we come to formulate the following notion.

    Suppose that we have the appropriate differential set-up on a topological space X (e.g., a Bianchi space, see Chapt. III; (3.17)). Then, the pair

    (5.126.1) $\qquad\qquad\qquad (X, \omega)$

is said to be a prequantizable symplectic (sheaf) space in the generalized sense, whenever

(5.126)   (5.126.2) ω is a closed 2-form on X supplying a 2-dimensional integral cohomology class of X, being also the curvature (field strength) of a Maxwell field $(\mathcal{L}, D)$ on X,

that is, the situation we met in (5.125), formulated here in abstracto.

The above fully explains our claim in (5.47), while it also leads us, finally, to the desired conclusion that

(5.127)   every (free) fermion is also prequantizable (as this happens already for (free) bosons as well; cf. (5.37)).

Accordingly, one arrives at the final statement, namely, that

every (free) elementary particle is prequantizable. That is,

(5.128)   (5.128.1)   every such physical system provides, by itself (canonically), a prequantizing line sheaf.

(See (5.124.1) and (5.124.3), as well as (5.123), in conjunction with (5.126).)

Scholium 5.3   In nowadays current physics we also usually assume that even gravity is a field theory. A. Einstein asserted it already, said author [1: p. 140], due, however, now to an elementary particle, that is, the quantum of the gravitational field, or else, the graviton, thus, by its very definition, a (spin-2) boson. Hence, in that respect, and in accordance with the standpoint of the preceding discussion, one has the following mathematical formulation of the previous claim, hence an equivalent statement, in view of the foregoing, about gravitons. That is, one concludes that

(5.129)   the gravitational field, being in fact a particular case of a Maxwell field, that is, by our hypothesis, a boson, is (cf. (5.37)) also prequantizable.

Concerning the above perspective, we refer the reader to Vol. II of this treatise (in particular, to Chapt. IV; Section 9), where we discuss general relativity as a gauge theory always within the point of view of the present abstract differential-geometric set-up. The appearance of this subject of our study in Part II of this account, devoted in principle to considerations about Yang–Mills fields from the standpoint of abstract differential geometry, is only a technical matter; namely, it is due to relevant necessary differential-geometric notions that are always formulated in the abstract sense adopted here, such as, for instance, Lorentz metrics for Yang–Mills fields. All these, however, are deferred for the pertinent place in Part II of the present study.

Note 5.1   The preceding material can be formulated within the appropriate Hermitian framework, provided, of course, one is given the pertinent $\mathbb{C}$-algebraized space

$(X, \mathcal{A})$. See, for instance, Section 3 in the foregoing. On the other hand, cf., for example, B. Kostant and J.-M. Souriau [1], concerning the classical counterpart.

Another relevant aspect to the preceding discussion, pertaining in particular to the notion of a symplectic (sheaf) space (cf. Section 2 above), is the consideration of a Hamiltonian mechanical (sheaf) system, or sheaf Hamiltonian system

$$(5.130) \qquad\qquad\qquad (\mathcal{A}, \omega, \alpha).$$

Here $(\mathcal{A}, \omega)$ stands, by definition, for a symplectic (sheaf) space (see Definition 1.1, or (5.126)), while we also assume that

$$(5.131) \qquad\qquad\qquad \alpha \in \mathcal{A}(X),$$

viz. we are given a global section of our structure sheaf $\mathcal{A}$. However, we are not going to discuss this material here. Concerning the classical case, we refer, for instance, to M. Puta [1: p. 28].

On the other hand, as a final remark to the previous considerations, we can refer to the significance of geometric (pre-)quantization for the classical theory. See, for instance, the relevant quotations at the beginning of this chapter, pertaining to the bypass of the classical correspondence principle and look instead at the elementary particles themselves directly; this is achieved in a (geometrical-)quantized dressing, in the sense that the elementary particles under consideration carry by themselves (canonically) a prequantizing line sheaf (/bundle). This mechanism is rooted, as we have seen, on the classification of elementary particles by their spin-structure, in conjunction with what we may call Selesnick's correspondence principle, as exhibited in Chapter II. Furthermore, the differential geometry we apply, motivated by the classical theory, is that of the abstract standpoint that dominates the present treatise.

References

R. Abraham – J.E. Marsden
1. *Foundations of Mechanics* (2nd ed.). Benjamin/Cummings, Reading, Mass., 1978.

W.A. Adkins – S.H. Weintraub
1. *Algebra. An Approach via Module Theory.* Springer-Verlag, New York, 1992.

M.F. Atiyah
1. *Complex analytic connections in fiber bundles.* Trans. Amer. Math. Soc. 85(1957), 181–207.

S.Y. Auyang
1. *How Is Quantum Field Theory Possible?* Oxford Univ. Press, Oxford, 1995.

S.J. Avis – C.J. Isham
1. *Quantum field theory and fiber bundles in a general space-time* in "Recent Developments in Gravitation," Cargèse, 1978 (eds. M. Lévy-S. Deser). Plenum Press, New York, 1979. pp. 347–401.

J. Baez – J.P. Muniain
1. *Gauge Fields, Knots and Gravity.* World Scientific, Singapore, 1994.

H. Bass
1. *Algebraic K-Theory.* W.A. Benjamin, New York, 1968.

J. Bernstein – V. Lunts
1. *Equivalent Sheaves and Functors.* Lecture Notes in Mathematics, No 1578. Springer-Verlag, Berlin, 1994.

J.D. Bjorken – S.D. Drell
1. *Relativistic Quantum Mechanics.* McGraw–Hill, New York, 1964.
2. *Relativistic Quantum Fields.* McGraw–Hill, New York, 1965.

R.J. Blattner
1. *On geometric quantization* in "Non-linear Partial Differential Operators and Quantization Procedures," Clausthal 1981 (S.I. Andersson and H.–D. Doebner, eds.). Lecture Notes in Mathematics, No 1037. Springer-Verlag, Berlin, 1983. pp. 209–241.

274   References

D. Bleecker

1. *Gauge Theory and Variational Principles.* Addison–Wesley, Reading, Mass., 1981.

N.N. Bogolubov – A.A. Logunov – I.T. Todorov

1. *Introduction to Axiomatic Quantum Field Theory.* Benjamin/Cummings Reading, Mass., 1975.

A. Böhm

1. *Quantum Mechanics.* Springer-Verlag, New York, 1979.

N. Bourbaki

1. *Théorie des ensembles, Chap. 1–2* (3me éd.). Hermann, Paris, 1966.
2. *Théorie des ensembles, Chap. 3* (2me éd.). Hermann, Paris, 1967.
3. *Algèbre, Chap. 1–3.* Hermann, Paris, 1970. p. 238,

P.J. Braam

1. *How empty is the vacuum?* in *"The Philosophy of Vacuum"* (S. Saunders and H.R. Brown, eds.). Clarendon Press, Oxford, 1991. pp. 279–285.
4. *Commutative Algebra.* Addison–Wesley, Reading, Mass., 1972.

G.E. Bredon

1. *Sheaf Theory* (2nd ed.) Springer, New York, 1997.

J.-L. Brylinski

1. *Loop Spaces, Characteristic Classes and Geometric Quantization.* Birkhäuser, Boston, 1993.

Y. Choquet-Bruat – C. DeWitt-Mozette – M. Dillard-Bleick

1. *Analysis, Manifolds and Physics* (rev. ed.). North–Holland, Amsterdam, 1982.

L. Conlon

1. *Differentiable Manifolds.* A First Course. Birkhäuser, Boston, 1993.

R.W.R. Darling

1. *Differential Forms and Connections.* Cambridge Univ. Press, Cambridge, 1994.

R. Deheuvels

1. *Formes quadratiques et groupes classiques.* Presses Univ. France, Paris, 1981.

P.A.M. Dirac

1. *The Principles of Quantum Mechanics* (4th ed.). Clarendon Press, Oxford, 1958.

C.T.J. Dodson – T. Poston

1. *Tensor Geometry. The Geometric Viewpoint and its Uses* (2nd ed.). Springer-Verlag, Berlin, 1991.

B.A. Dubrovin – A.T. Fomenko – S.P. Novikov

1. *Modern Geometry–Methods and Applications,* I. Springer-Verlag, New York, 1984.

A. Einstein

1. *The Meaning of Relativity.* Princeton Univ. Press, Princeton, N.J., 1988.

A. Einstein – N. Rosen

1. *The particle problem in the general theory of relativity.* Phys. Rev. 48(1935), 73–77.

2. *On the Ether* in *"The Philosophy of Vacuum"* (S. Saunders-H.R. Brown, eds.). Clarendon Press, Oxford, 1991. pp. 13–20. [The original article, under the title, *"Über den Äther,"* was published in 1924].

D. Eisenbud

1. *Commutative Algebra with a View Toward Algebraic Geometry.* Springer-Verlag, New York, 1995.

G.G. Emch

1. *Mathematical and Conceptual Foundations of 20th-Century Physics.* North–Holland, Amsterdam, 1984.

D.R. Finkelstein

1. *Matter, space and logic* in *"The Logico-Algebraic Approach to Quantum Mechanics, Volume II: Contemporary Consolidation"* (C.A. Hooker, ed.). D. Reidel Publ. Co., Dordrecht, 1979. pp. 123–139.
2. *Quantum Relativity. A Synthesis of the Ideas of Einstein and Heisenberg.* Springer, Berlin, 1997.

O. Forster

1. *Funktionentheoretische Hilfsmittel in der Theorie der kommutativen Banach-Algebren.* Jber. Deutsch. Math.-Verein. 76(1974), 1–17.

H. Fritzsch

1. *Quarks. The Stuff of Matter.* Penguin Books, London, 1983.

I.M. Gel'fand – R.A. Minlos – Z.Ya. Shapiro

1. *Representations of the Rotation and Lorentz Groups and Their Applications.* Pergamon Press, London, 1963.

R. Godement

1. *Topologie algébrique et théorie des faisceaux* (3ème éd.). Hermann, Paris, 1973.

H. Goldstein

1. *Classical Mechanics.* Addison-Wesley, Reading, Mass., 1950.

H.Grauert – R. Remmert

1. *Coherent Analytic Sheaves.* Springer-Verlag, Berlin, 1984.

A. Grothendieck

1. *A General Theory of Fiber Spaces with Structure Sheaf* (2nd ed.). Univ. of Kansas, Lawrence, Kansas, 1958.

A. Grothendieck – J.A. Dieudonné

1. *Eléments de Géométrie Algébrique I.* Springer-Verlag, Berlin, 1971.

V. Guillemin – S. Sternberg

1. *Symplectic Techniques in Physics.* Cambridge Univ. Press, Cambridge, 1984.

R.C. Gunning

1. *Lectures on Riemann Surfaces.* Princeton Univ. Press, Princeton, N.J., 1966.

W. Heisenberg

1. *The Physical Principles of the Quantum Theory.* Dover.

F. Hirzebruch

1. *Topological Methods in Algebraic Geometry.* Springer-Verlag, Berlin, 1978.

D. Husemoller

1. *Fibre Bundles* (2nd ed.). Springer-Verlag, New York, 1975.

H. Inassaridze
1. *Algebraic K -Theory*. Kluwer, Dordrecht, 1995.

K. Jänich
1. *Topology*. Springer-Verlag, New York, 1984.

M. Karoubi
1. *K -Theory. An Introduction*. Springer-Verlag, Berlin, 1978.

A.A. Kirillov
1. *Elements of the Theory of Representations*. Springer-Verlag, Berlin, 1976.

S. Kobayashi – K. Nomizu
1. *Foundations of Differential Geometry*, I. John Wiley, New York, 1963.

B. Kostant
1. *Quantization and Unitary Representations in "Lectures in Modern Analysis and Applications, II."* Lecture Notes in Mathematics, No 170. Springer-Verlag, Berlin, 1970. pp. 87–208.

A.I. Kostrikin – Yu.I. Manin
1. *Linear Algebra and Geometry*. Gordon and Breach, Amsterdam, 1997.

S.E. Landsburg
1. *Algebraic fiber bundles*. Trans. Amer. Math. Soc. 266(1981), 259–273.

S. Lang
1. *Algebra* (2nd ed.). Addison-Wesley, Reading, Mass., 1984.

R. Levi Setti
1. *Elementary Particles*. The Univ. of Chicago Press, Chicago, 1963.

K. Lønsted
1. *An algebraization of vector bundles on compact manifolds*. J. Pure Appl. Algebra 2(1972), 193–207.

S. Mac Lane
1. *Mathematics: Form and Function*. Springer-Verlag, Berlin, 1986.

A. Mallios
1. *On a convenient category of topological algebras, II: Applications*. Prakt. Akad. Athēnōn 51(1976), 245–263.
2. *Vector bundles and K -theory over topological algebras*. J. Math. Anal. Appl. 92(1983), 452–506.
3. *Homotopy invariants of the spectrum of a topological algebra*. J. Math. Anal. Appl. 101(1984), 297–307.
4. *Topological Algebras. Selected Topics*. North-Holland, Amsterdam, 1986. [For convenience, this item is also noted, throughout the text by TA.]
5. *Continuous vector bundles over topological algebras, II*. J. Math. Anal. Appl. 132(1988), 401–423.
6. *On Karoubi's density theorem in (non-normed) topological algebras* (manuscript; to appear).
7. *On an abstract form of Weil's integrality theorem*. Note Mat. 12(1992), 167–202.
8. *On an axiomatic approach to geometric prequantization: A classification scheme à la Kostant-Souriau-Kirillov*. J. Math. Sci. (New York), 95(1999), 2648–2668.

9. *Geometry of Vector Sheaves. An Axiomatic Approach to Differential Geometry*, Vols. I (Chapts I–V), II (Chapts VI–XI). Kluwer, Dordrecht, 1998. [This item is still noted, for convenience, throughout the text simply by VS.] There is a Russian edition of this 2-volume book, by MIR Publs, Moscow (Vol. I, 2000 and Vol. II, 2001).

10. *An axiomatic treatment of differential geometry via vector sheaves. Applications*. Math. Japonica (Intern. Plaza) 48(1998), 93–180.

11. *K-theory of topological algebras and second quantization*. Proc. Intern. Conf. on "*Topological Algebras and Applications,*" Oulu (Finland), 2001. Acta Univ. Ouluensis A 408(2004). Oulu Univ. Press, Oulu, 2004. pp. 145–160.

12. *On localizing topological algebras*. Contemp. Math. 341(2004), 79–95.

13. *Localization and Extension of Topological Algebras* (book, in preparation).

Yu.I. Manin

1. *Gauge Field Theory and Complex Geometry*. Springer-Verlag, Berlin, 1988.

2. *Strings*. The Math. Intelligencer 11(2)(1989), 59–65: p. 60.

M. Manoliu

1. *Abelian Chern-Simons theory. I. A topological quantum field theory*. J. Math. Phys. 39 (1998), no. 1, 170–206.

2. *Abelian Chern-Simons theory. II. A functional integral approach*. J. Math. Phys. 39 (1998), no. 1, 207–217.

K.B. Marathe – G. Martucci

1. *The Mathematical Foundations of Gauge Theories*. North-Holland, Amsterdam, 1992.

C.J. Mulvey

1. *A generalization of Swan's Theorem*. Math. Z. 151(1976), 57–70.

J.R. Munkres

1. *Elementary Differential Topology*. Princeton Univ. Press, Princeton, N.J., 1966.

M.A. Naĭmark

1. *Normed Algebras*. Wolters-Noordhoff, Groninger, 1972.

M. Nakahara

1. *Geometry, Topology and Physics*. Adam Hilger, Bristol, 1990.

C. Nash

1. *Differential Topology and Quantum Field Theory*. Academic Press, New York, 1991.

M.H. Papatriantafillou

1. *The category of differential triads*. Bull. Greek Math. Soc. 44(2000), 129–141.

2. *Initial and final differential structures* in Proc. Intern. Conf. on "*Topological Algebras and Applications,*" Rabat (Morocco), 2000. École Normale Supérieure, Takaddoum, Rabat, 2004, pp. 115–123.

W. Pauli

1. *General Principles of Quantum Mechanics*. Springer-Verlag, Berlin, 1980.

2. *Writings on Physics and Philosophy*. Springer-Verlag, Berlin, 1994.

W.A. Poor

1. *Differential Geometric Structures*. McGraw-Hill, New York, 1981.

M. Postnikov

1. *Leçons de géométie. Géométrie différentielle.* Éditions MIR, Moscou, 1990.

E. Prugovečki

1. *Quantum Mechanics in Hilbert Space* (2nd ed.). Academic Press, New York, 1981.

M. Puta

1. *Hamiltonian Mechanical Systems and Geometric Quantization.* Kluwer, Dordrecht, 1993.

A. Robert

1. *Introduction to the Representation Theory of Compact and Locally Compact Groups.* Cambridge Univ. Press, Cambridge, 1983.

J. Rosenberg

1. *Algebraic K-Theory and its Applications.* Springer-Verlag, New York, 1994.

J.J. Rotman

1. *The Theory of Groups. An Introduction* (2nd ed.). Allyn and Bacon, Boston, 1978.

2. *An Introduction to Homological Algebra.* Academic Press, New York, 1979.

S.A. Selesnick

1. *Second quantizations, projective modules, and local gauge invariance.* Intern. J. Theor. Phys. 22(1983), 29–53.

2. *Correspondence principle for the quantum net.* Intern. J. Theor. Phys. 30(1991), 1273–1292.

J.–P. Serre

1. *Modules projectifs et espaces fibrés à fibre vectorielle.* Sém. Dubreil-Pisot (1957/1958); exposé 23(1958), 1–18.

D.J. Simms – N.M.J. Woodhouse

1. *Lecture on Geometric Quantization.* Lecture Notes in Physics, No 53. Springer-Verlag, Berlin, 1976.

J.–M. Souriau

1. *Structure of Dynamical Systems. A Symplectic View of Physics.* Birkhäuser, Boston, 1997.

N. Steenrod

1. *The Topology of Fiber Bundles.* Princeton Univ. Press, Princeton, N.J., 1951.

R.F. Streater – A.S. Wightman

1. *PCT, Spin and Statistics and All That.* W.A. Benjamin, New York, 1964.

F. Strocchi

1. *Elements of Quantum Mechanics of Infinite Systems.* World Scientific, Singapore, 1985.

R.G. Swan

1. *Vector bundles and projective modules.* Trans. Amer. Math. Soc. 105(1962), 264–277.

R.M. Switzer

1. *Algebraic Topology-Homotopy and Homology.* Springer-Verlag, Berlin, 1975.

G. 't Hooft

1. *Obstacles on the way towards the quantization of space, time and matter.* Spin-2000/20.

B.R. Tennison
1. *Sheaf Theory*. Cambridge Univ. Press, Cambridge, 1975.

Ph. Tondeur
1. *Introduction to Lie Groups and Transformation Groups*. Lecture Notes in Mathematics No 7. Springer-Verlag, Berlin, 1969.

L.N. Vaserstein
1. *Vector bundles and projective modules*. Trans. Amer. Math. Soc. 294(1986), 749–755.

E. Vassiliou
1. *Geometry of Principal Sheaves*. Springer, Dordrecht, 2005.

M.E. Verona
1. *A de Rham theorem for generalized manifolds*. Proc. Edinburg Math. Soc. 22(1979), 127–135.

A. Weil
1. *Introduction à l' étude des variétés kählériennes* (nouv. éd.). Hermann, Paris, 1971.

S. Weinberg
1. *The Quantum Theory of Fields. Volume I: Foundations*. Cambridge Univ. Press, Cambridge, 1995.

A. Weinstein
1. *Symplectic geometry*. Bull. Amer. Math. Soc. 5(1981), 1–13.

C. von Westenholtz
1. *Differential Forms in Mathematical Physics*. North-Holland, Amsterdam, 1981.

N.M.J. Woodhouse
1. *Geometric Quantization* (2nd ed.). Oxford Univ. Press, Oxford, 1991.

L. Wittgenstein
1. *Culture and Value*. B. Blackwell, 1980.

J.M. Ziman
1. *Elements of Advanced Quantum Theory*. Cambridge Univ. Press, Cambridge, 1969.

E. Zafiris
1. *Dynamics of quantum observables algebras* (manuscript).
2. *Boolean coverings for quantum observables structures: a setting for an abstract differential geometric mechanism*. J. Geom. Phys. 50(2004), 99–114.

Index of Notation

$\mathbb{C} \underset{\varepsilon}{\subseteq} \mathcal{A}, 5$

$\varepsilon(\lambda) := \lambda \cdot \mathbf{1}_\mathcal{A}, 5$

$\partial : \mathcal{A} \longrightarrow \mathcal{E}, 5$

$(\mathcal{A}, \partial, \Omega), 6$

"dx", 6

$\mathcal{A} \equiv {}^{\mathbb{C}}\mathcal{C}_X^\infty, 6$

$\bar{\partial}(\alpha) \equiv \partial((\alpha_{ij})) := (\partial(\alpha_{ij})), 6$

$\bar{\partial} \equiv M_n(\partial), 6$

$\tilde{\partial} : \mathcal{A}^{\bullet} \longrightarrow \Omega, 7$

$\tilde{\partial} : \mathcal{GL}(n, \mathcal{A}) := M_n(\mathcal{A}^{\bullet}) \longrightarrow M_n(\Omega), 8$

$\tilde{\partial}(\alpha^{-1}) = -Ad(\alpha) \cdot \tilde{\partial}(\alpha), 8$

$D : \mathcal{E} \longrightarrow \mathcal{E} \otimes_\mathcal{A} \Omega \cong \Omega \otimes_\mathcal{A} \mathcal{E} \equiv \Omega(\mathcal{E}), 9$

$\partial : \mathcal{A} \longrightarrow \mathcal{A} \otimes_\mathcal{A} \Omega = \Omega \equiv \Omega(\mathcal{A}), 10$

$\Omega_X^1 := \mathcal{S}(\Gamma({}^{\mathbb{C}}T^*(X))), 10$

$(\mathcal{C}_X^\infty, d, \Omega_X^1), 10$

$\Omega_X^1 := ({}^{\mathbb{C}}T(X))^* \equiv \mathcal{E}^*, 12$

$\mathcal{E}|_U = \mathcal{A}^n|_U, \ n \in \mathbb{N}, 13$

$e^U \equiv \{U; (e_i)_{1 \le i \le n}\}, 14$

$D = \partial + \omega, 16$

$D|_U \longleftrightarrow \omega \equiv (\omega_{ij}) \in M_n(\Omega(U)), 16$

$\omega_{\mathcal{U}} \equiv \omega \equiv (\omega^{(\alpha)}) \in C^0(\mathcal{U}, M_n(\Omega))$, 17

$\omega^{(\alpha)} \equiv (\omega_{ij}^{(\alpha)}) \in M_n(\Omega(U_\alpha))$, $\alpha \in I$, 17

$\omega^{(\beta)} = Ad(g_{\alpha\beta}^{-1})\omega^{(\alpha)} + \tilde{\partial}(g_{\alpha\beta})$, 17

$(g_{\alpha\beta}) \in Z^1(\mathcal{U}, \mathcal{GL}(n, \mathcal{A}))$, 17

$Ad(g_{\alpha\beta}^{-1}) \cdot \omega^{(\alpha)} := g_{\alpha\beta}^{-1} w^{(\alpha)} g_{\alpha\beta}$, 17

$\delta(\omega^{(\alpha)}) := \omega^{(\beta)} - Ad(g_{\alpha\beta}^{-1})\omega^{(\alpha)}$, 20

$\delta(\omega^{(\alpha)}) := \tilde{\partial}(g_{\alpha\beta})$, 20

$D_{\mathcal{E} \otimes_A \mathcal{F}} := (D_{\mathcal{E}} \otimes 1_{\mathcal{F}}) + (1_{\mathcal{E}} \otimes D_{\mathcal{F}}) \equiv D \otimes 1 + 1 \otimes D'$, 22

$D_{\mathcal{Hom}_A(\mathcal{E},\mathcal{F})}(\phi) := D_{\mathcal{F}} \circ \phi - (\phi \otimes 1_\Omega) \circ D_{\mathcal{E}} \equiv D' \circ \phi - (\phi \otimes 1) \circ D$, 22

$D_{\mathcal{End}\mathcal{E}}(\phi) = D \circ \phi - (\phi \otimes 1) \circ D \equiv D \circ \phi - \phi \circ D \equiv [D, \phi] \equiv L_D(\phi)$, 22

$\mathcal{E}^* := \mathcal{H}_A(\mathcal{E}, \mathcal{A})$, 23

$\partial(u(s)) = u(D(s)) + D^*(u)(s)$, 24

$\omega^* \equiv (\omega_{ij}^*) = -{}^t\omega \equiv (-\omega_{ji}) = -(\omega_{ji}) \in M_n(\Omega(U))$, 24, 25

$D|_U = i_U^*(D)$, 26

$\mathcal{A} \equiv \mathcal{O}_X$, 27

$(D_\alpha) \in C^0(\mathcal{U}, \mathcal{Hom}_{\mathbb{C}}(\mathcal{E}, \Omega(\mathcal{E})))$, 28

$\mathfrak{a}(\mathcal{E}) := [\delta(D_\alpha)] = [(\tilde{\partial}(g_{\alpha\beta}))] \in H^1(X, M_n(\Omega))$, 29

$H^1(X, M_n(\Omega))$, 29

$\mathcal{Conn}_A(\mathcal{E})$, 30

$\mathcal{Conn}_A(\mathcal{E}) = D + \Omega(\mathcal{End}\mathcal{E})(X)$, 30

$D_{\mathcal{F}} \circ \phi = (\phi \otimes 1_\Omega) \circ D_{\mathcal{E}}$, 33

$D_{\mathcal{F}} = Ad(\phi) \cdot D_{\mathcal{E}}$, 34

$D_{\mathcal{End}\mathcal{E}} = Ad(\theta) \cdot D_{\mathcal{E} \otimes_A \mathcal{E}^*}$, 34

$\mathcal{E} \otimes_A \mathcal{E}^* = \mathcal{Hom}_A(\mathcal{E}, \mathcal{E}) \equiv \mathcal{End}\mathcal{E}$, 34

$\mathcal{Aut}_A(\mathcal{E}) \equiv \mathcal{Aut}\mathcal{E} := \mathcal{Isom}_A(\mathcal{E}, \mathcal{E})$, 34

$\mathcal{Hom}_A(\mathcal{E}, \mathcal{E}) \equiv \mathcal{End}\mathcal{E}$, 34

$\mathcal{Aut}\mathcal{E} = (\mathcal{End}\mathcal{E})^{\cdot}$, 34

$\mathcal{E}^{\cdot}|_U = (\mathcal{E}|_U)^{\cdot}$, 35

$Aut\mathcal{E} := (\mathcal{Aut}\mathcal{E})(X)$, 35

$^t\omega = -\bar{\omega}, 60$

$\mathcal{J} \in \mathcal{H}om_A(\mathcal{E}, \mathcal{E}) \equiv \mathcal{E}nd\mathcal{E}, 61$

$\mathcal{J}^2 = -id_{\mathcal{E}} \equiv -1, 61$

$D(\mathcal{J}) \equiv D_{\mathcal{E}nd\mathcal{E}}(\mathcal{J}) = 0, 62$

$\mathcal{R}ic(\mathcal{E}) = \alpha \cdot \rho, 63$

$\omega = \tilde{s}_1 \wedge \ldots \wedge \tilde{s}_n \in (\det(\mathcal{A}^n))(X) = \mathcal{A}(X), 65$

$\omega := \sqrt{|\tilde{\rho}|} \cdot \varepsilon_1 \wedge \ldots \wedge \varepsilon_n, 65$

$* : \bigwedge^p \mathcal{E}^* \longrightarrow \bigwedge^{n-p} \mathcal{E}^*, \ 1 \leq p \leqq n, 65$

$(*\alpha)(\beta) := \omega \cdot (\alpha \wedge \beta^\#) \equiv \ <\alpha \wedge \beta^\#, \omega> \ \in \mathcal{A}(X), 65$

$\# := \bigwedge^{n-p} \tilde{\rho}, \ 1 \leq p \leq n, 65$

$* \in \mathcal{A}ut_A(\bigwedge \mathcal{E}^*) \cong \mathcal{A}ut_A(\bigwedge \mathcal{E}), 66$

$\check{H}_{phys} = \check{H}_{bare} \oplus \check{H}_{etc}, 73$

$\mathcal{E}(X) = \mathcal{F}(\mathcal{C}(X)), 82$

$E(\xi) \equiv \xi = (E, \pi, X), 86$

$Vect_{\mathbb{C}}^n(X) = \Phi_{\mathcal{A}}^n(X), 86$

$\Phi_{\mathcal{A}}^n(X) = H^1(X, \mathcal{GL}(n, \mathcal{A})), 94$

$H^1(X, \ \mathcal{GL}(n, \mathcal{A})) := \varinjlim_{\mathcal{U}} H^1(\mathcal{U}, \mathcal{GL}(n, \mathcal{A}))$

$$= \bigcup_{\mathcal{U}} H^1(\mathcal{U}, \mathcal{GL}(n, \mathcal{A})) = \sum_{\mathcal{U}} H^1(\mathcal{U}, \mathcal{GL}(n, \mathcal{A})), 95$$

$\Phi_{\mathcal{A}}^1(X) = H^1(X, \mathcal{A}^\cdot), 95$

$Pic(X) = H^1(X, \mathcal{A}^\cdot), 96$

$\det \mathcal{E} \equiv [\det \mathcal{E}] \in \Phi_{\mathcal{A}}^1(X), 96$

$\mathcal{E} = (\mathcal{F} \underset{X}{\times} \mathcal{A}^n)/\mathcal{G}, 108$

$\mathcal{E} = (\mathcal{I}som_A(\mathcal{E}, \mathcal{A}^n) \underset{X}{\times} \mathcal{A}^n)/\mathcal{GL}(n, \mathcal{A}), 108$

$\mathcal{M}_X, 115$

$(\mathcal{L}, D) \underset{\phi}{\sim} (\mathcal{L}', D'), 116$

$[(\mathcal{L}, D)] \in \Phi_{\mathcal{A}}^1(X)^\nabla, 117$

$(\mathcal{U}, \mathcal{A}^\cdot), 120$

$(\theta_\alpha) \in C^0(\mathcal{U}, \Omega^1)$, 120

$\delta(\theta_\alpha) = \tilde{\partial}(g_{\alpha\beta})$, 121

$(\mathcal{L}, D) \longleftrightarrow ((g_{\alpha\beta}), (\theta_\alpha))$, 123

$\delta(\theta_\alpha) = \tilde{\partial}(g_{\alpha\beta})$, 123

$\tau : \Phi^1_\mathcal{A}(X)^\nabla \longrightarrow \Omega^2(X)$, 130

$\Phi^1_\mathcal{A}(X)^\nabla = \sum_{R \in im\tau} \Phi^1_\mathcal{A}(X)^\nabla_R$, 130

$\Phi^1_\mathcal{A}(X)^\nabla_R := \tau^{-1}(R), \quad R \in im\tau$, 130

$\tau : \Phi^1_\mathcal{A}(X)^\nabla \longrightarrow \Omega^2(X)_{cl} \subseteq \Omega^2(X)$, 131

$\Omega^2(X)_{cl} = \ker d^2_X \equiv \ker(d^2)$, 131

$\mathbb{Z} \underset{\longrightarrow i}{\subset} \mathbb{C}$, 134

$i^* : H^p(X, \mathbb{Z}) \to H^p(X, \mathbb{C})$, 134

$z \in im(i^*) \equiv im(H^p(X, \mathbb{Z}) \to H^p(X, \mathbb{C}))$, 134

$R(D) \equiv R \in im(H^2(X, \mathbb{Z}) \to H^2(X, \mathbb{C}))$, 134

$H^1(X, \mathcal{A}^\cdot) = H^2(X, \mathbb{Z})$, 137

$im\tau \equiv \tau(\Phi^1_\mathcal{A}(X)^\nabla) = \Omega^2(X)^{int}_{cl}$, 139

$\tau^{-1}(R) \equiv \Phi^1_\mathcal{A}(S)^\nabla_R$, 140

$\theta'_\alpha = \theta_\alpha + \tilde{\partial}(t_\alpha^{-1})$, 144

$g'_{\alpha\beta} := \delta(t_\alpha^{-1}) \cdot g_{\alpha\beta}$, 144

$\Omega^2(X)^{int}_{cl}$, 149

$Conn_\mathcal{A}(\mathcal{L})/\mathcal{A}^\cdot$, 150

$Conn_\mathcal{A}(\mathcal{L})/Aut(\mathcal{L})$, 150

$\mathbb{C}^\cdot \subseteq \mathbb{C} \underset{\longrightarrow \varepsilon}{\subset} \mathcal{A}$, 152

$\mathbb{C}^\cdot \underset{\longrightarrow \varepsilon}{\subset} \mathcal{A}^\cdot$, 152

$\Phi^1_\mathcal{A}(X)_{her}$, 171

$\Phi^1_\mathcal{A}(X)_{her} = H^1(X, \mathcal{SU}(1))$, 171

$\Phi^1_\mathcal{A}(X)^{\nabla_{her}}$, 173

$\Phi^1_\mathcal{A}(X)^{\nabla_{her}} < \Phi^1_\mathcal{A}(X)^\nabla$, 173

$\theta + \bar{\theta} = \tilde{\partial}(\rho)$, 174

$\mathcal{E}^{\bullet} : 0 \longrightarrow \mathcal{A}^{\bullet} \xrightarrow{\tilde{\partial}} \Omega^1 \longrightarrow 0 \longrightarrow \cdots, 216$

$\check{\mathbb{H}}^1(\mathcal{U}, \mathcal{E}^{\bullet}) := \ker D^1 / im D^0, 218$

$\mathcal{SU}(1) \equiv \mathcal{SU}(1, \mathcal{A}) \lhd \mathcal{A}^{\bullet}, 231$

$\Phi_{\mathcal{A}}^1(X)_{\omega}^{\nabla_{her}} = \check{H}^1(X, S^1) \cdot [(\mathcal{L}, D^{her})], 248$

$(\mathcal{L}, D)_{her} \equiv (\mathcal{L}, D^{her}), 248$

$[(\mathcal{L}, D^{her})], 248$

$\Phi_{\mathcal{A}}^1(X)_{\omega}^{\nabla_{her}} = \pi_1(X)^* \cdot [(\mathcal{L}, D^{her})], 249$

Index

Made in the USA
Monee, IL
07 July 2026